A Modern Theory of Evolution

Carl J. Becker

iUniverse, Inc.
New York Bloomington

A Modern Theory of Evolution

Copyright © 2010 by Carl J. Becker

All rights reserved. No part of this book may be used or reproduced by any means, graphic, electronic, or mechanical, including photocopying, recording, taping or by any information storage retrieval system without the written permission of the publisher except in the case of brief quotations embodied in critical articles and reviews.

iUniverse books may be ordered through booksellers or by contacting:

iUniverse
1663 Liberty Drive
Bloomington, IN 47403
www.iuniverse.com
1-800-Authors (1-800-288-4677)

Because of the dynamic nature of the Internet, any Web addresses or links contained in this book may have changed since publication and may no longer be valid. The views expressed in this work are solely those of the author and do not necessarily reflect the views of the publisher, and the publisher hereby disclaims any responsibility for them.

ISBN: 978-1-4502-2449-9 (sc)
ISBN: 978-1-4502-2450-5 (ebook)

Printed in the United States of America

iUniverse rev. date: 04/05/10

In this dedication I would like to remember my grandfathers, Carl Lotus Becker and Marion Sylvester Dooley, a professor of history at Cornell and a professor of medicine at Syracuse, respectively. They set a very high bar of intellectual achievement. I would like to remember Walter J. Freeman for picking up a graduate student at Berkeley who had been cut loose by another department for being involved in an unfortunate and unwanted discovery. In that situation I learned the practical limits of politically correct science. From Freeman I would discover the existence of a level of intellectual achievement above and beyond the Nobel Prize level. I would also like to remember Robert Macey, who introduced me to the amazing world of physiology, and Sidney Brenner, who initiated me into the secrets of DNA. I am deeply indebted to Robert Treese who gave me a friendly introduction to the Bible. Over the years I have communicated with and have been encouraged by Fred Hoyle (astronomy), Donald Griffin (ethology), Luis Alvarez (physics), Roland Fischer (philosophy), and Mario Vaneechoutte (microbiology), among many others. In the course of educating myself over the decades I would have been lost without the used-book sellers of this world now, sadly, few and far between. They stocked a library that is devoted to the question, *What is the human condition and how did we get into this mess?* Finally I would like to acknowledge the use of Google and Wikipedia as my fact checker of last resort.

Contents

Introduction	1
Chapter 1: Typhon and the Iron Age	3
Chapter 2: The English Enlightenment	26
Chapter 3: Charles Darwin and Victorian Culture	50
Chapter 4: Science on the Continent	75
Chapter 5: Evolution Comes to America	95
Chapter 6: Modern Geology	119
Chapter 7: The Fitness of Darwinism	143
Chapter 8: In the Beginning …	165
Chapter 9: Microbiological Evolution	185
Chapter 10: The Shapes of Things to Come	208
Chapter 11: Plant/Insect Evolution	230
Chapter 12: The Last of the Dinosaurs	256
Chapter 13: The Age of the Mammals	281

Chapter 14: Human Evolution	305
Chapter 15: Last Thoughts	329
Notes	333
Bibliography	339

Introduction

ENGLISH UPPER-CLASS CULTURE AND the Anglican Church were ready for a reformation of the Old Testament theory of Genesis by the nineteenth century. It was an active member of the Church, Charles Lyell, who would propose an old-earth theory of change called Uniformitarianism. This was a replacement for the young-earth theory of change associated with biblical literalists and the likes of Bishop Ussher, Isaac Newton, Johannes Kepler, and the Venerable Bede. Lyell proposed that all changes that have ever occurred on earth over the long periods of time of his new geology were caused by the same factors we have seen in our own little age of reason: changes wrought by the sun, rain, wind, and cold—miniscule causes that resulted in big changes over a long period of time. Lyell's theory made the rationalistic argument that the geological record only appears to show violent changes in the past. He believed that when the missing pages were retrieved, Uniformitarianism would be confirmed.

Lyell continued to hold with the traditional Christian view of an original creation with no subsequent evolution of life: he held the old-earth, single-creation theory. Lyell was influenced by the bishop of Wearmouth, William Paley. Paley was a skilled naturalist who said that manifested creation was simply too marvelously complex to have been anything less than the creation of a divine watchmaker. The biological counterpart to Uniformitarian geology, built on the plinth of Lyell's geological theory, would now follow. Charles Darwin, a candidate for the Anglican ministry converted to biology, took Lyell's geology and combined it with the population theory of Thomas Malthus, an Anglican cleric, to propose an old-earth, gradual-change theory of biological evolution. As far as Darwin was concerned, anyone who could not believe in Lyell could not believe in his theory of evolution.

There was a brief flurry of excitement, but a year after Darwin published *The Origin of the Species*, future Archbishop of Canterbury Fredrick Temple and several other young clerics would affirm Charles Darwin's thesis. Darwinian evolution was compatible with the Anglican Church, especially in contrast to another variation on the theme of evolution associated with Jean-Baptiste Lamarck and Karl Marx. In less than a generation the hopeful gospel of Uniformitarianism became the operating system for geology, paleontology, archeology, and history. Combined with Adam Smith, who had provided the moral justification for English mercantilism in the previous century, Lyell and Darwin became the holy trinity of Victorian science. The invisible hand of Smith's economics, the missing pages of Lyell's geology, and Charles Darwin's imperceptible biological changes were the three-legged stool that supported Victorian science. Revolutionary change was already outlawed in the social culture. Now it was outlawed in Nature itself. Prophecies of catastrophe were declared the fearful delusions of the superstitious, uneducated mind.

Numerous irritations, exceptions, and anomalies to Victorian Uniformitarian science have always existed for anyone who cared to look, but the terminal moment for Lyell's theory came in 1976 when Nobel laureate Luis Alvarez presented incontrovertible evidence for a major meteor impact around 65 million years ago. It was coincident with the end of the age of dinosaurs. All of a sudden a flood of information on catastrophic events became available to the discipline of geology. Of course, the evidence was there all along. In spite of much backing and filling, as well as outright attempts to suppress what was happening, the Alvarez Impact was eventually accepted. Miraculously, Darwin's theory of evolution was not affected. Darwinism was saved. Darwinism had become too big to fail; yet the only opponent that Darwinism can now still hope to contend with is the Creationism of William Paley.

The first half of this book will be a history of how and why we find ourselves worshiping the secular dogma of Darwinism. The second half will be an attempt to put together a picture of evolution from scratch derived from the latest findings in science without concern for what is politically correct.

CHAPTER 1

Typhon and the Iron Age

HOMO SAPIENS EVOLUTION HAS been broadly segregated into three major periods: a Stone Age, a Bronze Age, and an Iron Age. The Stone Age was associated with the gradual cooling of the planet. Around 50 million years ago the planet was tropically forested from pole to pole. The average climate was warm and humid. It didn't vary much. Around 5 million years ago the tropic forests had been reduced to a narrow band centered around the equator. At that time there were still a hundred or so different species of apes living in the tropical forest canopy. By the Pleistocene epoch the planet was significantly desiccated, colder, and hotter, as well as dry. Occasional impact events followed by horrendous glaciations began to show up dramatically in the geological record because they were followed by short warm periods. The planet became green for a while. By the time *Homo habilis* moved out of the cradle of life on two legs and into the savannahs of Africa and then to the rest of the world around 2 million years ago, there were only a handful of apes left. *Habilis* used stone tools. The peak of the Stone Age came with Cro-Magnon in Ice Age Europe with their famous cave paintings. As revealed by Alexander Marshack at Harvard, they survived by the use of calendar bones and had a primitive knowledge of the seed. By this knowledge they were able to time conception so that parturition would occur in the spring of the year. They could not have survived long in Ice Age Europe without this knowledge.

Richard Firestone, Allen West, and Simon Warwick-Smith at Berkeley have revealed the geologic strata that show evidence of major impact events at 13,000 years ago and 9,000 years ago. An iridium spike, the marker for sizable meteor impacts, identifies them both. The cause of the extinctions of numerous species of megafauna, the sudden creation of massive glaciers that wiped out Western Europe and a broad area of North America, and the reason for a sudden revolution in human culture are now on a solid footing. We will discuss our geological denial and confusion on this matter in greater detail in Chapter 6. The initiation of agriculture and the domestication of animals that followed these events were accompanied by a switch from stone tools to the use of copper and bronze tools, marking the beginning of the Bronze Age.

The shift to iron tools began around 3,500 years ago and it formed the basis for the Iron Age culture that we now live in. From discoveries made in the last quarter century we are in a position to know that this too was correlated to a significant impact with space junk. The gradual decline in the environment of the planet and the shattering impact events at 13,000, 9,000, and 3,500 years ago have been the primary cause of our recent biological and cultural evolution. We will begin our story with the last impact since it has shaped how we think about such matters today. Once we unravel the sociological reasons for the way we think, or refuse to think, it will be easier to proceed to a discussion of what is now possible for us to know from the database of science.

A group of scientists called the Holocene Impact Group is specifically looking for impacts in the last ten thousand years. They estimate five to ten significant impacts have occurred during this period. Operating out of an academic tradition befogged by Sir Charles Lyell's hopeful Uniformitarian geology, their search is rather like looking for a needle in a haystack when it comes to finding such events. So far they have missed the most important impact of all. The largest Holocene impact is inextricably linked with stories of the Flood and with the myth of Atlantis.

The trail to that impact time and location begins with Dr. Alan H. Kelso de Montigny, a freethinking American anthropologist brought back to light by Murry Hope in her book *Atlantis: Myth or Reality?* After studying depth charts of the eastern Caribbean seafloor, de Montigny concluded that a recent impact had created the Lesser Antilles. The impact had occurred west of that partial ring of seventeen active volcanic islands. Otto Muck, a German engineer, expanded this thesis but correlated the impact with two deep wounds inflicted on the Earth's crust under the Atlantic Ocean just north of Puerto Rico. He associated this with the Carolina Bays impact sites scattered up and down the southeastern seaboard of North America. He connected this event with the story of Atlantis. He also associated the event with the termination of the Stone Age, around 10,000 years ago. He was probably correct about the impact site, but incorrect about the time or the Carolina Bays. They are part of the catastrophic events described by Firestone et al.

The traditional understanding of the formation of the Lesser Antilles is contained in the monograph by W. M. Davis of Harvard University published in the Proceedings of the National Academy of Sciences in June 1924. It is the standard Uniformitarian explanation for geological change. According to Davis the Lesser Antilles volcanoes describe the eastern edge of the Caribbean Plate. Both the North American and the South American Plates are moving toward the west and have been thrust under the Caribbean Plate, creating the volcanic activity. In addition, the western edge of the Caribbean Plate is riding up over the Cocos Plate to the west. This creates the Central American Volcanic Arc through Guatemala, El Salvador, Nicaragua, and Costa Rica. Uniformitarian tradition has no particular explanation for the energy input needed to arouse such a movement, activity that is still going on to the present day. Muck's impact site on the North American Plate with a low southeast incoming vector does the trick. It also explains the evidence for a recent drowning of part of the North American Plate off the coast of Florida and Cuba, leaving the Bahamian Islands.

Another important piece of the geological puzzle comes from Jack Morelock and his associates at the Department of Marine Sciences, University of Puerto Rico, who have cored the corral reef off the east coast of Puerto Rico. This core covers the whole of the Holocene period. It is located just south of Muck's impact site and just north of the Lesser Antilles. There is one outstanding anomaly in the core. Called backstepping, it is essentially a major die-off of the coral colony followed by a recovery. By radiocarbon dating, the die-off occurred around 6,000 years ago. The reef came back to life around 4,000 years ago. It is the only major event to mark this coral reef for the past 10,000 years.

Pleistocene oceans are too cold to support corral reefs. All of the world's major corral reefs were formed around 10,000 years ago; I presume they formed after the Firestone impact events heated up the oceans. Corral reefs are the tropical rain forest of the ocean and support a great array of life forms. They are a sensitive measure of the cooling of the ocean. The conventional explanation for the formation of these reefs is that it was due to glacial meltdown, which flooded shallow continental shelves, but glacial water and corral reefs don't coexist. A more reasonable scenario is that the Puerto Rico reef was an indicator of the massive infusion of cold water to the Atlantic Ocean that occurred in the glacial meltdown beginning around 7,500 years ago. The cooling of the ocean also precipitated a serious downturn in the weather worldwide and is marked worldwide from glacier cores. I presume the reef came back to life again with the reheating of the ocean by local marine volcanic activity around 4,000 years ago. As dating with radiocarbon in this period is very imprecise, it would be easy to think that the date of 4,000 years ago was really 3,500 years or so. With these meteorological possibilities in place, we are now prepared to look at mythological creation stories worldwide.

Many Atlantis seekers have sought to place the time of that myth tradition at the termination of the Stone Age. Jürgen Spanuth is one of few modern Atlantis seekers that have seen the evidence of an impact causing the demise of that idealized culture around 3,500 years ago. This North German cleric's book is called *Atlantis of the North*. The myth of Atlantis came from Egyptian priests to Greece by way of Solon. Solon gave the story of Atlantis to Plato, who preserved it in *Timaeus* and *Critias*. The date given by Solon of 11,000 years ago, or 9000 BC, for the flooding of Atlantis was obviously wrong according to Spanuth, since that would place Bronze-Age Atlantis in the Paleolithic age. There simply are no Bronze-Age remains to be found from that time. A major Neolithic Stone Age site at Çatal Hüyük, whose earliest date was 7500 BC, was abandoned well before the Bronze Age.

Spanuth says that Solon misinterpreted the Egyptian priests concerning the time of the event. Egyptian priests operated on a lunar calendar. He suggests they were talking about 9,000 lunations, not years. When this correction is made, the date of the flooding of Atlantis according to the Egyptian priests was around 1230 BC. Egyptian culture just barely managed to hang on through the tragedy—it marked the change from the Old Kingdom to the New Kingdom—yet it appears to be an unbroken first hand account. It is probably the most accurate date we will ever get. Radiocarbon dating is notoriously suspect and all too readily converted to Uniformitarian prejudices. The theory assumes a constant level of carbon dioxide in the atmosphere, which is certainly not the case when there is a dramatic increase in volcanic activity.

Any individual, tribal, or national memory can be easily disregarded as a phantasmagoric perception of a superstitious mind struggling with limited knowledge. Academic specialists will guard their narrow intellectual domains to the detriment of a broader understanding. A summation of human experience, however, augmented with scientific details appropriately interpreted, makes an ironclad case for an impact event. The description of the event from human reports around the globe is as clear a description an impact event as one could possibly imagine from its first sighting in the nighttime sky, its appearance in the day light as it got within days of impact, and the impact event itself as it burned through the upper atmosphere of the Indian subcontinent and buried itself in the Atlantic Ocean north of the Caribbean

Plate followed by the physical consequences that went on for days, years, and even centuries.

A meteor or a comet on a direct impact course with Earth will only appear in the sky a week or two ahead of time. Many of the world's peoples, including the Chinese, personified the approach of a foreign body in the night sky as a struggle between Mars and Venus, with Venus being out of place. This identifies the sector in the night sky that was the origin of the impacting body. As it got closer and became visible during the day, it was seen as a struggle between two suns in some stories. It is safe to presume that there was a considerable amount of anxiety in the days preceding the impact, and that most people on the planet had their eyes glued on the sky. All around the world Flood stories are connected to the setting of the Pleiades, which marks the beginning of winter in the northern hemisphere. The peoples of the Indian subcontinent saw the comet as it first entered the atmosphere. They described it as a flaming lingam. It was personified as the "feared one" or "red one," Shiva. It appeared during the night and came in at a low angle to the horizon. We know this because people in China, who were also in the dark at the time but in the northern hemisphere, only experienced the results of the impact. This they attributed to the Black Dragon. Its low angle of entry to the atmosphere allowed it to be seen by the peoples of India, North Africa, and the Mediterranean area before it impacted.

It is entirely possible that at least one human sacrifice was being conducted on Crete just as Typhon was passing over. Such a sacrifice was frozen-mid stride in a collapsed Minoan Temple, the legendary tomb of the Cretan Zeus, at the base of Mount Juktas. It is likely that the comet caused earthly tremors below it as it passed over. The people of the Mediterranean area saw the flaming lingam as it was breaking up and left descriptions of it in florid and terrifying detail. One name given to it by the Greeks was Typhon, which means "stupefying smoke." It was daylight in the Mediterranean when the Greeks saw Typhon coming from the direction of Crete. Typhon was described as the most grotesque of all creatures that ever lived, having a hundred serpent heads.

Many peoples in many places saw the same event and struggled to find metaphors for the terrible event. Only later would they be gathered, often scrambled together, by storytellers. A more distant second-hand personification of the event is contained in Ovid's *Metamorphoses*, the story of Phaëton, who drove the chariot of the sun too close to the Earth so that the vegetation dried and burned, most of Africa was turned into a desert, and the skin of the Ethiopians was turned black. There is reason to suppose that the myth of the Medusa was an attempt to describe the impossible. The Medusa was an offspring of Typhon. With teeth like a wild boar, her black tongue protruding and too large for her mouth, her hands like brazen claws, and her wings changed into serpents, her gaze alone turned men to stone. The Medusa was a dragon. It wouldn't be until fairly recently that we would discover that the Medusa was a North African serpent-goddess, the destroyer aspect of a triple-headed goddess named Neith, Anath, or Athene. In the cycle of birth, death, and rebirth she destroys in order to rebalance life. This myth puts the event back into the context of a regular process of nature.

After first hiding in terror in a cave, according to the Greek myth of Typhon, Zeus battled with and defeated the monster with a terrible barrage of lightning bolts. The comet created conditions for a massive display of static electrical discharge. From the Greek perspective the monster was thrown to the ground in Sicily in the Western Mediterranean, there to be entombed under Mount Etna, which would soon become very active. Massive volcanic clouds also create stunning lightning displays. Even in calmer times lightning storms carry with them the halo of terror as human myths provide the basis for catastrophic possibilities. In a later time the Greeks would invent a cartoonish after thought. Zeus as superhero had to intervene, it was said, striking the runaway solar chariot with a lightening bolt. In reality the comet crossed the Atlantic.

According to the *Annals of Quauhtitlan*, Quetzalcoatl immolated himself on the shores of the eastern sea, and from his ashes rose birds with shining feathers, symbolizing the souls of dead warriors rising to heaven. The Feathered Serpent was associated with the planet Venus. Native folk tales speak of the old moon falling to Earth. According to the priests accompanying Hernando Cortez thousands of years later, Moctezuma heard the story of the Second Coming of Jesus Christ and believed that Cortez had something to do with the second coming of Quetzalcoatl. The Aztecs hoped to discourage the return of Quetzalcoatl, the celestial snake that devoured the Pleiades in a great fire and deluge. This is implicit in the human sacrifices they were still performing, sacrifices whose purpose was to keep the Earth from going haywire again. They were unsure as to whether they should worship this newly arrived god or exterminate it.

When Europeans occupied North America they collected Turtle Island stories from the Cherokee in the southeast of North America. The Cherokee had a belief in a race of snakes that were as huge as a tree trunk. These snakes were horned with a bright, blazing crest. Even to see one was death in this Medusa-like myth. This story is not a borrowing from other Native American traditions, nor of European traditions; it is a story that is a part of the original, worldwide experience of an impact event. Stories like this acquired secondary functions such as instructional tools: don't do such and such, or the terrible snake will strike you dead. Enlightenment Europeans with a poetic sensibility would eventually turn the images of Turtle Island into metaphors, reducing their power to the civilized level of *sturm und drang*.

On impact observers in China saw the stars appeared to fall from the sky as though one of the pillars of the heavens had collapsed. To observers in the Mediterranean area, Typhon caused the sun to appear momentarily to stop its movement. All bodies of water on the planet were jolted out of their beds. Low-lying landmasses facing large bodies of water to the northwest were devastated by mega tsunamis within minutes. The Indus civilization, already seriously weakened by centuries of drought, was finished off by a tsunami. The Tigris and Euphrates valley suffered relatively less from the Typhon impact as far as flooding was concerned, since it was protected by the highlands of the Eastern Mediterranean. Only the rivers were temporarily knocked out of their beds. The Mesopotamian civilizations rebounded fairly quickly. It was a different story for the new cities remembered by the name of Sodom and Gomorrah, drowned by the Dead Sea and burned by bitumen and natural gas deposits that flared off nearby.

In Egypt the Semitic Hyksos from their capitol at Avaris in the Nile delta had dominated the African dynasty of Upper Nile that had been in decline for centuries. This was the result of a serious drought, plague, and famine in the time of Joseph as recounted in the Bible. The tsunami triggered by the Typhon impact destroyed the Hyksos and actually allowed for a restoration of traditional African culture. This was the flood of Misphragmuthosis that marked the restoration of the Eighteenth Dynasty under Ahmose. *Mose, mosis, moses*, and *musa* are suffixes that mean "born of." By poetic license with regard to the breaking of the waters at birth, both Ahmose and Moses were born with the flood to become great leaders of new nations. It is safe to assume that at least one army of an Egyptian king was drowned, crushed, or otherwise obliterated as recorded in the Bible. Jews remember this event as their liberation from slavery during the celebration of Passover.

The low-lying areas of Northwestern Europe (Atlantis) were largely wiped out by flood, whereas the Scandinavian highlands preserved a small group of fair-haired, blue-eyed survivors. From the poem called the Vafthrudnismal we hear:

> *The sons of Bör slew the Jötun Ymir, but when he fell there flowed so much blood from his wounds that it drowned the whole race of the Hrim-Thursar, except one who escaped with his household. Him the Jötnar called Bergelmir; he and his wife went on board his ark, and thus saved themselves; from them are descended a new race of Hrim-Thursar.*

The myth of Atlantis collected by Solon from Egyptian priests was correlated by Jurgen Spanuth with Homer. Homer's *Iliad* is about a typical human military venture that began at the height of the glory that was Mycenae in Greece, the Bronze Age, later remembered as the Golden Age. After the victory at Troy the *Odyssey* is a tale that goes through the impact of Typhon creating the long return of Odysseus to Ithaca, and it gives us a vision of an impoverished world reduced to beggary and Stone Age rusticity. It is the classic post-apocalyptic travelogue. Spanuth locates Homer's Scylla and Charybdis as the Straits of Gibraltar with the Atlantic Ocean roaring through; Ogygia he identifies as the Azores that were on the Mycenaean, Bronze-Age, tin route to Northern Europe. Phaeacia he places in the area of Heligoland in modern Denmark. Phaeacia is identified by Spanuth as the capital of Atlantean culture.

There is considerable evidence that the Atlantean culture had colonies in the New World, probably along with the Phoenicians. The drowned portions off the Bahama Islands seem to preserve remains of one of these colonies. Native folk tales also preserve memories of this transatlantic connection. The impact of Typhon severed the contact between distant cousins for nearly three thousand years. Although the Atlantis myth was handed down orally for many centuries, thereby picking up red herrings and other anachronisms of a later age, there remains a solid core of history. There is a cottage industry devoted to identifying the places mentioned in Homer or to explain the story of Atlantis, but none of them have anywhere near the wide reach and internal coherence of Spanuth's explanations and dates along with Muck's impact site north of Puerto Rico.

With striking detail the apocalypse was riveted to founding stories of virtually all tribes and nations. Along the western shores of the great Pacific Ocean are astonishing glimpses of the ocean disappearing as mega tsunamis with monstrous waves rush toward North and South America, only to return again with great destruction. The Kojiki in Japanese tradition contains memories of the event. Chinese stories confirm what archeology has discovered: habitations all along their Pacific coast were wiped out. The Aborigines of eastern Australia tell the story of a giant frog that sucked all the water out of the oceans, but then he was tickled by an eel and caused to laugh, belching out the water again and drowning almost everyone. On the main island of Tonga there are seven giant coral boulders that have been ripped from an offshore reef. The scientists who have studied them suggest they were transported by a mega-tsunami not more than a few thousand years ago. All of the Pacific islands were seriously damaged or wiped clean by Typhon's floods, including the Galapagos Islands.

Earthquake and volcanic activity was dramatically increased worldwide.

After Quetzalcoatl immolated himself on the shores of the eastern sea, his heart became the Morning Star, wandering for eight days in the underworld before it ascended in splendor. In this tradition the volcanic cloud obscured the sun for eight days. The period of darkness remembered in legend the world around created the terrible Fimbulwinter of the Germanic Eddas. The Bog Oaks preserved in peat fields in Ireland show the absence of a growing season for eighteen years. There was a drop in temperature. An impact of this sort opens up volcanic hot spots and drives plates of earth down under overbearing plates. This was the cause for increased volcanic activity for centuries into the future. The island of Thera, however, exploded in the eastern Mediterranean around 400 years before the impact of Typhon, causing great damage to the brilliant Bronze-Age culture of Crete nearby. The impact of Typhon finished the culture that had limped through the first catastrophe.

The cosmic impact of Typhon raised a question in the mind of humanity: has God forsaken us? The stories that speak to a resolution of this issue break into two types: the stories carried by the tribal peoples by oral tradition and the stories carried by the highly socialized peoples with a literate tradition. The oral traditions were allegories hinged to nature: coyote the trickster was the cause of it. They are generally alike in that they come to view the impact as a natural part of life and a reflection of the profound power of the divine. The tribal, stone-age peoples of the time recovered as rapidly as the rest of life on the planet would allow. The literate traditions of the high civilizations that survived tended toward detailed allegories with a cultural theme. Commonly the creator god was disgusted with the imperfect nature of his creation and had decided to start over. Often these traditions often contain a memory of the time before Typhon. In Greek tradition it is the Titans who occupied Bronze Age Greece. The Iron Age Dorians who occupied the Greek peninsula after it was devastated praised Prometheus, a Titan, for bringing light to humanity but cursed him for tricking Zeus, their chief god, with the Titans' corrupt rituals. In Chinese culture it was the vices of the last Hsia emperor that brought on the trouble. After a lapse of two hundred years the Shang reorganized the people and took power.

Religions the world around would eventually show the imprint of the transition from the Bronze Age to the Iron Age with the advent of the fully invested male-dominated or all-male theology that blames old-time religion/science. Soon enough the knowledge of the event and its explanation became the special domain of high-priest intellectuals or of brotherhoods living in hermetically isolated monasteries. The high priests alone knew the secret name of God and the rituals necessary to appease the almighty. Cultures conjured up the power of the divine by their ferocious ritual appeasements. Human sacrifice was common to the great civilizations in the Americas, in

China, and in the Near East. Considering the costs that are paid when the gods go berserk and the world goes off its hinges, it seemed a small price to pay. It was certainly not worth the risk of doing without and the gods losing their temper again.

Many peoples became migrants with the impact of Typhon, especially around the Mediterranean. They followed lightning gods such as Zeus, Jehovah, and Indra. It was lightning bolts that were presumed to have defeated Typhon. Nowhere else on Earth do lightning gods assume such preeminence as here. Nowhere else are dragons visions of evil omen. These lightning gods were angry as hell. They were the only ones capable of leading the people out of the valley of the shadow of death. The character of these gods was the character of their people: angry, irrational, looking for revenge, and implacable. People and gods that were timid, depressed, and infertile did not survive.

On the one hand, the destruction caused by Typhon was truly massive, of sufficient size to cause regional devastation and years without summer around the world. Coming in at a severe, acute angle, Typhon jolted the oceans, seas, lakes, and rivers out of their beds temporarily and drove the crustal plates down and under one another. On the other hand, it was considerably smaller than the impact that brought the Stone Age to an end; that impact had been capable of triggering a major glaciation event, warming up the oceans, and greening the planet. In any case modern language simply cannot convey the magnitude of change such as this since all the disaster descriptives in the modern language have been usurped for comparatively minor events. One courageous, modern commentator who has come to terms with the reality of what happened 3,500 years ago has guesstimated that had Typhon not come our way, television would have been discovered by the Age of Christ.

Archeology marks the impact of Typhon as the transition of the Bronze Age to the beginning of the Iron Age. Archeologists don't recognize the cause, but they record the dark age that followed. In Greece the collapse of the Mycenaean civilization was followed by a dark age up to 800 BC. It was during this period that the value of a sharper blade became obvious and necessary. In the fury of a tragic birth the Iron Age was forged in the Near East, Iran, India, and Greece, the most important axis of cultural exchange on the planet. Even in the Classical Age around 300 BC the Greeks would still remember the Mycenaean Bronze Age as a Golden Age now long past. In the *Timaeus* Plato implied that ancient wisdom was still being lost. Egyptian priests were telling them that they didn't know their own history.

As we have already mentioned Typhon was the cause of a major gender change in heaven: the accession of the male to the complete power and responsibility. In the earliest, darkest days a female voice was sometimes heard. Pandora still had hope in her box even after the release of famines and plagues, but it was a fanatic kind of hope enforced by wars engaged in by the chosen people. Soon enough there was a fundamental shift in the gender roles in the human society. Males had already assumed extremely important roles in the Bronze Age required by the demands of specialization in high civilization. Now, as male gods settled into more powerful roles in heaven, males also begin to assume a more powerful role in the human family. As is the microcosm below, so is the macrocosm above is the ancient metaphysical axiom.

The most extreme case of theological reform stemming from the crushing impact of Typhon occurred in areas where its passage was observed and in those made homeless migrants as a result. A radical reduction and simplification of theology is the normal response of people in a state of diaspora at any time. Many who survived the collapse of Atlantis moved to the East. Zeus led his people out of Europe into Eastern Europe and the Crimea, down into the Greek peninsula where they would be called Dorians, and down into the Italic peninsula. Indra led his Central-Asian, Indo-European-speaking people into the Indian subcontinent. The Aryans occupied the territory in the Indus River system that had been the homeland for the Indus civilization. The native Indus civilization, which had been completely devastated, offered no resistance to this invasion. The Europeans would eventually mix with that ancient, native race.

Among the Indo-European-speaking cultures it was the Aryans that recovered first. High Indo-Aryan civilization arose well before time had had a chance to efface the physical and especially psychological effects of Typhon. The earliest roots of Vedic tradition have been traced to 1200 BC. When northwestern Europeans were still mostly living in wattle and daub huts these, Aryans had reformed their alphabet so that it replicated all the sounds that they spoke and produced a fully rationalized grammar in order to preserve the purity of their sacred texts. The West has still not achieved that. In the fully rationalized cosmology of Vedic tradition we find that the full cycle for an age is called a *Maha Yuga*, the Great *Yuga*. A *Maha Yuga* is 4,320,000 years by human reckoning but merely a single day or *kalpa* of Brahma. Every *kalpa* was subdivided into fourteen *Manu* intervals, each of these were comprised of seventy-one and a fraction *Mahayugas*, each of which terminates with a deluge. We recognize in these mathematical constructions the work of a highly rationalistic academic institution and an Indian mathematical mind that would anticipate western European intellect by thousands of years. These rationalistic calculations never seem to have become the basis for an apocalyptic movement in India such as we will see in northern Europe.

A relatively benevolent view of life was heralded around 600 BC with the appearance of Lao Tzu in China and Siddhartha Gautama in India. The weather was good; the sun was shining. Memory of a physical reality to connect the fantastical stories of Typhon's passing had settled into just-so stories suitable for the instruction and children. It was time to turn a critical eye on the abuses of high civilization. It was time to return to pacific Mother Nature to meditate and get one's life in order for a higher purpose. Buddhism would reform Vedic tradition and eventually create a religion devoted to redemption. While northwestern Europeans were still following their herds, Indian culture had reached an apogee and was counting the heavy costs. India developed the most profound tradition of self-realization and illumination of any culture by this point.

The Jewish religion born out of the impact of Typhon came to be the most complete masculinization of religious principle. The creation of the world is considered to have taken place 3760 and three months BC in the Jewish calendar. The pre-impact chapter of the story sets the stage for those Kurdish or mountain people of Jehovah to blame the polluted science and religion of Babylon for the impact of Typhon, thereby to inspire purification of its own base impulses, in preparation to embrace the return of the Messiah, and thereby to receive redemption as a people. The snake is emblematic of the Hermetic scientific tradition of Sumer. By the snake's seduction of Eve, it is said, the feminine principle became polluted.

The gods of the mountains are not the same as the gods of the valleys, and this is dramatically illustrated in the story of the offerings that Cain and Able gave to their father, Adam. In spite of their social differences, the historian of religion Charles Guignebert compares the day of Yahweh (Jehovah), when traditional idolatrous enemies were to be destroyed, with the elaborate End of All Days scenarios of Babylon and Persia. The Jews were not the only ones given to allegorical extremes; the flame of revenge ignited by Judaism would make the Torah the talking head for Christianity. That persecuted minority would have no interest in the more philosophical and even-tempered Talmud. They would build their message of peace and love on the sacrifice of a Jewish messiah that the Jews themselves ignored.

Christian academic theologians are part and parcel of the contemporary Uniformitarian worldview. To them the impact of Typhon never happened. They take the catastrophic aspects of the Bible, especially the Old Testament, as the work of the undeveloped mind: a primitive stage of their religion that has evolved up to our elevated ideal of a quiet, well-ordered universe. They associate the vengeful aspects of the charismatic church with a rustic level of spiritual and intellectual development. As we know from several cycles of civilization that have already transpired it has been the Old Testament that sustains Judaism and Christianity when high civilization falls, its well-ordered kingdoms broken and buried again. The perfumed view of life lasts only as long as the climate is moderate and the larder is full. Only under those circumstances do we focus on these words from Isaiah 65:17:

> *For behold, I create new heavens and a new earth: and the former shall not be remembered, nor come into mind. But be ye glad and rejoice forever in that which I create: for behold, I create Jerusalem a rejoicing, and her people a joy.*
>
> *The wolf and the lamb shall feed together, and the lion shall eat straw like the bullock: and dust shall be the serpent's meat. They shall not hurt nor destroy in all my holy mountain, saith the LORD.*

From the perspective of the First Jewish Kingdom, Jehovah, like Indra and Zeus, no longer strode the earth carousing and gaming with his people. He was safely retired to the Holy Temple on Temple Mount in the old city of Jerusalem. It was here and only here that the terrible, sacred name of the Lord could be uttered. Evolution in the story of Genesis has a timeless quality reflecting the great distance from those terrible days centuries earlier. It has the quality of being politically correct. The unseemly events or tribal rusticities of oral tradition are either edited out or seen as a metaphor. The Hebrew word for "day" that was used to define the different stages of evolution could also mean "a long time." It is an evolution that is as long as it needs to be. God's days are indeterminate. It is a Uniformitarian or high civilization story of evolution. In his book *The Book of J* Harold Bloom critiques this earliest level of Jewish written tradition and suggests that the editor/writer may well have been a woman, he interprets everything as metaphor as a good Uniformitarian must, and he reminds us that the correct name for the Lord God is Yahweh, not Jehovah the destroyer of nations.

The First Jewish Kingdom was short lived, however, and in the disorder that followed the reign of Solomon, the northern Kingdom of Samaria fell to the sword of Shalmaneser of Assyria and the southern Kingdom of Judea fell to the sword of Nebuchadrezzar of Babylon. The cream of the Jewish religious and intellectual upper class went into captivity. During the Babylonian captivity visions of apocalypse would flare up again in the Books of Daniel and Ezekiel. The Israelites were released from captivity by Cyrus the Great, founder of the Persian civilization. Following this, by around 500 BC, the Israelites would eschew human sacrifice by injunction in Leviticus and substitute circumcision as a mark of obedience to God. Maybe the failure of that profound act of appeasement to God to save them from the swords of Shalmaneser and Nebuchadrezzar had something to do with it.

In Greece the old thunder-god Zeus was cut down to size, humiliated, and put out to pasture by a culture devoted to theatrical tragedy, tragedy meaning "goat song." It was hubris on the part of Bronze Age Greeks that irritated the Titanic gods who brought on the flood of Deucalion dated at around 1400 BC according to this sardonic point of view. Apollo, the sun god, was the new hero of Olympus. Apollo was reasonable and organized. Aristotle would formally define a static, crystalline universe in which the rule was that nothing untoward ever fell out of the sky. He claimed that Plato invented the story of Atlantis as a metaphor for the invisible hand that renders evil forms of government. His universe was pure Uniformitarian.

The upper class of the Second Jewish State as well as the surrounding Gentile culture was comfortably in denial about the impact of Typhon as we approach the time of Jeshua. We find the Jewish god in the Book of Jubilees is a thoroughly decent, law-abiding god, and rather unapproachable. The lower classes, on the other hand, were still actively engrossed with a literal interpretation of the Jewish tradition. In the century or so before Jeshua, or as the Christians would know him, Jesus, full-blown Messianism had gripped a large portion of the Israelite people. Disgust with liberal Greek culture introduced forcibly by Alexander, fear of the new colossus of the West, Rome, and internal feuding threatened to undo the Jewish State. The prognostications of Daniel and Enoch seemed to be very relevant.

The Book of Enoch was of late authorship, but it marks an older tradition maintained by a celibate brotherhood. Enoch was the seventh patriarch from creation modeled after the seventh antediluvian mythical king of Babylonia, Enmeduranki. Enoch's prophecy stems from an older theory of history that emphasizes return. This is the principle of the ever-returning or reborn god, the seasonal god of agricultural peoples, but also the vengeful god rendering justice for oppressed peoples. Enoch represents a fundamentalist tradition. Jewish tradition through the influence of Enoch rejected most of the common methods of divination and omen-seeking, relying instead upon direct divine revelation. The prophet basically states what will come on the basis of what has come before. As reviewed by John J. Collins in the collection of works entitled *The Sage in Israel and the Ancient Near East*, it was through Enoch that the belief in reward and punishment after death first appears in Judaism. Enoch was decidedly apocalyptic.

The Sadducees were gathering together the canon of the Jewish Torah at the time of Enoch. Although the tradition of Enoch started in an urban, intellectualized upper class, the tables were now turned. Jewish high-class culture was not interested in returning gods. The Sadducees were Hellenized priests to the Jewish upper class. They left the Book of Enoch out of the canon, but they also left out the Book of Jubilees. With the climate of social violence increasing in the Second Jewish State, as the age of Jeshua approached there was a need for Yahweh, the irrational berserker of old, to be on call. It was a very fine balance. The Book of Enoch and the Book of Jubilees would not be accepted as a part of either the Jewish or the Christian canon a few centuries later, and they are stashed away along with other so-called pseudepigrapha. The compilers of the canon would take the middle road.

For the Jewish commons the hopeful fervor of the apocalyptic literature raised their spirits after the holocaust of the defeat of the Second Jewish State by the Romans. There was actually a greater range of difference within religions between the upper class and the lower class than there was between similar classes of different races. These apocalyptic ideas would also play an important role in the formation of Christian beliefs, about the resurrection and the Second Coming of Jesus the Christ in parallel with several classical myths of rebirth prevalent in pagan traditions. For the poor and the dispossessed the need for redemption would be personified by Jeshua, meaning "Jehovah is salvation."

The three wise men that set out to find the birth of a new god or the Son of God were triggered by an unusually close conjunction between Jupiter and Venus that resonated with the old stories about the fight in the night sky between two planets that preceded the impact of Typhon. One of those planets, of course, was actually Typhon. In addition, Earth's orbit was passing through the Taurid stream from 500 BC up to AD 0. There was considerable activity in the night sky. The wise men found Jesus of Nazareth, who was a country boy from an Aramaic-speaking family. When Jesus came of age in the cosmopolitan setting of Jerusalem he was obviously influenced by the stories of Enoch. According to the Synoptic Gospels (Matthew, Mark, Luke, and John) Jesus seems to have believed that the coming of the Kingdom of God and the time of Judgment were near. Little about his ministry attended to the building of a religious institution. Repentance was the theme of his ministry, as was a reform from the states of mind and material necessities that are schooled by and for the business of civilization. Jesus' statements, as reported in Mark include, beginning with 13:7, visions of actual catastrophe and are a chillingly realistic report of a cosmic impact event:

> *And when ye shall hear of wars, and rumours of wars, be ye not troubled: for such things must needs be; but the end shall not be yet.*
>
> *For nation shall rise against nation, and kingdom against kingdom: and there shall be earthquakes in divers places, and there shall be famines, and troubles: these are the beginnings of sorrows.*

But wo to them that are with child, and to them that give suck in those days!

And pray ye that your flight [to the mountains] be not in the winter.

But in those days, after that tribulation, the sun shall be darkened, and the moon shall not give her light, And the stars of heaven shall fall, and the powers that are in heaven shall be shaken.

Watch ye therefore: for ye know not when the master of the house cometh, at even, or at midnight, or at the cock-crowing, or in the morning: Lest coming suddenly, he find you sleeping.

After the ghastly crucifixion on Golgotha, astrological anomalies including Halley's comet came and went without an appearance of a messiah. Within a generation the basic message of Jesus was beginning to be adjusted to fit the requirements of a gradually unfolding history. The Gospel of Thomas is earlier than the four gospels of the traditional canon and, theologians surmise, contains the actual words of Jesus. The Gospel of Thomas uses the simple parables that are presumed to have been very close to what Jesus actually said, without any explanation. The Gospel of Mark is dated as falling within the generation following the earliest possibility for a Second Coming, and it is the earliest of the Synoptic Gospels. Mark subtly adjusted the parables into allegories, giving the stories told by Jesus a prophetic dimension that would predict what would subsequently happen to him in his own life.

One of the most important parables, found in Thomas and all the Synoptic Gospels, was the Parable of the Sower. It is reported that Jesus said, echoing the Book of Isaiah, that those who were initiated into the mysteries of the kingdom of God may see and may hear directly, but those outside of the mystery, particularly the Pharisees, would not. The Parable of the Sower described the sower sowing his seeds and how it came to pass that some seeds fell by the wayside and the fowls came by and devoured them up. Some fell on stony ground, where there was not much earth; these sprang up, but because they were not deeply rooted they soon withered away. Only those with ears to hear will hear; only those fitted to survive would know the kingdom of God.

As a result of the stunning act of self-sacrifice that the crucifixion represented, his words would survive. Primitive Christianity spread around the Mediterranean like wildfire and would evolve into the institution of the Christian Church that would carry the stories down to us, if not the messianic state of mind.

In the second book of his *Old Testament Theology*, theologian Gerhard von Rad proposed the thesis that the Jewish apocalyptic literature was the child of wisdom rather than prophecy, thereby collapsing the familiar dialectic of science versus religion. Prophecy is actually at the root of the human wisdom tradition that included the single most important discovery humans have ever made: the calendar. However, von Rad's thesis, that apocalyptic literature might be reality-based, is controversial among educated theologians and largely rejected or ignored the same way that the catastrophists in the discipline of science are generally rejected or ignored by the academy. The High Church follows the secular academy supposing that belief in meteor or comet impacts is prima facie evidence of heathenish superstition. These theologians portray Jesus as a healing sage or Soter, Enoch as a prophet of apocalypse. Religious fundamentalists and common folks share a belief in the literal possibility of ancient catastrophes, but they occur in the service of an allegory in which the chosen few will be saved at the Second Coming.

The rebirth of Greek intellectual culture after the dark ages following Typhon occurred in Ionia from 700 to 500 BC. Ionia was a confederation of Greek city-states lining the Anatolian coast of the Adriatic Sea. Today we demean this Oriental Greek culture as fatalistic. Democritus summed up the Ionian thesis of Atomism. Democritus believed in the evolution of human society. He believed that at one time humans lived in caves without clothing, houses, fire, or domesticated animals. At that time Homeric song men were still reciting the old stories from memory in country villages, in the market places after business had finished for the day. The stories were of Odysseus's encounters with peoples reduced to Stone Age conditions, survivors of Typhon. These astonishing tales filled the night air with wonder as goatskin bota bags filled with retsina were passed around.

Democritus was one of the most widely traveled men of his time. He traveled throughout the Middle East, India, and Egypt. He was very impressed by the Egyptian mathematicians he encountered. When he moved to Athens to continue his meditations on the human condition he was eclipsed by the local superstars. Unlike Plato or Aristotle, who operated from presuppositions, *purpose, prime mover,* and *final cause,* Democritus was a strict determinist and believed in the operation of natural laws. He asked about the cause of events, not about the purpose for events. Democritus supposed that originally the universe was nothing but tiny atoms churning in chaos. They collided to form larger units including the Earth and everything on it. He supposed that the Earth was round. He surmised that the universe held many worlds, some growing, some decaying, some with no sun or moon, some with several. He held that every world had a beginning and an end, and that worlds could be destroyed by collision with other worlds. This, of course, undercut religion and state-sponsored rationalistic science. The Athenian philosophers that the

West remembers, with notable exception of Socrates, tuned their theories to the ear of their patrons, the wealthy class. Democritus and his Ionian predecessors are still little known today in the West.

The hero for lower-class Greeks in the Golden Age was popular throughout the Mediterranean area—Heracles. Heracles was much beloved for defeating such chthonic monsters as the Learnaean hydra, a Medusa-like monster. The hydra was supposed to be the offspring of Typhon. The existence of two Medusa-like monsters comes from isolated traditions whose experience of a common event was collected centuries later. Heracles suffered from his association with the gods of Olympus and their kind. After Heracles killed the Cithaeron lion for the king of Thespiae, the king offered Heracles his fifty daughters for the night. The man of mighty deeds performed his duty and impregnated them all. This is sometimes called his thirteenth labor. Impressive as this might have been, the commoners would turn to the Jewish prophet Jeshua as their new hero.

With Hercules's greatest accomplishments in the past, Jesus offered hope for the future by a Second Coming. The savage reality and fear of a cosmic impact was transubstantiated into a future miracle. Paul's Christian Church attracted the oppressed and the lower middle class. Greek culture did not survive long on its highly skilled art and drama, its finely tuned philosophy or material wealth. Greek culture would survive as a social entity by absorbing a version of Jewish religion reformulated around its own traditions. Christian tradition would eventually inherit the remarkable speculative philosophy of Socrates, Plato, and Aristotle a thousand years later and mark the beginning of science with the Golden Age of Athenian Greece.

The theme of class was a central issue during the period when the Christian canon was being assembled following the conversion of the Roman Empire to Christianity by Constantine. The center of Constantine's new Holy Roman Empire was in the East. Rome was left at the altar when he established the capital at Constantinople in 330. He built the new Church of the Holy Apostles on the site of a temple to Aphrodite. Among the books under consideration for the Christian canon was The Book of Revelations of St. John the Divine, the epic poem of John of Patmos. Revelation fairly dripped with apocalyptic, occult, "End of All Days" imagery. And it was full of mathematical harmonies, so it was relatively believable. Revelation resolves on a note of transcendent peace and beauty for the true believer in the bejeweled, heavenly city of Yerushalayim. No rational philosophy of life has so sweet an ending, but Revelation's prophecy of violent destruction of the Evil Empire from above was not a suitable theme for high civilization in its days of wine and roses.

The Book of Revelations would not be included in the Greek Orthodox

Bible. In fact the Christian Roman Empire at Constantinople would opt for the Code of Diocletian, written in AD 300 during the last days of Imperial Rome. It was part of the most massive persecution ever directed at Christians by pagan Rome. The Code decreed the death penalty for astrologers and prophets of hell and damnation as being a danger to the state. Those prophets of doom were largely Christians. The Book of Revelations would be included in the Latin Bible, Revelations becoming its last chapter. This occurred over the protests of the first great Latin intellectual, Aurelius Augustinus, but Revelations gave impoverished Romans hope for revenge for their lost prerogatives.

St. Augustine was an admirer of Islamic mysticism before he converted to Christianity. This prepared him to be receptive of Plato's concept of essences behind the appearance of things. He would decry a literal translation of Ovid's myth in *Metamorphoses* of Phaëton, who drove the chariot of the sun too close to the Earth, let alone the seven-headed, ten-horned, Red Dragon in Revelation whose tail drew one third the stars of heaven and cast them to the earth. He would claim that the only metamorphosis worth considering was the subtle magic of transubstantiation: the change of bread and wine into the body and blood of Christ. He saw the horrific prophecies of Revelation as an allegory. St. Augustine would agree with Aristotle that atmospheric phenomena such as meteors merely hung in the air and were signs to be interpreted by scholarly astrologers. This would remain the position of educated Christians all the way to the twentieth century.

St. Augustine would break with the ancient theory of return and the relevance of prophecy and by that decision plant the seed of linear history and reasonable progress. St. Augustine supposed that a symbolic concordance existed between the events in the Old Testament and the process of Christianity from Christ to the establishment of the Kingdom of Heaven on Earth. He created a formal grammar of past, present, and future for high Christian culture. St. Augustine marks the beginning of a secularized, intellectual Christian culture. The highest goal of Augustine's theological posture was personal salvation: the capacity to feel the love of God. St. Augustine is one of the few saints that are common to both the Latin and the Orthodox Churches.

Waves of re-occupation had started to come back into devastated western Europe following the impact of Typhon as early as 1200 BC, beginning with the Celts from Bohemia. Celtic Europe experienced a halcyon age in their rustic simplicity and tribal vainglory until the more civilized Romans came north beginning around 100 BC to civilize them where they could. New waves of Germanic barbarians from further east burst across the Rhine in 406 to bring on a dark age for Celtic culture. Only Ireland escaped this holocaust. Rome, in its reduced and humiliated state, and Europe in general were teeming with Germanic barbarians from the East. The early Latin Church would outlaw science. Individuals did what they had to do to survive. It was Eastern Byzantine Christian culture that would flourish with creativity and invention as a high civilization in the first centuries after Christ.

A peak of Orthodox culture in Constantinople was reached with Justinian around 500, which coincided with a nadir in the west that occurred when Justinian's general Belisarius defeated the Goths over the bleeding carcass of the old Imperial city. Rome, once a city of a million, was reduced to 40,000, half of whom were paupers. Rome was in dire need of supernatural rebirth. While Justinian revived the Code of Diocletian, which outlawed prophecy as a pernicious and false enterprise, the hammer of revenge as portrayed in Revelations was a lively theme in the Latin Church. With no actual history to muddy the issue for the religious fundamentalist, every catastrophe was a revelation that the return of the Savior was nigh. This was the good news of Christianity energetically prophesied by the wandering prophet. The vision of apocalypse would supply the Latin Church with its missionary zeal, and it would eventually make the Latin Church the stronger and more widely spread of the two churches that were built upon the rock of St. Paul.

A serious revelation would occur in 535 when Krakatoa exploded in the Sunda Strait between Java and Sumatra. This was considerably bigger than the well-known Krakatoa explosion of 1883. The sulfate ash created a reflective layer in the atmosphere and a significant climatic aberration reflected in tree rings around the globe. It was the most severe short-term episode of cooling over the past 2,000 years. David Keys chronicles the results in his book *Catastrophe: A Quest for the Origins of the Modern World*. The bad weather went on for decades; the effects on human culture lasted for centuries. The Plague of Justinian brought about by the cold weather started the decline of the Eastern Roman Empire. Lack of forage forced the Avars to migrate to the west with devastating results. The collapse of the opulent Yemenite civilization in southern Arabia caused population to shift north to Mecca and Medina, motivating the apocalyptic rise of Islam.

In the fifth century the Celtic and British chieftains had been abandoned by Rome. According to Keys, the Celts had continued to trade with Rome. From the Romans they got the plague, and they were incapable of defending themselves from pagan Angles and Saxons. The Celtic magus behind the throne of Arthur, Merlin, would render an existential, apocalyptic prophecy that reflected the times. As described by Norma Lorre Goodrich in *Merlin*, Merlin's prophecy took no sides and contained no hope. The listener to this prophecy was not coerced into any belief, so no disappointment could occur. Removed from any fixed associations of person, place, and time, the prophecy works at any time, and it strikes a sympathetic chord as one approaches the end of one's lifetime. The main theme of western Christianity through this period would be that the inhumanity and brutality of life mattered little except as a test of faith and virtue whose reward was eternal life in paradise.

The flowering of the Indo-European–speaking peoples had already occurred and gone to seed two or three times by now in the Indian subcontinent. The Indo-Iranian moment in the sun in Persia had occurred during and following the collapse of the First Jewish State. The Persians met their comeuppance at the hands of their language cousins the Greeks as they began to return to the level of Bronze-Age Mycenae. The complete collapse of the Persian Empire followed the eruption of Krakatoa. While the Mediterranean peninsular cultures of southern Europe were flowering, northwestern Indo-Europeans were still living in wattle and daub huts. But the glory that was Rome had given those people a taste of high civilization, they were learning to fight fire with fire, and the weather was getting better again after the setback caused by Krakatoa. As we know, to everything there is a season, and a time to every purpose under heaven.

It was with Charlemagne in the ninth century that northern Indo-Europeans were finally beginning their metamorphosis and western European history passed from its Dark Ages to its Middle Ages. Germanic Roman Catholic culture had a glorious flowering. Magnificent cathedrals were constructed to the greater glory of God. In the Royal Portal at Chartres Cathedral in France were sculpted images of Euclid, Cicero, Aristotle, Boethius, Ptolemy, either Donatus or Priscian, and Pythagoras quietly playing on his tintinnabulum. The rendering of these pagan intellects, especially Pythagoras, on the sacred edifice from an earlier southern Mediterranean cultural flowering reflected the sophistication of this high culture.

Pythagoras, that sixth century Ionian Greek mathematician, was a bellwether for the state of mind of the intellectual institution. His ideas were remarkably inventive, but irritating. Pythagoras limited attendance to his school to those who he thought were capable of adopting and living by his precepts. The Pythagoreans followed practical suggestions for lifestyle as to diet, exercise, and thoughtful living, and women were treated as equals. With the climate warming and hay in the haylofts along the Danube in the Middle Ages, a plump and happy baby Jesus rolling in the fulsome lap of Mary was emphasized, not the tragic, crucified Christ hanging on the cross with his crown of thorns. This European culture tolerated its own giants of spirit and intellect such as Hildegarde von Bingen, Joachim of Flora, Meister Eckhart, and the greatest Latin intellect since Augustine, Thomas Aquinas. Aquinas held that there was an objective moral order to life that reason could understand, divine revelation was not necessary, and that we had the free will to follow it or not.

But the good times were not good enough. They never are. Irish philosopher Johannes Scotus Eriugena distilled Irish Catholic religion into the word Nature: Nature being a synonym for reality, and reality being a continuum of all of God's creatures that are theophanies of God himself. In 1225 Pope Honorius III ordered all copies of John Scotus's gnostic *De Divisione Naturae* to be burned. At the heart of the Latin Church is the definition of the Trinity. This is not explicitly defined anywhere in the Bible; it comes through church fathers, such as Origen and Augustine, who were influenced by Plato. Plato declared that man is a trichotomy consisting of body, mind, and soul. The Trinitarian concept of Christianity applies only to the divinity that is a coequal, co-eternal, and con-substantial nature to the Father, Son, and the Holy Ghost. This triune divinity is separate from the Aristotle's celestial sphere and acts upon it from the outside.

With the modest overpopulation occasioned by the good weather, there followed the idea of crusades to retake the Holy Land from the infidel. The center of the universe was still Jerusalem. The feedback from these adventures caused Europe to open up to high Islamic culture. This enthusiasm led to local outrages. Innocent III launched a Crusade against the refined Jewish-Moorish-Christian, Albigensian culture in the Langue d'Oc region of southern France. The Albigensians practiced a heretical version of Christianity called Catherism that did not believe in the divinity of Jesus. This was the definition of the Trinity held by the Orthodox Church. This Crusade killed three birds with one stone: a Christian heresy, a flourishing Jewish community, and a community heavily influenced by North African Islam. The extermination of this culture was accompanied by the formal institutionalization of the Inquisition, whose role would soon be expanded with a change in the weather.

The warmer climate had increased the melting of the glaciers in Greenland. The ninth through the eleventh century also saw the outburst of Viking Norse peoples from the far north of Scandinavia. Looking for more land, the Vikings introduced settlements at the southern fringes of that glacially impacted island around 1000. But the melting of the glaciers eventually caused the oceans to cool. Lloyd Keigwin at Wood's Hole has found evidence from ocean cores taken from the sea floor plateau called the Bermuda Rise of ocean cooling reflecting a drop of four degrees Fahrenheit correlated with the period now called the Little Ice Age. All hell was about to break loose.

An "aerolite" fell in 1492 at Ensisheim, in the Upper Rhine, in the presence of Maximilian I, the Hapsburg Holy Roman Emperor. In a culture imbued with the expectation of a Second Coming, the church and Maximilian himself considered it a miraculous event. Although the details have been forgotten and the halo has faded from the Ensisheim Donneraxten, a plaque affixed to it contains a statement that is still relevant: "Concerning this stone many have said much, everybody something, no one enough." Although this was a singular violation of Aristotle's rule that nothing ever fell out of the sky there were more serious issues to attend to at the time. Of greater import was the climatic catastrophe that had been going on for nearly two hundred years: a precipitous downward slump in the climate that started roughly around 1315. The Little Ice Age provoked a sudden rash in European culture. It precipitated a religious reformation of the Catholic Church, it motivated the Age of Discovery, including the voyage led by Christopher Columbus to the New World, and it would eventually give birth to a new science not beholden to classical models.

CHAPTER 2
THE ENGLISH ENLIGHTENMENT

BEGINNING WITH THE CROP failure and famine of 1315 the bloom was off for Europe. For the next four hundred years there was nothing little about The Little Ice Age for the people who lived through it. According to historians such as Barbara Tuchman, the so-called "Little Ice Age" did as much damage as an all-out atomic war would have done. In one striking example of the change, during the age of Elizabeth I, Raleigh, and Shakespeare the River Thames would freeze over most years. Dirty sea coal that had been banned from use in London on the pain of death in the Middle Ages was now tolerated. This created the killing London Fog.

Brian Fagan points out in a recent review that it was not simply a case of deep freeze that so disrupted Europe. Frigid winters gave way to torrid summers and storms, with the climate veering from floods to droughts. This general instability was reflected by waves of famine, disease, and war. Despair was followed by exuberance, only to be crushed again. The Italian Renaissance was a very short-lived, manic phase of the manic-depressive cycles that now gripped European culture. Philip VI of France and Edward II of England would continue the nasty little secular conflicts that had been going on between the French and the English. The conflict begun in 1338 would drag on mindlessly for a century and become a religious war, for only God could justify it. It would be most remembered for the tragic appearance of Joan of Arc.

Ice Age Europe looked like an Old Testament–style condemnation of the

Roman Church by God: crops lay blighted in the field; arms and legs withered and dropped off from St. Anthony's Fire; rats swarmed everywhere looking for food, their fleas carrying a bacterium that infected them; human beings marked with the tell-tale buboes of the plague were nailed up in their houses and burned alive; flagellants swarmed through the streets whipping the air into a red haze of blood in penitence; and the armies of God preyed upon the survivors. And people could not pay their taxes. The plague was no respecter of power, status, or education.

All civilizations face the problem of unwonted knowledge. The Jewish banker, the wandering pig castrator, the barber surgeon, and the gypsy midwives who provided opiates to ease the trials of birthing in Medieval Europe all performed profane but necessary chores. They were outside genteel culture. Their work fell under the heading of black magic. These black magicians had filled in behind the scenes where the polite society dared not go. It was thought by some that these practitioners were the impurity for which God was punishing all society. Those who operated under the cover of darkness were persecuted. Witches were burnt at the stake along with their cats, shrieking imprecations or pleading for forgiveness. Having no inkling of the cause of the plague—rat fleas—many pointed the finger of blame in the direction of cats. But the sheer magnitude of the fall called for a bigger scapegoat. Martin Luther was an Augustinian monk. But instead of appreciating the pagan culture of Classical Greece, as did St. Thomas Aquinas during finer weather, he turned to the Old Testament. The Old or Hebrew Testament that had blamed profane Gentile culture for the collapse of life after the impact of Typhon now became the script turned against the high, Latin, Christian Church. To Luther the decay of the Church had begun in the ninth century when the Pope acquired the secular power of a landowner, thereby giving him the power of a king. In the original Christian state of Constantine there had been a clean separation between church and state. The pagan state had carried on the profane business of secular life, the ugly business of civilization. The Roman Church of Luther's day was struggling to survive the worst ecological calamity in at least a thousand years. Among its numerous attempts to raise money, Rome was selling indulgences. Grace came to the highest bidder. Luther also argued that reason was so corrupted by the Tomist church that pure blind faith in the divine was humanity's only hope for salvation.

In the underground of eternal discontent, common folks were roiling over the prophecies of Merlinus Ambrosius. The tradition of Merlin goes back to the holocaust of Germanic tribes invading western Europe and defeating the aboriginal Celtic peoples during the Dark Ages. The update of that aggrieved tradition resulted from thirteenth-century Italian politics. At the Council of Trent the Catholic Church put the dire prophecy of Merlinus Ambrosius on the forbidden list. Privy to such excitement, Martin Luther decided to reform religion the way Jesus had once attempted to when he threw the moneylenders out of the Temple, but Luther's most revolutionary act was to translate the Bible into vernacular German. Luther finished translating the Bible in 1534. The Old Testament and the Book of Revelations came out of a state of relative dormancy through the power of oral recitation that the people could understand. It would give the Puritan Reformation its cleansing moral authority and its courage. The Catholic Church now found itself to be of object of Old Testament-like prophecies.

In both the Protestant tradition and in the Reformed Catholic tradition, the enlightened period that produced Chartres Cathedral would be remembered as the Gothic period. The high hopes and civilized culture reflected by Chartres would be suppressed or simply forgotten. There would be no more Crusades, or at least they were smaller undertakings. In the far north where conditions were the worst, Christian sects were mirroring the ruthless, seemingly random destruction, proclaiming that it was simply impossible to know who the chosen people were and that good works were a vain pretence of no consequence. You were either born among the chosen or you were not. It is very hard for us to imagine the conditions that produced such a state of affairs.

There was a steep decline in education and in the educated, but out of the desperation and curiosity of a brave few the received wisdom of Greek philosophy began to be superseded. The first inklings of a modern scientific method began to appear in the alchemical methods borrowed from Islam. Medicine would look furtively over the shoulder of Andreas Vesalius and his friends the grave robbers. William Harvey solved the mystery of the heart that had so perplexed antiquity. The use of coal as fuel created the need for better flues and a higher grade of cast iron for grates and stoves. Cutlers who hand-finished knives from start to finish began to give way to blacksmiths who specialized in interchangeable parts mass-produced. The Industrial Age is officially considered to have started in nineteenth century England, but its inchoate beginnings are to be found laboring over forges during the Little Ice Age.

The Polish cleric Nicolaus Copernicus took on the chore of attempting to establish the length of the year and the date of Easter. It appeared that the seasonal rituals were not timed properly. Among the books he read was the work of the Greek astronomer Aristarchus, who proposed a sun-centered universe in the third century. He would throw out Ptolemy's Aristotelian Earth-centered cosmology for a solar-centered system in order to refine the calendar. On the page where he depicted the new arrangement of the solar system, in his *De revolutionibus* published in 1543, there is a paean to sun worship. Copernicus would invoke Hermes Trismegistus, the Greek version of the Egyptian god Thoth, as the origin for his insights. Renaissance scholars identified Thoth with ancient theology—*prisci theologi*. Copernicus wrote in Latin, the private language of European academics. Knowing that Aristarchus was chastised for his impiety by the Latin academy, Copernicus also expected scorn and derision and waited until he lay on this deathbed to publish.

Copernicus's work might have buried as a mathematical correction that improved the calendar, but these were revolutionary times. On the evening of November 11, 1572. Tycho Brahe was on the road, on the way to a dinner. He happened to look up to see an object with the brightness of Venus in the constellation Cassiopeia. He asked his servant if he saw this apparition. He did. He asked some other travelers. They did. For a few weeks this new star was bright enough to see during daylight. At the time every stargazer on the planet noticed this celestial event. The stargazers in Europe, fed on a diet of Aristotle or his follower Ptolemy, knew that changes in the heavens of this sort were supposed to be impossible. The Aristotelian conception of a celestial sphere—immutable, incorruptible, and superior to the terrestrial region—was the law of established aristocracy and of the High Church.

Concerning this event, Benjamin Woolley states in *The Queen's Conjurer*, "The history of modern science tends to be portrayed as a series of brilliant ideas that formed more or less spontaneously in the capacious minds of individual men of genius." He goes on, "Usually, however, developments emerge in a far more haphazard fashion, their full significance remaining historically invisible for decades or even centuries."[1] In addition to this insight we also note that cultures, faced with certain hypothetical conditions, tend to behave in very stereotypical ways that are directly attributable to their traditions. In a single stroke a cosmic flare had oriented all of the astrologers of that time in the charged environment of Little Ice Age Europe. The transformation of astrology into astronomy received a kick in the ass.

The work of Copernicus would be embraced and enlarged upon by Giordano Bruno. Bruno was a Dominican who became a Hermetic philosopher, a poet, and firebrand in the name of social change. He had written of an infinite universe that left no room for that greater infinite conception which is called God. He could not conceive that God and Nature could be separate and distinct entities as visualized in Genesis, as taught by Aristotle and adopted by the Church. He encouraged the study of Nature's footprints in the pursuit of knowledge rather than importing received wisdom. Bruno, Copernicus, and Galileo mark the beginning of a rebirth of the tradition of Greek philosophy that had been edited out by high culture—the Ionian tradition of Democritus. Bruno spoke of a heroic love in proportion to the heroic suffering that had come from the separation from God in life. This was too visionary and revolutionary for normal consumption. He was not, as history generally records, burned at the stake for advocating the heliocentric planetary system of Copernicus. The Catholic Church had no official stand on the question at the time. He was burned at the stake in Rome in 1600 for denying the divinity of Jesus Christ. His God was too big for the church of the sixteenth century.

In 1604 another new star flared up in the constellation of Sagittarius. Galileo Galilei was teaching astrology to medical students at Padua at the time. Not only was he familiar with Vesalius's affront to Galen's received wisdom, he also knew of Copernicus's challenge to Aristotle's astronomy. With the newly developed astrological instrument the telescope, Galileo pierced the veil of ideology and saw the machinery of a different cosmos in the phases of Venus, the craters of the Moon, the moons of Jupiter, and the spots or clouds that disfigured the perfection of the sun. Galileo was no outsider. He was a much-celebrated philosopher in the house of Cosimo II of Florence. He was a jewel in the crown of the Italian Renaissance. During the papacy of Paul V, the Holy Office of the Inquisition instructed Galileo to attend upon Lord Cardinal Bellarmino because of his writings. Bellarmino admonished Galileo to cease and desist from treating the sun-centered universe as a natural law. He could, as a private citizen, only propose a thesis. There is nothing new about this. The state has always maintained its prerogatives over individuals down to the present day on matters deemed hostile to national security.

Although the intellectual excitement of discovery was intense and the emotional satisfaction of belittling academic pedants was intoxicating, Galileo was no Bruno. He would limit himself to quiet observation. Galileo makes it clear in a letter to the Grand Duchess Christina of Tuscany in 1615 where the bulk of opposition to his ideas came from:

> *Some years ago, as Your Serene Highness well knows, I discovered in the heavens many things that had not been seen before our own age. The novelty of these things, as well as some consequences which followed from them in contradiction to the physical notions commonly held among academic philosophers, stirred up against me no small number of professors—as if I had placed these things in the sky with my own hands in order to upset nature and overturn the sciences. They seemed to forget that the increase of known truths stimulates the investigation, establishment, and growth of the arts; not their diminution or destruction.*

Intellectuals generally conform to the formalities of their age, intellectual ideologies are generally justifications of their culture, and intellectual institutions are very jealous of these traditions. Intellectuals had carried on the tradition of Aristotle for 2,000 years even through it was riddled with obvious flaws. It was Aristotle's Earth-centered universe translated through Ptolemy that was the basis for astrology. In those days the well-to-do could hardly move without casting an astrological chart. There was a steady and lucrative business in chart writing among academic philosophers. Galileo also flouted academic tradition by writing in Italian. "I wrote in the colloquial tongue because I must have everyone able to read it," he wrote. "… I want them to see that just as Nature has given to them, as well as philosophers, eyes with which to see her works, so she has also given them brains capable of penetrating and understanding them."[2]

Generally missed by historians is a third offense committed by Galileo. Galileo cited the Greek philosopher Pythagoras in a positive manner. With the change in the weather Pythagoras now stood out as a symbol for the intellectual radical. Now it was remembered that the followers of Pythagoras were in direct competition with early Christians for the hearts of the poor and oppressed. Father Riccardi, in charge of reading Galileo's *Dialogue*, described the sun-centered universe of Copernicus and Galileo as Pythagorean arguments. Pythagoras was among those who believed in the Earth-centered universe, but that is not the point. It's who your friends are that is the point. Down to the present day in academia Aristotle is still the godfather of normative science, Plato is on the cusp of rejection, and Pythagoras is so suppressed that he is nearly forgot except for the Pythagorean theorem.

In Tuscany where Galileo lay at Bellosguardo the clouds of oppression and fear dispersed momentarily when his friend and fellow stargazer Cardinal Barberini became Pope Urban VIII in 1623. Upon his accession Urban had immediately begun to feel the greater burden of public responsibility that is borne by a pope. Urban was under enormous pressure to back the powerful Spanish Catholics against the newly minted Protestant Germans. Urban had studied under the Jesuits at Collegio Romano and admired the style with which Galileo skewered conventional scholars on his similes, such as *I believe that good philosophers fly alone, like eagles and not in flocks like starlings*. Galileo would return to the issue of the solar system with his *Dialogue*. Galileo was allowed to offer public proof for the Copernican theory of the sun-centered planetary system by his protector and friend on the condition that he presented convincing counter-arguments. It was assumed that this would lead to the impossibility of establishing physical proof for any particular cosmological system. The pope assumed that there could be no physical evidence that would ever establish Galileo's thesis. This would leave tradition safely in place by default, and intellectual freedom would appear to have been served.

Galileo would, however, violate this trust. Galileo's argument for the conventional Aristotelian universe in the *Dialogue* was put in the mouth of Simplicis. Simplicis was a pompous ass. Also, again, he would write in Italian. In the words of one of his Jesuit inquisitors, Melchior Inchofer, "he writes in Italian, certainly not to extend the hand to foreigners or other learned men, but rather to entice to that view common people in whom errors very easily take root."[3] This was the final undoing of Urban's tolerance. The silent killer, plague, was walking the streets in northern Italy again. Copies of the *Dialogue* could not even be delivered directly to Rome without being dismantled and fumigated. It was no time for new ideas. Galileo was forced to recant his defense of Copernicus publicly. His ideas had remained the same, but the times had changed and he had lost the protection of the Pope.

At no time was the basic, upper-class, Uniformitarian perspective of Aristotle—a fixed, mathematically defined, divinely ordered—challenged by Galileo. In the debate over the passage in Joshua 10:12–14 where the sun is described as halting in the middle of the sky, Galileo argued that the sun ceased to rotate and that this caused the Earth to stop rotating by the will of God. Galileo Galilei was devout Catholic.

In France, René Descartes followed the trial of Galileo with great interest. Given the result, he would decide to shelve a book project of his own on the cosmos. Burning at the stake was a real possibility, or worse, exile to Sweden. Descartes would fashion a scientific philosophy that fitted with the requirements of the Church with the assistance of his mentor, the Jesuit Father Marin Mersenne. The triune divinity, we recall, is separate from the cosmos and acted upon it. Man's connection to this divinity is conditional and limited. Man's relation to this divinity stems from Adam and is sanctioned by the Church. The rest of the manifested universe is outside this limited blessing. Descartes supposed that mind was spirit, which is to say godlike, but not coequal, co-eternal, or co-substantial with God. By separating mind-spirit (*res cogitans*) from matter (*res extensa*), Descartes separated himself from the Hermetic tradition—represented by secret brotherhoods such as the Rosicrucians and the Freemasons—and the Cabalist tradition of Judaism. These traditions sidestepped the Trinitarian axiom of the Christian Church. The Hermetic tradition viewed matter and spirit as in a state of coextension and inseparable throughout creation.

By dividing the world into matter (the domain of science) and spirit-mind (the domain of the church) Descartes was acceptable to the Catholic Church. With Galileo under house arrest in 1637, Descartes published his *Discourse on the Method*, where he would say, without fear of condemnation, *cogito ergo sum*: "I am a thing that thinks". The God-given mind-spirit conceived of by Descartes would use logical induction to reveal the mechanical workings of the body. Control of the flesh by the spirit was an old theme of the Church, which preached that the mind-spirit lived on after the death of the body. This was the basis for the Christian formulation of resurrection on Judgment Day following the Second Coming. Descartes was, therefore, a Christian in line for redemption. Having received his security clearance, Descartes would take up the motivational vision of the Rosicrucians and proclaim a rebirth of human culture based upon scientific-mathematical principles.

Rosicrucians, as well as those who were spirit mediums, such as conjurers and astrologers or alchemists bent on changing the essence of lead to gold and even distilling the spark of life itself, had to work by the dark of the moon. They were tolerated or even quietly encouraged by two reigning monarchs, Elizabeth I and Rudolph II. The Protestant followers of Martin Luther frowned upon these doings. The Catholics bound such devil-worshipers over to the flames of the Inquisition, yet the fact of the matter is that the Jesuits and the Rosicrucians were almost mirror images of each other. The Jesuits led the Counter-Reformation movement for intellectual reform. As Frances A. Yates writes in her book *The Rosicrucian Enlightenment,* which covers a little known area of European history, the Jesuits followed Hermetic tradition closely without giving up the Trinitarian basics of the Church.

Descartes is often called the father of modern philosophy. He was the leading intellectual spirit of the Counter Reformation. He proposed a static universe that would have satisfied Aristotle. Geometry was his philosophical algorithm. Evolution played no role in his cosmology. He designated a pecking order of the scientific disciplines: the hard or inflexible sciences came first; the social sciences, such as they were in his day, were mere pretenders because they had not been reduced to mathematical formulae. This was the Counter Reformation. It would become acceptable to the Puritans, and it is the scientific hierarchy followed to the present day. The most famous mathematical innovation of this Church philosopher is the Cartesian coordinate graph with its X- and Y-axes. This gave Aristotle's dialectical Rule of the Excluded Middle a mathematical expression. It is still the most common method of demonstrating truth in the modern academy. It is a process whereby imperfect data is rationalized so that a straight line can be drawn to the truth.

Dr. John Dee was an advisor to Elizabeth I of England. He was another follower of Hermetic traditions. He converted the divinatory astrology of Nostradamus to a natural astrology based upon mathematics. Dee introduced the Copernican universe to the English. English professors were no more interested in new ideas than Italian professors, according to Benjamin Woolley in his biography of Dee. Dee assured Elizabeth that a major London earthquake that rang all the church bells in unison and other disturbing astral events were only portents of her glorious reign. He knew better than to be the messenger of unhappy portents. In his *General and rare memorials pertayning to the Perfect Arte of Navigation* he promoted the idea of a "Brytish Impire." In his *Mathematicall Praeface to The Elements of Geomitrie of Euclid* he endorsed the observation that two bodies of a different weight fall at the same rate. This observation is generally cited as proof of Galileo's pioneering role in scientific inquiry.

Dee was associated with occult practices, however, especially in communication with the spirit of Uriel, the angel who revealed the astrological secrets of the heavenly luminaries to Enoch. Dee attempted to channel the Book of Enoch through his spirit guide Edward Kelly, written in the language taught in Paradise to Adam. Such activities were widely believed in and feared at the time. In these undertakings Dee finally did offer his soul as a pawn for knowledge. A play based on Dee's life, *The Tragical History of Doctor Faustus*, was written by Christopher Marlowe. It deals with the hubris of the human attempt to understand its existence. Puritan polemicist William Prynne reported years later that two devils actually appeared on stage at a performance of the play.

Dee's decline from a position of high regard in his own day resulted in part from his association with traditional astrology and the predictions made by his fellow astrologers that the conjunction of Saturn and Jupiter in 1583 would culminate in the *annus horribilis* of 1588: the much anticipated catastrophic collapse of Elizabeth's reign. Elizabeth had always been shadowed by dark mutterings due to her tolerance for heterodox points of view and her unconventional rule. With her piratical navel captains using unconventional methods of naval warfare, Elizabeth defeated the Spanish Armada in 1588 with a large assist from the weather. As it turned out, 1588 was one of the most brilliant years in a long and brilliant reign for Elizabeth. As John Lyly would put it, "He [the astrologer] told me a long tale of Octogessimus octavus, and the meeting of the Conjunctions & Planets, and in the meantime he fell backward himself into a pond."[4] Dee died in poverty and obscurity. The modern academy is very sentimental about its received traditions. Individuals such as Woolley may be able to restore Dee to his important role in the evolution of the English intellect, but the modern intellectual institution already has the story it likes.

The death of Elizabeth, and shortly after of Rudolph, created a crisis for the Rosicrucian societies and other free thinkers. In the upset over the loss of balance of power caused by the death of Rudolph, the Thirty Years War began. This brutal war was motivated by the now well-established religious dialectic of Catholic versus Protestant. The Thirty Years War would drain the energy for further ideological conflict, but leave in its place a Protestant northern Europe and a Catholic southern Europe. The various states of Europe could now turn to consolidation. There had been some adaptation to the cold weather, some easing of the ecological situation, but free thinking remained a dangerous undertaking everywhere.

Francis Bacon in England was Lord Chancellor to Elizabeth. He made popular the first principles of experimental science, that it should free itself from the traditional biases of interpretation: the Idols of the Tribe, the Idols of the Cave, the Idols of the Marketplace, and the Idols of the Theater. Under James I, who followed Elizabeth, he was more circumspect, even though James was a Freemason. Bacon and did not allow the publication of his *Utopia*, basically a Rosicrucian vision of the possibilities for a scientific society, until after his death. Among the English educated and upper class the interest in violent prophecies, such as the "End of All Days," was gradually ebbing. This would leave the theological conundrum of rationalizing the depiction of such violence in the Old Testament.

Even the Puritans were beginning to thaw. It was decided that although it was uncertain who the chosen people were, it was likely that those who achieved material success in this life were those people. The Puritans began to resonate to the siren call of Democracy, One God over All. The fire of Puritanism brought on the English Civil War and the beheading of Charles I in the name of the one true religion. This was a bloody contest between High Church Anglicanism and Middle Church Puritanism. The Puritans won, but the country was exhausted after twenty years of the rule of Oliver Cromwell, and the English restored their regnal head with Charles II. With a restored aristocracy and the Great Fire of London in 1665, Freemasonry had a rebirth. Forgetting much of its metaphysical, non-Christian traditions, the brotherhood turned to the geometry of good works. Sir Christopher Wren orchestrated the rebuilding of London on Freemasonry principles. London, rebuilt in stone instead of wood, was thought to be the New Jerusalem. At least the rats were cleaned out and would no longer plague the capital city.

The English Civil War had fundamentally discredited all shades of Christian ideology. English culture was thoroughly chastised and fully prepared to let tradition rule with all its faults in order to avoid the vendettas of religious intolerance. The culture was prepared to be tolerant by default. There followed a dark, existential conception of life, a view focused by the philosophy of Thomas Hobbes. In the *Leviathan*, Hobbes concluded that the people's lives are poor, nasty, brutish, and short. One venue for crushed idealism, beside the old standbys of sex and drugs, was a renewed interest in the minutia of existence. With the glories of providence having fallen on hard times, the probabilities for finding a quotidian truth or two came to the fore. John Locke struggled with the fundamental definition of meaning and the elemental circumstances of causation. From his work came empiricism, the primary tool of taxonomy. These were necessary preconditions for there to be even a modest remodeling of the culture based upon science. The existing database was far too narrow to either affirm or to deny any exposition of Natural Law.

Yet there was still one more firecracker in the old school locker. From Dee, the mathematical astrologer to Elizabeth, to Francis Bacon who visualized a modern Utopia, we come to Isaac Newton, the mathematical astronomer from Cambridge. Newton discovered a law of gravitation in 1664. This law explained the motions of the planets and comets from the behavior of objects dropped to the ground on Earth. Furthermore he invented a mathematical calculus that could neatly describe the influence of this law of gravity. He was very discreet about his work. It was only twenty years later when the astronomer Edmond Halley approached him that his stunning insights became public.

It wasn't until some of Newton's private writings came to light in the twentieth century that we discovered where he got his inspiration. These papers were purchased at public auction by the economist John Maynard Keynes. Keynes reported to a stunned Royal Society that Newton was not the first eminence of the age of reason; he was the last of the magicians, the last to practice the Hermetic tradition going back to Sumer. Newton was last great mind to look out on the world with the same eyes as those who began to build our intellectual inheritance thousands of years ago. Newton stated explicitly that the law of gravitation had been fully understood in ancient times. He stated that Thoth, the Egyptian god of science, was a believer in a Copernican system. Newton believed that the detailed instructions for the building of the Arc of the Covenant had been a cryptogram of the universe and contained a floor plan of the Temple of Solomon that would be built to contain it. He was impressed with the worldwide tradition of a deluge among ancient peoples, and the pious Newton supposed that Noah was indeed the common ancestor to all humanity. It was Newton's silence on his sources during his lifetime that left a clean slate for his discoveries in astronomy, mathematics, and physics to be accepted.

The area of spirit was still in the hands of the Church. The issue of catastrophic change depended on moods created by the weather as much as anything, and the weather was getting better. The River Thames had not frozen over for a century or so. As the Rosicrucian movement was driven underground and eventually lost any social influence, the Jesuits too were disbanded by the Catholic Church. The radical movements for reform and counter-reform died out together. The modern academy, born out of and after the intellectual persecutions of the seventeenth century, suffers from a general amnesia about such conforming influences at work at its difficult birth.

Through a combination of an ignorance of history and a deeply engrained superstition for any but its own concocted traditions, the rise in modern science is identified with the insights of Copernicus, Galileo, and Newton, who seem to have appeared miraculously, like Athena fully formed out of the head of Zeus. They are seen as the result of miraculous conception.

Genesis is the first theory of evolution in Judeo-Christian tradition. The birth of history, which is the social application of the principle of evolution, follows naturally from the progression of the Old Testament to the New Testament. This was extended by Augustine's theory of change that flattened the cyclical or seasonal view of change that was the root of prophecy. A principle of biological evolution began to suggest itself after the pronounced changes that occurred with the Little Ice Age. Henry of Hesse (1325–1397) supposed that plants and even animals might evolve naturally out of older species. This insight was aided by the rudimentary herbals and manuals of husbandry that were available at the time, but there was no way to expand on the idea. There was not enough data. The question of evolution became a lively possibility in western European culture during the eighteenth century. The examples of exotic flora and fauna avidly collected worldwide on the heels of the Age of Discovery intensified curiosity over biological relationships in space and time. The systematic documentation of a biological taxonomy had begun.

Some, most notably the Swede Carl Linneaus, fended off the intimations of evolution even though it fairly leapt off the page in front of them. Genesis in the Bible was a suitable theory of evolution to his way of thinking, even though the vernacular translation of his Bible used the word "day" to describe its seven stages. The word no longer had an alternative meaning "a long time" that it had in the Torah. James Ussher, the Anglican Archbishop of Armagh, had already figured out that a literal seven-day evolution required that the time and date of creation be the night preceding the twenty third of October 4004 BC. The God of his culture performed his creation beginning with the simplest aspects of the physical cosmos, moving on up to the higher forms of life, the great chain of being. Even the most sophisticated modern theory of evolution still respects this order of things. The religious axiom about the immutability of the species stemming from Genesis and the relationship of those species to man and God was at the heart of Christian ethics. The ethical axioms that would necessarily come into question made change far too vexing.

Sociological changes concerning English class structure from the seventeenth century were also very important to the eventual implementation of English evolutionary theory. Up to this period, royalty had been supported by a landed aristocracy. Charles I had actually dynamited the spinning factory of an upstart Capitalist at the beginning of this period. Factory production of yard goods threatened the quality standards guaranteed by Royal imprimatur for centuries and impinged upon the sources of wealth of landed aristocrat. The English Civil War saw a recrudescence of middle-class aspirations, and out of this came the development of a mercantile class. This would continue with the restoration of the crown.

Adam Smith would baptize this social movement by supposing that if the new business class were allowed to compete, without government control or tax, they would create an industry that the invisible hand of the free market would guide to the result of the necessities of life at a price the poor could afford: the economies of scale. Selfishness would be conformed to a Christian morality; business success was confirmation of a Christian life. Moneymaking and money lending had always been a paradox in a Christian culture. Jesus, after all, had chased the moneylenders out of the Temple. In the past, this function was often performed in Christian culture by outsiders: the Roman civil class, Jewish bankers, and famous bankers and venture capitalists, such as the Fuggers, the Welsers, and the Höchstetters, who didn't seem to be bothered by the sin of handling money. Smith's clever moral justification for making money by lending money was quickly embraced. A new breed of swashbuckling, Christian Capitalists was born. And the money stayed at home.

Benjamin Franklin soon pointed out the other scale of this ethical balancing act: an excess of labor tended to suppress wages. It was the economies of scale again, but now working against the ethical principle of Smith's Capitalism. Of course the most egregious abuse of power and wealth is to condemn people to greatest economy of scale, slavery. English merchants and bankers underwrote the importation of slaves by the cheapest possible methods to their plantation colonies in North America to work under severe conditions of deprivation. Capitalism quickly proved itself to be a winner take all, zero-sum game in practice. A few people were getting all commodities that they could possibly imagine, while many other people were being worked to death. Capitalism was an important early intellectual motivation of the industrializing culture, and it would profoundly change English society. The landed aristocracy would be replaced by a mercantile aristocracy. With the forges roaring, the Iron Age was beginning to hit its full stride.

Educated Puritans would provide a well-ordered, highly moral, professional middle class. This merchant class was already replacing the landed gentry as the main support for the crown before the French Revolution. It was the mercantile class, in the main, that would fund the French War at the end of the eighteenth century. The Puritan ethic would be revived in a workingman's version courtesy of two brothers who were Anglican clerics. John and Charles

Wesley would create Methodism in the eighteenth century. Unlike the Calvinists John Wesley, believed that each person could be saved by faith in God. Predestination was on the wane. Reflecting his education he would constantly emphasize the "gradual work" of God sanctifying human beings and all of creation. A natural extension of gradualism for Wesley was the concept of the *anima mundi* found in Virgil's *Aeneid*: "the all-informing soul, that fills, pervades and actuates the whole." This was an attempt to counter the quick, emotional, and commonly flawed expectation of an imminent Second Coming.

Subsequent generations of Methodist ministers have not been in a habit of quoting Virgil, but they generally avoid the extreme anthropomorphic simplicities of their more charismatic brethren. Methodism offered such hope as could be gathered by a thoughtful life without abusing the senses, as the Buddhists would put it: no smoking, no drinking, no whoring. Singing also was one of the cornerstones of the Wesley brother's ministry. Methodism helped to give the English a hardworking, chaste, stiff upper lip, God-fearing, working class and the Empire would fill English coffers and by the trickle-down principle provide well-stocked larders in the middle class. And the removal of fire and brimstone from the Methodist pulpit was a preview for a complete Uniformitarian overhaul to the Anglican Church in the next century.

Since the reformation Protestant education had lagged way behind the Catholic education offered by the Jesuits. Oxford and Cambridge at the beginning of the eighteenth century were basically Anglican finishing schools for the landed gentry. Reference to classical Greece and Rome marked the outer boundaries of intellectual speculation at Oxbridge. Saint Thomas Aquinas had already determined what could and what could not be appealed to in the classics. Within these limitations Aristotle's theory of evolution was available for consideration: inorganic forms evolving into simpler organic life forms, to the pinnacle of biological creation—Man. The geometry of this inverted evolutionary pyramid was close enough to Genesis not to be offensive. The English Enlightenment would occur outside the crenellated walls of its main intellectual institutions.

Ever since England burst out of its rustic isolation as a marine culture under Elizabeth I, they were serious collectors and catalogers of the works of nature. The countryside was well stocked with botanizing clerics, such as Gilbert White; sea fairing naturalists, such as James Cook; and collecting squires, like Robert Darwin, who would thoroughly delineate the native flora and fauna right down to its local variations. Robert Darwin had trained as a lawyer but decided to retire to the countryside to raise a family and play the role of country squire. He belonged to a gathering called the Spaulding

Gentlemen's Society that included in its membership Newton, Bentley, Pope, and Gay. Isaac Newton, who tamed the movements of the heavenly bodies with his dynamic calculus, lived ten miles away at Woolsthorp. A wider conception of space is frequently the necessary handmaiden to the deeper consideration of time. As unlikely as it may appear, it was in the geographical center of England's rolling countryside, in a typical quiet English village, that we find the beginning of the Darwin family's confrontation with biological evolution.

It was the God-given immutability of the species, stemming from a seven-day act of creation, that was the challenge for the traditional Christian story of evolution. Dogs were being bred into new species within people's memories, but the most troubling contravention was fossils. Although the true nature of fossils was understood by Xenophanes, Pythagoras, Herodotus, and others, it had not become a serious stumbling block to anthropomorphic reproductions of creation until Europeans started to circle the globe. In search of raw materials to supplement the stressed resource base of Europe, naturalists were also onboard, curious to fill in the gaps in the available bestiaries and herbals. It began to become apparent that some bones had no living representatives. Some bones belonged to marine animals found in unusual circumstances. The word 'fossil' comes into English usage in the middle of the sixteenth century. It referred to a fossilized fish found, and at first believed to have lived, underground.

Robert Darwin, the great-grandfather of Charles Darwin, discovered a fossil ichthyosaur in the rector's garden just across the road from the family homestead at Elston Hall. Robert brought it to the attention of the antiquarian William Stukeley, thinking it was a human skeleton impressed into stone. It was given wider publicity by Stukeley who identified it as a fossil from the biblical flood that had occurred, he supposed, around 3,000 years earlier. The ichthyosaur fossil did much to fertilize the idea of evolution in the Darwin household, although the subject is pursued, as far as the historical record is concerned, until the next generation of Darwins enters the scene.

Robert Darwin's wife Elizabeth gave birth to seven children in seven years without destroying her health. Robert and Elizabeth Darwin succeeded in bringing all seven of their children through childhood even though the infant mortality rate in Nottingham at that time was about 40 percent. Elizabeth lived to the ripe old age of ninety-five. She is said to have been a woman of strong character, capable, practical, and scholarly. She was clearly an exceptional woman. Erasmus Darwin was the seventh and youngest child. He was born in 1731. The name "Erasmus" had Puritan origins in the Darwin family tree. There was an Erasmus Earle, sergeant-at-law to Oliver Cromwell, whose daughter Anne married William Darwin in 1653. The name is so unusual that one suspects that it may be a reference to the great humanitarian Desiderius Erasmus, who attempted to convert the Catholic Church from within, rather from the outside as in the case of Martin Luther. Erasmus Darwin would follow in the footsteps of Desiderius Erasmus. He would be an advocate for the abolition of slavery, the very engine of English wealth and power, and he would be a proponent of natural religion rather than revealed religion as a well-to-do country squire.

Natural religion was the religion that was revealed through God's works, especially in nature. As Erasmus put it, "That He influences us by a particular providence, is not so evident. The probability, according to my notion, is against it, since general laws seem sufficient for that end."[5] Such a posture was tolerated at the time as long as its was not accompanied by public advocacy. Erasmus retained the secular persona of the Puritan. With a sort of hybrid vigor he was temperate in his vices, producing fourteen children from two marriages and only two, possibly three, children out of wedlock; he was passionate in his causes; and he was well ordered and astonishingly industrious. In 1756 Erasmus took his medical degree at Edinburgh, where David Hume was the heart and soul of Scottish Enlightenment. There was no such program at Oxford or Cambridge. Oxford was stuffed to the gills with Greek and Latin studies and well-aged port. Cambridge, although a little less clerical, was content to rest on the laurels of its cult of Isaac Newton.

The medical model learned by Darwin was directly descended from the mechanical physiology, *res extensa*, that Descartes had wrested from the domain of the church. Where he was allowed Descartes had moved beyond Aristotle, Plato, and the High Church and their emphasis upon *purpose*, *prime mover*, and *final cause*. It was Descartes who made cause-and-effect analogies of the lungs to a bellows, the heart to a pump, the eye to a lens, and so forth, creating a human simulacrum. It has been one of the most productive intellectual inspirations of any time. It is a highlight of the cause of empirical science in response to the complete failure of the teleology of European religion to anticipate, change, or in any way to understand the nature of the massive ecological collapse caused by the Little Ice Age.

By the eighteenth century it had reached the point where a species of existence was imparted to animate, inert matter in mechanical inventions such as Jacques de Vaucanson's digesting, shitting duck; Wolfgang von Kempelen's chess automaton; and the Jaquet-Droz' writing automaton that would scrawl: I do not think … do I therefore not exist? In these Promethean productions, illusion and slight-of-hand were present to a lesser or greater extent. Von Kempelen's chess automaton had a human wedged into the machine actually guiding it. Even though it was exposed dozens of times, the public refused to see it. The chess automaton was a cynical parody of the

Descartian conception of life, but there was no one there to laugh. These automatons had become the holy relics of the new age. The machine in all its marvelous manifestations would become the scientific Trojan horse that would breach the defenses of a fundamentally religious population. The serious application of inductive reasoning to the invention of machines and then, by analogy, to the questions of life itself saw no greater champion than Erasmus Darwin, and his achievements were real.

Although located in the quiet English countryside, the Darwins were also close to the area endowed with cottage industry, a population gifted with mechanical instinct that was discovered by the new industrialists. English law was still overburdened by a pre-Capitalistic code that reflected medieval society, but a majority of its proscriptive labor laws were simply bypassed in the new, industrial cities of Birmingham and Wolverhampton. Erasmus Darwin lived the old cathedral city of Lichfield, "mother of the Midlands." Lichfield was less than twenty miles distant from those new, burgeoning industrial cities to the south. The spirit of invention and change intermixed with the old was intoxicating. One could actually measure progress, and even see evolution in operation.

In the year 1763 Erasmus was experimenting with gas laws. While casually enunciating the gas laws that are now credited to J. A. C. Charles and John Dalton, he was also busy improving existing steam engines and mounting them to new designs of carriages to make his rounds as a doctor more comfortable, thereby inventing a rude steam carriage. He was also involved in promoting the project of a twenty-mile canal from Lichfield to the River Trent. He was an inventor of imagination and skill; his poetry influenced Wordsworth, Coleridge, Shelly, and Keats; he was widely claimed to be the finest doctor of his time in England. George III invited him down to the sprawling metropolis of London to be one of his doctors, but Erasmus politely declined. The possibilities for creative expression and intellectual freedom of thought would have been severely curtailed. By this incident alone we may measure his wisdom.

With enormous enthusiasm, as well as impressive size, Erasmus was the heart and soul of the gathering of intellects called the Lunar Society, which was the chief intellectual force behind the English Industrial Revolution in the eighteenth century. Among the members of the Lunar Society was Matthew Boulton the manufacturer, James Watt the inventor, Josiah Wedgwood the potter, Joseph Priestley the English discoverer of oxygen, and an occasional American visitor, Benjamin Franklin. Erasmus shared with Franklin a fascination for electricity. He showed to Franklin the results of his experiments with electricity as a cure for an intractable gall bladder. When all conventional remedies had failed, the gallstones were passed following a series of electric shocks. Franklin shared his experience of drawing down lightning with a kite. In this he was tampering with the domain of the church and there were rumblings from the pews. But his discovery led to the practical invention of iron lightning rods. With lightning rods the house that God built would be saved from inadvertent acts of God.

Darwin acclaimed Franklin as the foremost experimental philosopher of his day in poetic form:

> *Led by the phosphor-light, with daring tread*
> *Immortal Franklin sought the fiery bed*
> *Where, nursed in night, incumbent Tempest shrouds*
> *The seeds of Thunder in circumfluent clouds,*
> *Besieged with iron points his airy cell,*
> *And pierced the monster slumbering in the shell.*[6]

Darwin was generally discreet about his religious and his political opinions in deference to his upper-class, medical trade, but in a letter to Franklin, the voice of the American Revolution to Europe, he could be more unstinting in his praise. He wrote: "Whilst I am writing to a Philosopher and a Friend, I can scarcely forget that I am also writing to the greatest Statesman of the present, or perhaps of any century, who spread the happy contagion of Liberty among his countrymen; and … deliver'd them from the house of bondage, and the scourge of oppression."[7]

Erasmus Darwin founded a botanical society in Lichfield in order to promote the new order in botanical taxonomy discovered by Carl Linnaeus. Linnaeus had taken the sex organs of plants as the key to distinguish them into various groups. English interest in botany and in Linnaeus took a great leap forward with the voyage of Captain Cook in the *Endeavour* that was completed in 1771. On that voyage the young and wealthy Joseph Banks acted as the ship's naturalist. Kew Gardens became the focal point of what has been called the century of imperial botany, imperial because of the interest and support given to it by the upper class. Banks became the president of the Royal Society, and it was he that had the collection at Kew arranged by the Linnaean system.

Dr. William Withering, who was a member of Darwin's Lunar Society, published an herbal arranged by the Linnaean principle in 1776 under the title *A Botanical Arrangement of all the Vegetables naturally growing in Great Britain*. Withering bowdlerized the sexual terms used by Linnaeus in order that they not give offense to the ladies. Stamen was changed to chive and pistil to pointal. This irritated Darwin so much that he launched his own translation of Linnaeus's encyclopedic work. It took him nine years, but with the help of Banks he completed this highly academic chore. Well founded in botany, Darwin now turned to his poetic talents to the issue of vegetable naughtiness. While Captain Bly was collecting breadfruits in Tahiti in 1789 Erasmus Darwin published *Loves of Plants*, which included the following lines:

> *With honey'd lips enamour'd Woodbines meet,*
> *Clasp with fond arms, and mix their kisses sweet.*
> *The fair Osmunda [a fern] seeks the silent dell,*
> *The ivy canopy, and dripping cell;*
> *There hid in shades clandestine rites approves,*
> *Till the green progeny betrays her loves.*

The sticky question of sex would entangle theories of botany and of evolution, in ways that are almost impossible to segregate, down to the present day.

For Erasmus Darwin, the year 1794 was monumental. He published a small book advocating a much more advanced education for women. This grew out of the experience he had educating his daughter Violetta by his third wife Elizabeth Pole. He had converted a pub, the Nag's Head, into a ladies' seminary. Erasmus preferred women who were educated in the sciences, and he thought they should be accomplished in the practical skills and well exercised to boot. The great event of that year, however, was the publication of a project twenty years in the making, *Zoonomia, or the Laws of Organic Life*. His aim in this work was "to reduce the facts belonging to Animal Life into classes, orders, genera, and species; and, by comparing them with each other, to unravel the theory of diseases."[8] It was in a fifty-five-page chapter on generation that he specified the processes, which we now call biological evolution, that almost certainly influenced Jean-Baptiste Pierre Antoine de Monet, Chevalier de Lamarck in France:

> *Would it be too bold to imagine, that in the great length of time since the Earth began to exist, perhaps millions of ages before the commencement of the history of mankind, would it be too bold to imagine, that all warm-blooded animals have arisen from one living filament, which THE GREAT FIRST CAUSE [sic] endued with animality, with the power of acquiring new parts, attended with new propensities, directed by irritations, sensations, volitions, and associations; and of delivering down those improvements by generation to its posterity, world without end?*[9]

The scale of Erasmus Darwin's vision came from his personal friendship with three of the best geologists of his time. It was his friend, the Scotsman James Hutton, who proposed that the selfsame forces that shape the present had shaped the past. Everything about the changes on the planet, seashells found on mountain tops for instance, could be explained by the microscopic processes of wind and rain, and the occasional earthquake that we see around us, if allowed to take their course over millions of years. Hutton would name his thesis Uniformitarianism. Hutton was born in Edinburgh, attended University of Edinburgh as a student of humanity, got a degree in medicine at the University of Leyden with a thesis on blood circulation, was a chemist, naturalist, and experimental farmer besides being a geologist. Hutton would attempt to professionalize the subject of geology by separating it from cosmology, which is to say the dictates of the Church. Hutton helped Darwin to move beyond the Bishop of Ussher's young Earth perspective enabling him to suppose that coal seams represented forests that had been buried long, long ago. Hutton and Darwin disagreed about the role of geology in the origin of life. Darwin saw an important role for geology in shaping life.

Although Darwin cared little for conventional wisdom or the Anglican Church, he was not able to free himself from the ball and chain of the prepotency of the male sex, the justifying theory for the patriarchal tradition. The Judeo-European prejudice concerning sex came right out of Genesis. Eve was born out of Adam's rib; she was an afterthought. Solomon believed it to be so. God's awkward demonstration of the power of fertility was a necessary, after-the-fact cover-up stemming from the assumption of an all-male, monotheistic religion. The Classical Greek sermon on the mount concerning the prepotency of the male sex was given by Aeschylus in *The Eumenides* through the mouth of Apollo: "The woman you call mother of the child is not the parent, just a nurse to the seed, the new-sown seed that grows and swells inside her. The man is the source of life—the one who mounts."[10] Aristotle in *Generation of Animals* saw woman as a weakness in the father's generative power or, perhaps, the result of some external factor such as a damp south wind. The creative element was semen that was foam-like and, according to Aristotle, "not unknown even in early days; the goddess who is supreme in matters of sexual intercourse was called after foam."[11] This was Aphrodite. Aphrodite was born when Cronus castrated his father, Uranus. The blood and seed that fell on the sea was the cause of the birth of Aphrodite out of the foam. The Church of the Apostles was built right on top of the temple to Aphrodite. The exception to this litany of self-congratulation was the great Roman natural historian Lucretius. He pointed out the flaws of his illuminated predecessors, but he cast his seed on sterile ground. Yet he confirms our suspicion that the prepotency of male sperm was not a thoughtful consideration of the evidence, but a politically correct thesis.

Erasmus Darwin supposed in the *Zoonomia* that in the fetus the male provided the active nucleus, female merely provided the food upon which it grows. The insufficiency of this idea is realized only a few pages later when pondering over mules from his practical experience as a farmer. Mules are the hybrids of a male donkey and a female horse. Mules are always sterile. From this he decides that the female must contribute equally to the make up of the progeny. Perhaps he had remembered Linnaeus, who had stated: "the offspring proceeds not from the egg alone, nor from the male sperm alone."[12] His ambiguity reflects the burden of circumstantial evidence suggesting equal contributions from the sexes weighed against the overwhelming dogma of the age. The time was not ripe for the discovery of the equality of the sexes. This insight would have to wait to the end of the next century. At this time the mood of social disapproval was tangible.

The excesses of the French Revolution brought an end to free thinking in England. Erasmus Darwin, who was a Deist, had usually been guarded about his comments on religion, yet in the *Zoonomia* he included under the heading of diseases of volition *spes religiosa*. Darwin defined *spes religiosa*, or superstitious hope, as a "maniacal hallucination" that in mild form produces merely "an agreeable reverie," but when given public support has "occasioned many enormities." Such an enormity had now engrossed the English people. Because of his subversive ideas, Darwin was brought down from his high place in English arts and letters in a matter of a few weeks by a crude parody on his dramatic style, published serially in 1798 in three numbers, called *The Loves of the Triangles*. In this rude send–up, a parabola, a blue-eyed wanton hyperbola, and an ellipse sigh for the love of a rectangle. The ideas that human beings have evolved from lower beings, that electricity will someday have practical applications, or that the mountains are older than the Bible are ridiculed. The smear was orchestrated by spymaster George Canning, under-secretary for foreign affairs. Canning increased the damage by implying that William Goodwin, whose *Political Justice* was subversive to all forms of government, was the author of *Triangles*. Darwin and Goodwin never met, but Goodwin was a great admirer of Darwin. On the whole this was a relatively light sentence, considering the sentence for political rebellion was disembowelment.

James Hutton published his geology text, *Theory of the Earth*, one year after Erasmus Darwin published his *Zoonomia* in 1794. It too contained a theory of evolution by means of natural selection. It was met with an equal amount of scorn and ridicule. If such a long-term view were to be adopted, then the miracle of the act of creation in Genesis and the awesome events described in the Old Testament would be put under a cloud of suspicion. The god of Genesis might begin to look like nothing more than a local chieftain of a rustic village a few thousand years ago. For most people, however, the issue of greatest import was descent from the apes suggested by evolution of species, and it was this to which people applied the contemptuous word "Darwinism."

In this paranoid environment Erasmus even considered moving to America. Perhaps he was dissuaded having been made aware that Benjamin Franklin had taken on protective coloration in America by becoming a member of a church. Some of the fire was drawn away from Franklin by Tom Paine, who gave Quaker philosophy an incendiary voice. Joseph Priestley, the English discoverer of oxygen and a member of the Lunar Society, was not so lucky. Priestley was a dissenting minister in Leeds and Birmingham and a religious materialist. He had publicly argued that a close examination of the second book of Genesis described the whole man as made from the dust of the ground. There was no mention of spirit in this alchemy. If there was no spirit there could be no polite separation between church and science as negotiated by Descartes, although nobody went so far as to work out that consequence. It was just wrong like a bad smell, like a minor key. Priestley was burned out of house and home by a mob because of his scientific and revolutionary sympathies. The mob howled, "No Philosophers!" He would flee to the state of Pennsylvania in the newly formed United States where he would live quietly for the remainder of his life.

Across the Channel in France intellectuals were in even graver danger. The French discoverer of oxygen, Lavoisier, was guillotined by the French Republic in 1794. The president of the court uttered this famous dismissal: "The Republic has no need of men of science." From this distance it is hard to see scientists such as Priestley or Lavoisier as a threat, but when the ship of state enters a period of bad weather even the mispronunciation of the name of God can be a mortal offense identifying the offender as alien to be thrown overboard. What was allowed free rein in wartime England was technology and industry, applied science and patriotism. Nationalistic rhetoric flourished; naturalistic science died. Erasmus's son Robert Waring Darwin, Charles Darwin's father, was a silent evolutionist along with many other academics who had read the *Zoonomia* and appreciated it. Erasmus was the great genius of the Darwin family. He was the great English genius of his century, yet for making some unseemly observations about the state religion, memory of Erasmus Darwin and the Lunar Society would be buried under a tide of reaction in England. Two myths would be spawned out of this environment that still influence us to the present day. One was by William Paley; the other was by Mary Wollstonecraft Shelly.

Mary Shelly's mother was the pioneering feminist Mary Wollstonecraft, who published *A Vindication of the Rights of Women* in 1792. Her mother died giving birth to her in 1797. Her father was the radical political writer William Godwin. Godwin remarried and started a bookstore that became a meeting place for writers, political activists, and natural philosophers. Mary was encouraged to read. Her father recommended the works of Erasmus Darwin, whom he greatly admired. Among the visitors to the Godwin household was Percy Bysshe Shelly. Mary and Percy had a flaming affair. Shelly also encouraged Mary to read Darwin, as his books had influenced his own writing as well as his scientific ideas.

Mary and Percy were living in a rented house, Villa Diodati, near Geneva during the summer of 1816. The previous spring Mount Tambora on Sumbawa Island, Indonesia, blew off, putting a reflective layer into the atmosphere. This was not much on the scale of geological or meteorological possibilities, but that winter was extremely cold, and the following year was known as The Year With No Summer. It was a temporary regression back into the Little Ice Age. Sitting around a blazing fire with storms raging outside, Lord Byron, whom they had befriended, suggested that they write ghost stories. On another evening there was a group conversation about the creation of life and Darwin's theory. With lightning flashing and thunder rumbling later that evening, in a half dream state Mary had a vision: "I saw the pale student of unhallowed arts kneeling beside the thing he had put together. I saw the hideous phantasm of a man stretched out and then, on the working of some powerful engine, show signs of life and stir with an uneasy half-viral motion." The story *Frankenstein, or The Modern Prometheus* was born. It didn't take much to stir up the specter of ageless terrors, in this case from a classical Greek source combined with science rather than a biblical source.

William Paley, rector of Bishopwearmouth, published *A View of the Evidences of Christianity* during the French War. It was a formal response to Erasmus Darwin, Tom Paine, and the French Philosophes. Reading this book became a prerequisite for admission to Cambridge until 1900. In the still more famous *Natural Theology*, 1802, Paley focused his attack on Darwin's theory of evolution. He argued for the existence of Supreme Intelligence by identifying evidences of design from natural history. Following the exemption allowed to Descartes by the Catholic Church and appealing to a mechanical analogy to life, Paley argued that an irreducible complexity in nature pointed to purpose in life and a divine watchmaker—God. Referring to the accidents of change suggested by Darwin, he says, "What does chance ever do for us? In the human body, for instance, chance, i.e. the operation of causes without design, may produce a wen, a wart, a mole, a pimple, but never an eye. Amongst inanimate substances, a clod, a pebble, a liquid drop might be; but never was a watch, a telescope, an organized body of any kind, answering a valuable purpose by a complicated mechanism, the effect of chance. In no assignable instance hath such a thing existed without intention somewhere."[13] His work showed a considerable knowledge of natural history, quite a bit more than the most modern scientists have as a matter of fact.

Ironically, the movement of the Anglican Church in the direction of an increasingly rationalistic or deist position was aided and abetted by Paley. He had clearly separated the Anglican Church from a fundamentalist interpretation of the Bible that advocates from scripture alone. In Paley's argument we sense Wesley's concept of the *anima mundi* by way of Virgil. He was implicitly taking the position that Galileo had advocated two centuries before, allowing that scripture alone was not up to the task of evaluating the details of the material world. The Anglican Church was becoming secularized in spite of itself. Erasmus Darwin had helped to shift the whole range of the discussion to the left even while English culture was closing down intellectually.

CHAPTER 3

Charles Darwin and Victorian Culture

THE FRENCH WAR HAD thrust enormous responsibilities on the English government. Reviving a peacetime economy was almost as serious a dislocation as the war had been. The king and Parliament were responsible for keeping the social fabric from disintegrating without the focusing motivation of an external enemy. The English laws that reflected an agricultural society, reigned over by an inherited aristocracy with its surrogate nobility, were now being overturned by the engine of industrialization and the wealthy mercantile class it was creating. Resource-poor England was dependent upon its far-flung empire to sustain itself at a high level of material refinement. It had launched its industrial economy on shipbuilding; it now gathered the materials to sustain that industry from diverse foreign ports with its merchant marine, protected by the world's most powerful navy. With wealth pouring in from the colonies, recovery was not in serios doubt however. English workshops would produce the finest products in the world in the nineteenth century. With mainland Europe now in a full-fledged reaction to the specter of Napoleon, the Romantic Age, English culture would become intoxicated with its brilliant achievement as well as the horror that is always attendant upon being a world power. London had replaced Jerusalem or Rome as the center of the universe in Western European culture.

From a mountaintop the English historian Thomas Carlyle would observe that "the whole Life of Society must now be carried on by drugs: doctor after doctor appears with his nostrum, of Cooperative Societies, Universal Suffrage, Cottage-and-Cow systems, Repression of Population, Vote by ballot. To such height has the dyspepsia of Society reached: as indeed the constant grinding internal pain, or from time to time the mad spasmodic throes, of all Society do otherwise too mournfully indicate." Carlyle restored the character of Oliver Cromwell to good odor, thereby sponsoring a romantic revival for the ethically simpler and purer days of the Puritan prophets. Carlyle was an entertaining and a perspicacious observer of his times. He was not a particularly good historian, although he did blunt the edge of hatchet jobs done by other not particularly good historians. His theory of history was commonplace: history is a compilation of a few great benefactors to humanity. His view of the commoners, a senseless herd that must be drilled, led, and punished in obedience, was typical of the view from above.

The new or experimental form of philosophy promoted by Sir Francis Bacon was changing from a suburban hobby and a pulpit for polymaths such as Bacon and Erasmus Darwin to an urban profession. The men's societies of the days of Charles Darwin's great-grandfather and grandfather, such as the Spaulding Club and the Lunar Society, had become low-ranking satellites blinded by the aura of the Royal Society of London. The Royal Society had been founded in 1660 with the restoration of Charles II by a goodly number of reformed Freemasons who had come out of the closet. They were well-meaning, wealthy do-gooders on the whole, dabblers in natural philosophy. They were probably more than a little disturbed by a few brothers who were involved with the Hell-Fire Club that engaged in wicked debaucheries at the expense of institutional religion. There were two astounding natural philosophers that were members of the Royal Society from the beginning, Isaac Newton and Robert Hooke.

Newton we know about because he kept his interest in Hermetic tradition secret. Hooke we hardly remember even though he was called the English Leonardo De Vinci at the time. He was the Erasmus Darwin of his century. This Renaissance man invented the spring control for the balance wheel of watches and a reflecting telescope. He proposed a theory of evolution based upon the existence of fossils and he deduced the wave theory of light. He made major contributions to the field of microscopy, identifying the cell as the basic unit of biology. Hooke was an architect of great renown who helped Christopher Wren rebuild London after the Great Fire in 1666. The last part of his life was dogged by ill health and by jealous intellectual disputes that account for our lack of knowledge of him.

In the eighteenth century the Royal Society had been reformed by Joseph Banks, the great naturalist. He replaced the enthusiastic amateurs with a certified group of published scientists and practical empiricists. In the nineteenth century the universe of knowledge began to be sectored off into learned societies: the Geological (1807), Astronomical (1820), Entomological (1826), Zoological (1826), and Geographical (1830) Societies. This reflected the increasing body of knowledge that would bring to an end the age of the great polymaths, natural historians or philosophers who could embrace the totality of learning. The nineteenth century would be the age of the specialist.

Charles Robert Darwin was born in 1809. He largely missed the rabid paranoia of the war years as a cosseted youth on his father's estate in Shrewsbury, in the Midlands. His father, Robert Darwin, had 'The Mount' constructed on high ground overlooking Shrewsbury and the River Severn. Like his father Erasmus, Robert Darwin was a doctor with a degree from Edinburgh and a yen for gardening. The Mount had numerous gardens, greenhouses, and a hot house for tropical plants. There were a well-stocked library and several workshops. Robert was a formidable presence weighing in at well over 300 pounds, he was an atheist and a believer in evolution.

We are first born into our mother's culture, so it was as a Unitarian that Charles Darwin began life. His mother Susannah was a Wedgwood, the daughter of his father's lifelong friend. When Charles was a world apart in Buenos Aires on his trip around the world, the early signs of spring brought his memory back to The Mount, as he wrote in a letter to his sister Caroline: "It is now Spring of the year, & every thing is budding & fresh: but how great a difference between this & the beautiful scenes of England. I often think of the Garden at home as a Paradise; on a fine summers evening, when the birds are singing how I should enjoy to appear, like a Ghost among you, whilst working with the flowers". As a boy Charles was an avid collector of butterflies, beetles, shells, stones, and fossils. This idyllic childhood came when he was eight when he suffered the death of his mother. He was boarded at Shrewsbury school, and then, as a teenager, he was exiled to cold and dank Edinburgh.

Charles disappointed his father as a student at Shrewsbury School, but when he reached his maturity his father directed him toward a medical career at Edinburgh to continue the family tradition. Edinburgh was still the best medical school in England at the time. Most English universities were still engrossed with a purely classical education. Oxford and Cambridge still did not have medical or legal schools. Medicine in that stronghold of Puritan culture, Edinburgh, was efficient, thorough, and mechanical. It avoided the emotional or the spiritual, leaving that to the church. This was how intellectuals had survived persecution since of the middle of the seventeenth century.

For religious and political reasons Scotland had a special relationship with France and French ideas. The French had conspired with two Scot pretenders to the English Throne, Mary Queen of Scots and Bonnie Prince Charlie. Jean-Baptiste Lamarck's school of Transformationism corresponded with the school of comparative anatomy at the University of Edinburgh that included Professor Robert Jameson. In a seminar given by Jameson, the sixteen-year old Charles watched as John James Audubon demonstrated his method for wiring up bird skins for the purpose of painting them. Jameson published an anonymous paper in which he praised Lamarck for explaining how the higher animals evolved from the lower animals. Charles found Jameson's lectures boring.

Charles Darwin's first conscious introduction to his grandfather's poem on evolution, *Zoonomia*, began with a tirade in support of evolution administered to the eighteen—year-old by Dr. Robert Edmond Grant at Edinburgh. Charles had undoubtedly made some literal reference to Genesis. Charles listened to Dr. Grant, obviously still under the influence of his grandfather, in astonished silence. Charles had never met his grandfather, who died seven years before his birth. Grant's tirade suggested to Charles that his grandfather's ideas may have lain passively in memory, part of the wallpaper of youth. "It is probable," he recalled later, "that the hearing rather early in life such views maintained and praised may have favoured my upholding them under a different form in my *Origin of Species*. At this time I admired greatly the *Zoonomia*."[1]

The English avatar of evolution was still an unmotivated medical student at Edinburgh and he would switch to Cambridge, where he would become a divinity student. This was a sign that the Darwin family had decided that Charles was not an intellectual and not going to follow in the family's footsteps. As anyone who has lived under the aura of an intellectual superstar can testify, the ordinary challenges of life are magnified and the bar for achievement is raised to an extra high level for the younger generation. Lack of application is often a response to such a situation. Professor Grant would go on to publish a well-known paper in 1826 declaring his belief that species were descended from other species, and that they became improved in the course of modification. English culture, however, was well practiced in fending off Scots-French innovations of this or any kind.

The sons of the new upper-middle class attending Oxford and Cambridge, such as Charles Darwin, expected to receive a non-scientific education, make contacts useful for later business ventures, and gain the polish and feistiness necessary for moving in high society. Darwin was typical of other students having a scullion, a laundress, a shoeblack, a hatter, a tailor, and a barber to attend to his needs, as well as a chimney sweep and a coalman to keep the fires going in his spacious rooms. Cambridge was no hotbed of scientific curiosity throughout the Victorian century. As A. E. E. McKenzie sums it up, "It is indeed surprising how small a part the universities played in the progress of British science in the nineteenth century."[2] In fact, they did very little original research even in the classics compared to what was coming out of the German universities. Religious instruction, including compulsory attendance at chapel, was the thing. Geologist Charles Lyell would criticize the compulsory attendance of chapel occasionally inflicted in some colleges as a penalty for academic misdemeanors. Formal attendance, he thought, had a tendency to weaken, rather than to exalt, the sentiment of true devotion.

Charles Darwin also drifted at Cambridge, at first. He had vague dreams of becoming a country parson that would have given him a socially acceptable occupation while allowing him to investigate natural history on the side in the tradition of Gilbert White, best known for his *The Natural History and Antiquities of Selborne* (1789). Then Charles met the biologist John Stevens Henslow at Cambridge. He became the favorite student of the devout Henslow. Suddenly motivated, he began studying the prelims that consisted of a test for the knowledge of the classics, the Old Testament, the New Testament, and the works of William Paley. His religious education at Cambridge brought him face-to-face with the anti-evolutionary argument of the rector of Bishop-Wearmouth: the manifested works of nature were an irreducible complexity that could not have occurred without intelligent design. Charles would conclude, "All high-minded Englishman accepted his premises; every Cambridge ordained swore by his conclusions. His cold, clear reasoning proved Christianity, made apologists of young gents, and underpinned the Anglican order."[3]

Charles Darwin actually became less of a divinity student and more of a biology student under the influence of Henslow. Given a chance to be the naturalist aboard the H.M.S. *Beagle*, largely because of the influence of Henslow, he leapt at the opportunity. Just before he boarded the *Beagle* he was handed a copy of the newly published book by Charles Lyell on the *Principles of Geology*. It was 1831. Lyell had been a student of Reverend William Buckland, the most influential English geologist at the time. Buckland thought the purpose of science was to show the facts developed by it were consistent with the accounts of the creation and the flood. Lyell would do the opposite. Lyell revived and expanded James Hutton's thesis of Uniformitarianism. This was the same thesis that had inspired Erasmus Darwin. By this time it would prove to be more salutary to conform the High Church to Lyell's geological thesis than it would be to force-fit the expanding horizon of geological knowledge into the Old Testament. This was in the tradition of St. Augustine who turned the Old Testament into a metaphor following Plato.

Charles returned from almost five grueling years of travel around the globe initiated into manhood, but also an ill man. Some have speculated that he contracted chagas disease from the kissing bug in South America. He vomited every afternoon and was often swept away by fits of hysterical weeping. Charles largely retired from public sight. He returned to his grandfather's *Zoonomia* after his *Beagle* trip and found that he was much disappointed in the proportion of speculation compared to the facts given being so large. Charles Darwin had finally stumbled into a life's work that suited his interests. Charles maintained a disciplined routine for the rest of his life, enlarging upon the family tradition of evolution. He became the central clearinghouse for hundreds of naturalists who were scouring the Earth for its various species of life. He would accept crates of stuffed birds and animals; jars of pickled fish, eels and echinoderms; insects mounted on boards; crates of shells and bones; and letters and envelopes containing page after page of descriptions of the world's varied geography. It would take him nearly thirty years to complete his magnum opus. His two main works, *The Origin of Species* and *The Descent of Man*, corrected the paucity of facts in his grandfather's works.

In 1839 Charles married his cousin Emma Wedgwood, whom he had known since childhood. Both of their grandfathers were larger-than-life characters in the previous century. Josiah Wedgwood was a staunch capitalist, but was also an abolitionist. Josiah rejected the notion of the Trinity and the divinity of Jesus. He was a Unitarian. Liberal Unitarians suppose that the purpose of life was revealed through reason, scholarship, science, philosophy, scripture, and other prophets and religions. They maintain only the most tenuous connection to mainstream Christianity. Erasmus called Unitarianism a featherbed to catch a falling Christian.

In her youth Emma spent time in Paris studying with Frédéric Chopin. She took the grand tour of Europe and was keen on outdoor sports. She helped overcome Robert Darwin's objections to his son's proposed trip on the Beagle. After the marriage Charles and Emma spent most of the rest of their lives in a big house in the rural village of Downe surrounded by family, servants, cats, and dogs. Charles and Emma had ten children, three of whom died young. Emma ran the household and nursed Charles through his ups and downs. Emma was the intellectual and emotional equal to Charles, the essential partner to his accomplishments, and the whole endeavor was supported by the Wedgwood fortune.

Until the age of forty Darwin remembers himself as a Christian, but his *Beagle* trip caused him to raise questions about his religion. From a broadened perspective he began to wonder if the Christian sacred books could "no more be trusted than the sacred books of the Hindoos or the beliefs of any other barbarian."[4] The validity of those works came to depend upon the interpretation of the metaphors and allegories we now place upon them. The greatest trial he faced in his quiet apostasy was with his beloved wife, Emma, who was a devoted Unitarian. A turning for Darwin came with the death of his father, who, in the eyes of the church, was now condemned to eternal damnation as an atheist. There was also the death of his beloved little Annie. He could not abide Emma's view that suffering and illness were meant to exalt our minds with hope for a future state.

What was special about the relationship between Charles and Emma was the honesty of their relationship and the profound affection that was maintained across such a deep divide. On the one hand, this loving relationship restrained Charles from going any further than he did in his philosophy; on the other hand, it gave Charles strength of character that was just as important as any of his ideas for success in the public arena. Charles feared the damage that might follow from a public exposure of his loss of faith and was generally silent on questions of religion. Emma and his daughters edited his written material in order to maintain that proper Victorian front. A brazen character such as Oscar Wilde, for instance, could never have authored an adjustment in Victorian culture, which was hypocritical and abusive on the issue of homosexuality. Wilde was tolerated like a highly trained monkey in a sideshow. Darwin too was portrayed as an ape in the public press and always on the cusp of wholesale ridicule.

While an acceptable theory of biological evolution was gestating in Charles Darwin's mind, geology was making great headway in creating a climate of opinion that was hospitable for its birth. It is the discipline of geology that brings up the question of the number of days of creation and stumbles over the reality of fossils. If the six days of evolution followed by a day of rest in Genesis was taken literally, then the time since creation estimated by the Bishop of Ussher at the year 4004 BC was close enough. Genesis does not say anything specific about the mutability of species, but the immutability of the species is a reasonable conclusion under these circumstances.

If the Lord God tested Job to prove to Satan that his faith was strong even in adversity, the Lord God could certainly create fossil dinosaurs to test our faith in Genesis. Understanding a creation that proceeds over millions of ages past is also extremely challenging. Old-Earth or long-day Genesis really begins for modern English culture with the fossil ichthyosaur in the rector's garden just across the road from the Darwin family homestead at Elston Hall. It continues with Erasmus Darwin, who was the personal friend James Hutton, who proposed that the selfsame forces that shape the present had shaped the past. Everything about the changes on the planet could be explained by the microscopic processes that we see around us, if allowed to take their course over millions of years.

With the French War everything was put on hold from 1789 to 1830. The resurrection of an old-Earth creation theory began after the war when the young Charles Lyell became excited by the ideas of Jean-Baptiste

Lamarck, whose evolutionary theory presupposed an old Earth. He would also come to know James Hutton's theory of Uniformitarian geology. In *Principles of Geology* he would begin by reviewing ancient traditions. He cites Vedic tradition, Egyptian priests, Chinese tradition, and the legends of the Peruvians. He cites the *Timaeus* of Plato, Aristotle, and Pythagorean tradition with great respect. From this very impressive survey of history he found ancient traditions of cosmology and he also found stories of such vehement disturbances as earthquakes that leveled cities, rivers that ran out of their courses with widespread destruction, and volcanic eruptions with similar effects. These infrequent and local tragedies were sufficient for the development of an idea. Citing the opinion of Aristotle, he says that the deluge of Deucalion was a local event arising from great inundations of rivers during rainy weather. In the absence of any knowledge of impacts with space junk, there was no cause sufficient to result in worldwide calamities. They were what we would call today the hundred year flood.

Lyell also made a comment that reveals much about the psychology of the times and explains why he would become the godfather of old-Earth geology in English culture. He said that it would simply be too dispiriting in the early stages of geological science to think that violent events such as those described in the Bible had actually occurred. While most geologists were still finding evidences of The Flood or of major volcanic activity of catastrophic proportion, Lyell supposed that what looked like catastrophic events in the geological record would be rectified when the missing pages were found and the record ironed out. What gave Lyell added credibility was that he was a devoted Anglican.

When the generation of Catastrophists died out the tide of opinion shifted rapidly. Uniformitarianism became the philosophy of hope. Uniformitarian geology would sew up the gaping wound of fear that was carried by the Old Testament and The Book of Revelations, which had been revived with the Little Ice Age and its famines, plagues, and wars, and even during the nightmare of the French Terror. In its mood and its method, Lyell's Uniformitarianism was similar to Aristotle's cosmology. It was an extension of Augustine's philosophy, and it completely suited the social equation of Christian, upper-class, English Empire in the nineteenth century. A thriving middle class is always ready to embrace the illusion of stability as well, always ready to build its house on the delusions of influence and power. Only the poor and the oppressed would retain their franchise for violent change through The Book of Revelation or, as was now happening in Europe, revolution.

Charles Darwin was well prepared for an old-Earth view of life offered by geology, having had an early interest in the subject, yet the young Charles Darwin during his trip on the H.M.S. *Beagle* saw things that could not be explained by the gradual processes detailed in Lyell's *Principles of Geology*, and said so at the time. After decades of hermitage as a semi-invalid, however, he forgot his youthful doubts and in his signature work declared that those who could not believe in Charles Lyell could believe in him. Lyell's geology would be as close to Aristotle's theory of no change as any cosmology could be. Uniformitarianism would sweep through the academy like a sigh of relief. The Anglican Church would register a brief complaint, but Uniformitarianism would become a cornerstone and calling card of the Victorian Age. So we have moved from the gradualism of John Wesley the Methodist reformer, to the Uniformitarianism of Lyell, the devoted Anglican, to the wit and wisdom of Lord Salibury: "Whatever happens will be for the worse, and therefore it is in our interests that as little should happen as possible."

Five years after the publication of the *Principles of Geology* the cyclic comet discovered by Edmund Halley was predicted to return. Several years before its appearance a common French song captured the deep-rooted, human fear of such an event:

> *We shall not escape this great impact*
> *I feel our planet crumbling already ...*
> *Go quickly to confession you timorous souls*
> *Let us be done with it, the world is old enough*
> *... The world is old enough.*
> *God is sending a comet against us*

Halley's Comet swept by again in spectacular fashion, but only the birth of Samuel Clemens in America was commemorated by it. Everything was changed with the advent of the new, rational, Victorian science. All previous human mythological traditions could be viewed as the monstrous absurdities that were a product of depraved, superstitious minds. It was High Church versus Low Church. Catastrophists were in league with religious fundamentalists. Mythology would become a ward of the English department and could be deconstructed for wit and style, not knitted together for the purpose of unveiling the human condition. License was give to writing a canon of history that would suit this contemporary usage.

The subject of biological evolution was broached to a wider audience in 1844. Robert Chambers published a book called *The Vestiges of the Natural History of Creation*. His vision was an attempt at unifying the various disciplines of science, using Laplace's Nebular Hypothesis as the origin of one, glorious ramification of progressive development. As Copernicus opened the academic debate on a solar system for Galileo, Robert Grant opened the academic debate on evolution in English culture. As Galileo opened the public debate on the cosmos, Chambers opened the public debate on evolution. Copernicus published posthumously, Galileo was forced to desist from publishing, and Chambers decided to publish anonymously. Chambers had the advantage of owning his own publishing house.

In the introduction to *The Origin of Species* Darwin would criticize the accuracy of the information in Chambers' work, but thought that it had "done excellent service in this country in calling attention to the subject, in removing prejudice, and thus preparing the ground for the reception of analogous views."[5] Robert Chamber's book was popular, but popularity usually does not influence the intellectual academy; in fact, the opposite may true. Academic institutions are usually impervious to ideas that do not originate from one of their own.

Uniformitarian geology created the plinth upon which biological evolution was erected for both Erasmus and Charles Darwin. For Charles Darwin economics would give the vision a general method, and that would begin with Thomas Malthus. It was a reading from the Old Testament that triggered Malthus to write *An Essay on the Principle of Population*. Malthus was an Anglican cleric at Cambridge. His father was a personal friend of Jean-Jacques Rousseau and the most famous English philosopher of his day, William Goodwin. Although born into a liberal family, Malthus did not share in the utopian enthusiasms of his father. Malthus was of the new age that held the principles of Natural Equality and the Rights of Man in comic contempt. Democratic governance was viewed as pernicious. Malthus's essay was published in the same year that Charles's grandfather was brought down by a satirical attack launched by George Canning in 1798. Because of the man he was Malthus's essay was received with due respect.

What gave Malthus his insight on population was a phrase in Ecclesiastes. In the Book of Ecclesiastes, 5:11, it says "When goods increase, they increase that eat them ..." In other words, if the food supply increases, it is soon annulled by the sexual proclivities of man, and, the natural fertility of the earth being limited, starvation is the result. There was no chance that good works would alter this law of nature. The vindication for starvation came from the overall good effects of a benevolent creator. Probably this would not have been such a mind-boggling idea to the farmer who deals with the paradox of agricultural success first hand; it was an insight to an intellectual schooled in the religion of Christian charity. The Book of Ecclesiastes is a curious anomaly in the Old Testament. It is not cut out of the mainstream of the Jewish culture. It makes only one reference to anything else in the religious canon, yet on this book the sun never sets, or at least the sun always rises again. It is quoted more often than any other part of the Bible. Ecclesiastes is the wisdom tradition of the world-weary sage.

In contrast to Ecclesiastes, the main thrust of Judeo-Christian tradition is not circular, not circumspect, not recursive; it is linear. Christianity follows a simple moral algorithm: do this, suffer that. Ecclesiastes speaks of cycles—to everything there is a season—but when Malthus quotes from Ecclesiastes he does not even cite the full sentence of the Ecclesiastian principle. He uses it out of context with linear, adversarial intent. He uses it as a dialectical response to his father's generation. The end of the sentence that influenced Malthus is "and what good is there to the owners [of goods] thereof, saving the beholding of them with their eyes." The ephemeral nature of all things in their present form and the constant recycling from dust to dust is the basic message of Ecclesiastes. Although the collapse of the English Empire was only a century and a half away this was certainly not part of the thinking of Malthus, or his religion, or his culture at that time.

Malthus was doing, in particular, what English culture was doing in general: practicing enlightened Capitalism. He was looking for a way to fix a social dilemma, or at least avoid an even worse problem. Thomas Malthus was secularizing religious tradition for practical usage, picking and choosing from the sacred canon that which fitted the time. The Anglican religion had been removed from the killing fields of religious reformation after the English Civil War. It was now largely constrained to the pulpit and good works through the agency of economic power. It was still powerful enough to intimidate the scientific academy, but only by habit and because it was needed by the Capitalists to redeem and justify the Empire. There was no wisdom tradition in Victorian England to speak of, no metaphysicians whispering in the ear of power. The legendary gnostic adventures of Richard Francis Burton were attempts to escape the stifling embrace of Victorian culture. Poetical prophet William Blake stood out like a tiger burning bright—an anomaly.

Malthus and his half remarkable ecclesiastical analogy would continue the integration of economics into religion that Adam Smith's principle of the invisible hand had begun. With Adam Smith pushing from behind, Malthus would point to an underlying dilemma in the business cycle. The Malthusian principle justified what the English were already doing: shipping their redundant overpopulation first to North America and then to Australia. His observation continued the process. The Poor Law Commissioners of the 1830s were impressed with Malthus. They supposed that charity perpetuated the poverty it was intended to relieve by stimulating the growth of population and further depressing earnings. The English poor laws of the eighteenth century were seen as a misguided paradox and were constrained.

The Malthusian equation was completed by David Ricardo. Ricardo argued that the freedom to accumulate capital would lead to rapid growth. Ricardo thought that labor was the most valuable resource, but that population growth would eventually push down wages until they were insufficient to support the people. This was Benjamin Franklin's retort to Smith's Capitalistic miracle in the previous century. Ricardo supposed that with population growth land rent would rise, profits would be reduced, capital would stop accumulating, and growth would stop. This was the business cycle. But Ricardo went on to suppose that if trade were to be spread throughout the world maximum efficiency would occur, which is to say the depressions in the English business cycle could be flattened. Not to put to fine a point on it, the depressions in the business cycle would be imported to the colonies. The motivating vision of Victorian Capitalism would be endless growth at home. The Third World was born to suffer the burden of the business cycle. In exchange the White Man's Burden was to bring civilization to a pagan world. English culture now had the intellectual and spiritual justification it needed to continue to dismember and digest the cultures of its far-flung Empire. Carlyle would apply a name to this form of economics: the dismal science.

Charles Darwin heard these economic speculations with the horrific potato famines going on periodically in Ireland and the ugly socialist revolutions burning across the continent in the background. He saw in the Malthusian mechanism the method whereby biological life would struggle over limited resources, and that those with small, random variations to the better would survive, slowly, so as not to disrupt his cozy idyll at Down House. This was, as he saw it, the manner by which species changed and evolved. In a letter to the American botanist Asa Gray he said, in his typical demurring fashion, that while selection by chance was difficult to believe, having a divine watchmaker involved with all the multitude of details of life was even less believable.

Although it was descent from apes that had most people fired up with fear and loathing since his grandfather's day, the issue of incest had the potential to be an even hotter concern if not handled delicately. Incest is a central consideration that any theory of biological evolution had to attend to with great care. Inbreeding is a normal process used by animal breeders to sustain slightly improved variations. Reference is made by Erasmus in his *Phytologia* to "the plan of Mr. Bakewell in England in respect to quadrupeds, who continued to improve his flocks and herds by the marriages of those in which the properties he wished to produce were most conspicuous."[6] The inbreeding of animals for improvement or the saving of seed from the best plants has been going on since the beginning of agriculture. It is the very blueprint of evolution through selection. In Charles's copy of this work this passage is double marked with the word 'good' in the margin.

The question of incest is such an emotional issue that it is virtually impossible to have a reasonable discussion at any level of culture or education. We all know we are against it, but there are differing definitions over time about what it is that properly becomes incestuous. In the Old Testament it is in the story of Jacob going to Laban for a wife that is the story that opens the discussion of incest for Judeo-Christian tradition. Laban, pruning his flocks of sports to improve his herd, attempts to take advantage of Jacob. He gives the rejects to Jacob as well as his less-favored daughter Leah. A rustic, wandering Hebrew tribe such as the tribe of Laban would have been relatively inbred. Jacob turns misfortune to fortune, and Laban is demonized. Out-breeding wins the day.

The story was further defined by urban intellectuals. As Philo Judaeus of Alexandria would elaborate, "for the sacred scriptures attributed to him [Laban] a flock devoid of all distinctive marks. And matter, without any distinctive characteristics, is without any marks in the universe, and so is in men the soul, which is destitute of learning and which has no instructors."[9] In Darwin's day consanguinity at the level of first cousins was allowed. The marriage between Charles Darwin and Emma Wedgwood was permitted even though Charles's mother was a Wedgwood. Darwin emphasized the theme of out-breeding for the purpose of increasing variety, but out-breeding too far was also frowned upon of course. One did not breed outside of one's class.

In notes written in 1838, Charles suggested that sexual generation kept in check variations that might be produced during the production of eggs and semen. Interbreeding, as he put it, was "to obliterate differences ... if animals became adapted to every minute change, they would not be fitted to the slow great changes really in progress."[7] However advantageous a significant variation might be, it could not be the source of a new species because it would be swamped breeding with the multitudinous individuals of the older type. Sex was seen as a hedge against saltatory change, too much variation.

Charles Darwin followed the ancient Judeo-Christian tradition going back to Adam of gender bias in its principle of the seed. He stated, "Woman makes bud, man puts primordial vivifying principle."[10] The male was responsible for the seed; the female was the ground in which it was planted. This being the common understanding, the function of sex was a matter of great confusion in many ways and at all levels of education in Victorian England. Some of this involved plants such as the algae, fungi, mosses, liverworts, ferns, and the like that Linnaeus had put into a twenty-fourth class, the Cryptogamia. These were plants that hid their sexuality, or so he thought. It was beginning to appear, however, these plants were reproducing without sex, that is, parthenogenetically. As early as the 1830s botanist William Hooker suspected that the cryptogams were asexual, but in general there was very little research being done on plant physiology in England.

In early notes Darwin speculated that asexual or coeval generation would be different from sexual generation in that it produced individuals that were completely like their parents. There was no possibility for evolution here he thought. Much later in life Darwin stated circumspectly in *The Various Contrivances by Which Orchids are Fertilized by Insects*, 1877, that most orchids are able "to resist the **evil** [sic] effects of long-continued self-fertilization."[8] This reveals the ambiguity that necessarily followed from his misunderstanding of sex as well as the incapacity to conceive of more than one biological method of change. Darwinism is limited by this perspective to the present day.

In *The Descent of Man*, 1859, Darwin made a distinction between natural selection—being the struggle for existence in competition with other life forms or in response to external conditions—and sexual selection—being a struggle between the individuals of one sex, the male, for the possession of the opposite sex. Secondary sexual characteristics, the tail feathers of the peacock, the antlers of the elk, were certainly the result of sexual competition he supposed. He was still unsettled on the utility of sex in 1861 when he opined, "We do not know why nature should thus strive after the inter crossing of distinct individuals. We do not in the least know the final cause of sexuality."[11] Natural selection and sexual selection are now collapsed into one category by Darwinists, which makes it difficult for them to understand Darwin and how he viewed the world, but the confusion over sex still remains, as we shall see.

When Darwin discovered that Alfred Russel Wallace was about to publish a theory of evolution that was essentially the same as his own that finally got him organized to publish his own thesis. Charles Lyell and Joseph Hooker arranged for both Darwin's and Wallace's theories to be presented to a meeting of the Linnaean Society on July 1 in 1858. The birth of Darwinism as we know it may be dated to this event.

The legend of Darwinism remembers its birth as violent and heroic like Samson pulling down the temple of the idolaters. It is one of the famous legends in the Anglo-Saxon history of science. However, in the annual report by the president it was stated that the year of 1858 had not been marked by any striking discoveries. It is difficult to make sense of this unless we realize that it is nearly impossible to imagine any English biologist at the time that hadn't heard or discussed some version of man's descent from the apes within the protective confines of a men's club. Darwin himself attests to this fact in his book that would be published in the next year. The popular *Vestiges of the Natural History of Creation* published by Chambers in 1844 was popular, continuing to outsell Darwin's soon to be published book until the end of the nineteenth century. The simple fact of the matter was that the subject of evolution was not new with Charles Darwin, else he would have cast his seed in vain.

Wallace was largely ignored in the process of sainthood that would now occur. Wallace was not the sort that the Royal Society recognized as a bona fide scientist. Wallace was a commoner who skulked around at the back of the hall listening to the lectures of Sir Charles Lyell. He was influenced by social reformers like Robert Owen and Thomas Paine, and he didn't wear a school tie. As to the matter of who would replace Aristotle as the great savant of natural history and who would turn Genesis into a metaphor, the bug collector from the Welsh village of Llanbadoc would have to take second place.

In November of 1859, *The Origin of Species* was published and Darwin was pilloried in the general public. The response was pretty much choreographed. Some of William Paley's generation of theologians were already mumbling under their breath about the secularization of the Church and the liberalization of theology. There was even a reactionary movement at Oxford whose intent was to reconsider the relationship of the Church of England with the Roman Catholic Church. The year after *The Origin of Species* appeared a collection of essays on religion by seven young, liberal Anglican clerics was published entitled *Essays and Reviews*. They backed Darwin's thesis. This stirred up a large debate, and *Essays and Reviews* sold many more copies than the *Origin*. One of those liberal Anglicans, Frederick Temple, went on to become the Archbishop of Canterbury. Darwinism was a sailboat that was being blown along by the winds of secularization in Victorian culture that had begun after the English Civil War two centuries earlier as well as the warming climate that had revived the culture in general.

In June 1860 the British Association met at Oxford for the purpose of discussing evolution. Darwin was too ill to attend. The Bishop of Oxford and Fellow of the Royal Society, Samuel Wilberforce, opened the discussion on Darwin's theory. Bishop Wilberforce was opposed to the theory, but he was not part of the reactionary Oxford Movement. As described by J. R. Lucas, the bishop criticized Darwin's theory on its own grounds, arguing that it was not supported by the evidence. This is the standard retort by entrenched intelligentsia to any new idea. Wilberforce also noted that the greatest names in science, including his mentors Sedgwick and Henslow, were opposed to the theory. The bishop could not avoid the clever oratorical jibe, no doubt learned as a student at Oriel College, Oxford, where he was a student. Addressing Thomas Henry Huxley, who was in the attendance, he asked whether it was from his father's side or his mother's side that he was descended from monkeys. A lady fainted and had to be carried out. Huxley's response was a tad self-righteous and not so entertaining. Admiral Fitzroy, Darwin's old commander on the H.M.S. *Beagle*, stood up shouting and waving a Bible in the air. While on the *Beagle* Darwin had quoted the Bible as if it was an unanswerable authority. Reports from the time indicate that most enjoyed the imbroglio and went cheerfully off to dinner at the conclusion.

The atmosphere was electric, but certainly not revolutionary. No one was even arrested. The general public would get into the act as reflected by the famous caricature showing Darwin's head on the body of a chimpanzee, but it certainly didn't hold a candle to the dangers faced by intellectuals in previous centuries. Whether the outrage was anything more or less than the dialectical vilification that is the standard fare for contemporary politics in the House of Commons or the daily assassination of character that occurs on every street corner to sell copy is hard to say. In any case, Wilberforce was caricatured in turn in *Vanity Fair* by someone signing himself Ape. In the cartoon Wilberforce is doing his familiar, compulsive, hand-washing motions during the Huxley-Wilberforce debate. This was a reference to his oily lecture style that had earned him the sobriquet "Soapy Sam."

Erasmus Darwin's biographer, Desmond King-Hele, finds a few minor ways in which Charles's evolution was different from his that of his grandfather. Even the core element in Charles's theory, natural selection, is to be found in Erasmus stated in much the same way. He actually thinks that Erasmus was slightly closer to the modern way of thinking than his grandson. As King-Hele sums it up, "The few differences between Erasmus and Charles arose from the different climates of opinion in which they lived; from Charles's ill-health and consequent diffidence; and from Erasmus's greater range of talents, particularly as a poet and inventor. Otherwise the two seem to offer an excellent example of hereditary likeness between a man and his grandfather."[12] It was about this time that the word "Darwinism" was used to describe both Darwins and their theories of evolution as if they were the same. Charles Darwin's role in introducing evolution to English culture almost seems to have been held for him like a squire his county seat.

King-Hele points to four reasons why Charles Darwin's theory of evolution became the English theory of evolution. First, intellectual opinion had been 'softened up' by so many earlier theories of a similar nature. Second, Darwin presents a mass of evidence modestly. His grandfather, by way of contrast, was the very image of dramatic advocacy. Third, he uses the very innovative approach of asking leading questions and then answering them. And fourth, he never gets into the issue of human descent from the apes.

We would enlarge King-Hele's fourth category to add Charles Darwin's agnosticism about *the first great cause* that was evoked in his grandfather's *Zoonomia*. Also, nowhere in the works of Charles Darwin do we find an enormity such as that committed by Erasmus Darwin, who defined *spes religiosa* as a superstitious hope, or a "maniacal hallucination" that in mild form produces "an agreeable reverie." Also, he was very discreet in the way in which he handled inbreeding. But in general it was indeed "The Orang Outang theology of the human race," as Coleridge put it, from which he seems to have benefited most by not mentioning.

The shrieking, anti-monkey mania of the English is common to northern peoples who don't live with monkeys. In lands where people do live with apes and monkeys there are very ancient myth traditions and basic linguistic features that show that descent from the apes or monkeys is a fairly obvious deduction. The Egyptian god Thoth—the one, the self-begotten, the self-produced—is credited with all the works of science. He appears as an ape in the underworld, which is to say the past. The Greeks equated him to their god Hermes. A rudimentary idea of evolution from the apes in tropical countries is likely older than caveman living with the bears in the north.

The first convert and most ferocious defender of Charles Darwin, Thomas Henry Huxley, coined the word for the philosophical posture of this new science—'agnosticism.' According to the *Oxford English Dictionary*, Huxley intended the meaning: "One who holds that the existence of anything beyond and behind material phenomena is unknown and, so far as can be judged, unknowable and especially that a First Cause and an unseen world are subjects of which we can know nothing." He took the construction from St. Paul's mention of the altar to "the unknown God." Dividing the world into matter (the domain of science) and spirit-mind (the domain of the church) was an acceptable compromise for the Anglican Church. This all flies over the head of most modern academics, where the word "agnostic" is most often used as a synonym for atheist.

The détente with the Church orchestrated by Huxley pruned away any lingering associations with alchemists, astrologers, herbalists, necromancers, and geomancers who advocated from a basis of older non-Christian ethical traditions. These none gnostic "scientists," a word newly coined by Whewell in 1840, would restrain themselves to empirical analysis. Science would be a service organization to Victorian culture. Agnostic scientists would use pagan Golden-Age Greece as their role model. Aristotle was their saint. They were in general accord with the great Catholic intellect St. Thomas Aquinas of the great days of the Latin Church before the Little Ice Age. In short, English science was embedded within a general epistemology that still presupposed a *purpose, prime mover*, and *final cause*. Darwin began to experience considerable notoriety and began to be consulted on all manner of issues like an oracle.

Charles Lyell, who created the stage for the drama to unfold, would remain equivocal, or should we say agnostic, on the issue of Darwin's evolution for the rest of his life. Lyell stated in the second volume of the *Principles of Geology*, 1832, "The testacea of the ocean existed first, until some of them by gradual evolution, were improved into those inhabiting land." Yet for him how the huge gulf between man and beast was bridged remained a profound mystery. It was one thing to demonstrate changes within species through breeding, as pigeon fanciers were wont to do; it was quite another thing to show the reality of natural selection by which simpler species changed into more complex species—apes into men—and no one had done that. He was open to the suggestion that there were different centers of creation to account for the diversity and territory of species. Furthermore, even if the Earth and its inhabitants had undergone a succession of revolutions and aqueous catastrophes interrupted by long intervals of tranquility, as some still supposed, he believed that we had no right to believe it was due to chance. Lyell was content to live with a Uniformitarian, old-Earth theory of geology and an original, immutable creation.

After the initial furor died down, after most of the older generation of Anglican theologians died out, Darwin's selection process came to be seen as a part of an orderly design in nature that had rewarded Christian English society. And it was agreed that by whatever method this evolution occurred, intelligence was the sole preserve of humans, because only humans have intelligence. Descartes established that with his accommodation to the Catholic Church over two hundred years earlier. Darwin was given an honorary degree from Cambridge, and he would be buried at Westminster Cathedral. It was the best of all possible worlds, especially if one held stock in the railroads. On the occasion of the two-hundred-year celebration of the birth of Darwin, Rev. Malcolm Brown, the head of the Anglican Church's public relations department, apologized for getting their first reaction wrong and thereby encouraging others to misunderstand.

As far as the business moguls of Victorian England were concerned, science was tolerated as long as it attended to practical matters and behaved in a proper manner. Thomas Henry Huxley described the attitude of the money men and industrialists toward science: "They were of the opinion that science is speculative rubbish: that the scientific habit of mind is an impediment rather than an aid in the conducting ordinary affairs."[13] A well-tempered and well-financed Christian religion and applied science would be their cohort in justifying the means and redeeming the ends of Victoriana. While this was going on Europe, especially Germany, was in a state of social and intellectual revolution. Carlyle supposed that the Dickensian mind was incapable of understanding German philosophy.

During the celebration of the two-hundred-year anniversary of the birth of Charles Darwin, a great to do was made about his abolitionist position. From his *Beagle* voyage notes we read, "I thank God, I shall never again visit a slave country. To this day, if I hear a distant scream, it recalls with painful vividness my feelings, when passing a house near Pernambuco, I heard the most pitiful moans, and could not but suspect that some poor slave was being tortured ..." We are moved by this section, but abolitionism had been a tradition in both the Darwin and the Wedgwood families for at least three generations. Abolitionism was also the position of John Wesley in the previous century when there was considerable opposition, it was the position the Creationist William Paley, and of Samuel Wilberforce who ridiculed Charles Darwin's theory of evolution.

In his grandfather's day it was a courageous position to take, but the British had outlawed the transatlantic slave trade in 1807 and slavery in the British Empire in 1833. Anti-slavery had become little more than a liberal position. Bringing Charles Darwin's abolitionist stance up at such a late date for the first time is really just a case of polishing up the handles on the big front door by and for people with little historical education. Far more impressive is the absence of Eurocentrism in Darwin's thinking. The preeminent nature of one's own class and race was a very common belief in his day and, for that matter, today. He was aware, for instance, that it was only a few centuries earlier that Turkish culture was dominant and superior anything cobbled together in the way of culture or military strength by the European cultures.

The outburst of European exploration and exploitation of the world at large was a response to the Little Ice Age. The planting of slave-driven colonies was always hazardous and always had it opponents. No European state thought to introduce the practice on their home soil. But the abolition of slavery in the British colonies did not mean that all was well on the labor front. Now that the famine, plague, and war of the Little Ice Age were no longer wiping out a third of the population with regularity, overpopulation was giving a free hand to the Malthusian principle as it applied to wages and working conditions. The Industrial Revolution was introducing abuses as profound as anything seen under the whip of classical slavery. In the real world no more than a few miles north of where Darwin took his late morning tea, in his comfortable parlor overlooking a fine green lawn and a greenhouse where he kept his orchid collection, there were labor and poverty issues provoking the formation of a new political power.

We recall that the Poor Law Commissioners of the 1830s believed, with Malthus, that relief perpetuated the poverty it was intended to relieve. Their thought process was described in a 1963 article by M. Blaug, The Myth of the Old Poor Law and the Making of the New," as making no attempt to be objective about the analysis of the evidence. The commissioners selected the facts and opinions that supported their view, "so as to impeach the existing administration on predetermined lines ... what evidence they did present consisted of little more than picturesque anecdotes of maladministration."[14]

Selected evidence and picturesque anecdotes sometimes do reveal a hard truth, but a hard-nosed response to the problem can be just as short-sighted and ineffective as charity, perhaps even more so.

At the time Darwin was publishing his magnum opus, one of the greatest sociological studies in English was being published in the popular press by Henry Mayhew. The cofounder of *Punch* magazine was a bohemian intellectual who had absolutely no standing with the Oxbridge crowd. *London Labour and London Poor* came out in its last editions in 1862. In the introduction to the Dover edition, John D. Rosenberg says, "The image of London that emerges from Mayhew's pages is that of a vast, ingeniously balanced mechanism in which each class subsists on the drippings and droppings of the stratum above, all the way to the rich, whom we scarcely glimpse, down to the deformed and starving, whom we see groping for bits of salvageable bone or decaying vegetables in the markets."[15]

Charles Dickens emerged from the drippings and droppings of English poverty to focus English self-absorption on the horrid London circus. Dickens personified the Irish Sweepstakes reality of poverty in the life of Tiny Tim who, improbably, does not starve to death and Ebenezer Scrooge, who, miraculously, was inspired by a blast of Christian charity to save Christmas. A search of the complete works of Charles Darwin online reveals no citations for Henry Mayhew or Charles Dickens. Darwin was imbued with the myth of the healthy English lives in the lower classes, a thesis he supported with picturesque anecdotes.

There was another slant on the Dickensian world taken by the son of a Capitalist family by the name of Friedrich Engels. Engels had to flee the Continent because of his Socialist activities. The revolutions on the Continent by the middle of the nineteenth century were the response to the reactionary climate surrounding autocratic government and the abuses of industrialization in western Europe. Engels would use his adopted country to write a classic work on *The Condition of the Working Class*, published in 1844. He was a supporter of his fellow Socialist refuge Karl Marx, who lived in extreme poverty in the heart of the richest city in the world—London. The conditions of his London impoverishment would ruin his health and shorten his life. With financial support from Engels, whose family business in the Midlands was thriving, Marx would write the economic rebuttal to the economic principles of Smith and Ricardo. This came to be the fourth book of *Capital*. Both Engels and Marx read *On the Origin of Species* and gave it high praise. It took teleology out of natural history as they saw it, and this was exactly what they were intending with their dialectical and materialist analysis of human social evolution. They were in no position to judge the reality of Darwin's

Uniformitarian selection process, but they were impressed by the stunning effect it had on them. It was a long way from Genesis.

It was from a Socialist perspective that more than a few adjustments in metaphor would be made to religious tradition and the Church would be attacked directly. The fact of the matter, however, is that the Church was no longer seen as the main culprit among those with revolutionary intent. During the revolutions of 1848 nobody was nailing maxims to the door of the cathedral. The incendiary ideals were free speech, parliaments, religious liberty, and jury trials: the ideals of the American and French Revolutions. The violence of the Reformation had already put the Church in a secondary role. Capitalism was now enemy number one. Marx sent Darwin a copy of his book *Capital*. Darwin replied in 1873 in his typical friendly and humble manner, saying that it was well beyond his competence, but he hoped that the general extension of knowledge would add to the general happiness of mankind.

The sociology that Charles Darwin was allied with was that of his friend Herbert Spencer. Spencer was the son of a Methodist schoolteacher in Derby who, as Erasmus Darwin might have described it, fell into the featherbed of Quakerism. His father had served as the secretary of the Derby Philosophical Society. This was the group that Erasmus Darwin founded in 1783 after he had moved to Derby from Lichfield. Erasmus invited his old friends, the Lunitics, to hold joint meetings, and his son Robert and Joshua Wedgwood joined. The young Herbert Spencer was taught empirical science by his father and was introduced to the ideas of biological evolution held by both Erasmus Darwin and Jean-Baptiste Lamarck by members of the Philosophical Society. This was three decades after the passing of Erasmus Darwin. Spencer was a polymath of extraordinary proportions. Many in his day described him as the English Aristotle.

Before Darwin published his great work, Spencer was engaged in organizing a System of Synthetic Philosophy that demonstrated the principle of evolution in biology, sociology, psychology, and morality. He had already published *Principles of Psychology*, in which he supposed that the repeated association of ideas caused changes in the brain that could be passed along to succeeding generations using the Lamarckian mechanism of use-inheritance. After the publication of Charles Darwin's great work, Spencer began to be swayed to the idea of natural selection. The weak and frequently ill Spencer would paraphrase the natural selection thesis of the chronically sick Charles Darwin with the familiar axiom, "survival of the fittest."

As Darwin's biological evolution was based upon Lyell's geology, Spencer's sociology would become based upon Darwin's biology. Spencer immediately saw an analogy between natural selection and the Parable of the Sower in the New Testament. As he would put it in *The Study of Sociology*, "The parable of the sower has its application to the progress of Science. Time after time new ideas are sown and do not germinate, or, having germinated, die for lack of fit environments, before they are at last sown under such conditions as to take root and flourish. Among other instances of this, one is supplied by the history of the truth here to be dwelt on—the dependence of Sociology on Biology."[16]

Almost no Victorian scientists were without an education in religion. So it was not unnatural for them to see an analogy between natural selection and the Parable of the Sower. We find little indication that natural selection was a great cause for concern in Victorian culture. Intellectuals of all stripes discussed the issue with differing opinions, with the Bible offering support to all by what was said as well as what was left unsaid. On the issue of natural selection or chance, John Milton had proclaimed two centuries earlier in *Paradise Lost*, "That led th' imbattelld Seraphim to Warr Under thy conduct, and in dreadful deeds Fearless, endanger'd Heav'ns perpetual King; And put to proof his high Supremacy, Whether upheld by strength, or Chance, or Fate …" Chance was just another face of the mysterious ways of the Lord, part of the irreducible complexity of life. Chance offered also offered a way to avoid unpleasant realities. Chance offered an excuse to allow such realities to be swept under the rug and ignored.

Almost no modern scientists or historians of science have any training or knowledge of religion. All they have in this area are superstitions. Historians of science often have very little interest, skill, or understanding of the sociology of culture. In Thomas S. Kuhn's *The Structure of Scientific Revolutions*, he supposes that the unteleological crap game of evolution proposed by Darwin was the greatest violation of all, but the crap game of evolution only becomes an issue after science is removed from the protective husk of theology and the protective overview of Victorian culture. Chance becomes an entirely different animal in the context of the secularized cultures as they committed their heinous bloodbath of the first half of the twentieth century. The projection of existential concerns onto the past as Kuhn has done has contributed to the legend of Darwinism that would evolve in the twentieth century.

The Uniformitarianism of Charles Lyell in geology and the Uniformitarianism of Charles Darwin in evolution would be continued in the sociology of Herbert Spencer. He takes note of the recent evolution of human society. In ages past the welfare of the king was everything, the welfare of the people nothing. Now he sees that the welfare of the nation is paramount, the king acting as a civil servant. Improvising on the theme of Adam Smith about the inherent ambiguity in the very concept of a Christian high civilization, he says, "Ethically considered, there has never been any warrant for the subjection of the many to the few, except that it has furthered the welfare of the many."¹⁷ This exception is enough, as he sees it, to warrant the high degree of class-subordination that existed in English culture. It was the duty of the individual to maintain and perform their role in society and to be happy under the circumstances. A stark contrast was made with the savage peoples and their ignorant and violent ways. This was an intellectual worldview that suited Charles Darwin.

In the progression toward a more just society Spenser would give a Uniformitarian definition of social evolution, he would give the state religion a Darwinian polish:

> *To maintain the required equilibrium amid the conflicting sympathies and antipathies which contemplation of religious beliefs inevitably generates, is difficult. In the presence of the theological thaw going on so fast on all sides, there is on the part of many a fear, and on the part of some a hope, that nothing will remain. But the hopes and the fears are alike groundless; and must be dissipated before balanced judgments in Social Science can be formed. Like the transformation now in progress is but an advance from a lower form, no longer fit, to a higher and fitter form; and neither will this transformation, nor kindred transformations to come hereafter, destroy that which is transformed, any more than past transformations have destroyed it.*[18]

Charles Darwin had been fully vetted as the spokesman for his age without raising a hand in his own defense. He had survived a life-long burden living under the shadow of his grandfather and of the utopian ideals of the Enlightenment. As he got older he gained in self-confidence and intellectual courage. He wrote a biography of Erasmus's life giving him the credit he was due. It was edited by his daughter Henrietta, who removed nearly everything favorable to Erasmus. The Victorian Henrietta was highly disapproving of her notorious great-grandfather and his raucous, freethinking age. Henrietta's attitude toward Erasmus has prevailed as the general attitude down to the present day.

In Darwin's later writings we find three ideas that were not part of his grandfather's thesis, and they were not minor as suggested by King-Hele. The first idea was that important changes came from the body of the parent organism in the form of gemmules that influenced the reproductive germ. He called this Lamarck-like hypothesis "pangenesis." It was borrowed from Aristotle and was fully explained in *The Variation of Animals and Plants under Domestication*. Pangenesis, if remembered at all, is called unfortunate by Darwinists. Lamarckian theories such as this would continue to plague Darwinism down to the present day.

Charles Darwin would concentrate on ethnology toward the end of his life. His second unique contribution was made in *The Expressions of the Emotions in Man and Animals*, 1872. He would look at communication in the animal world and come to the conclusion that singing came before language in man. The academy prefers the theory that in the beginning was the word, and that is that. Darwin's suggestion, if remembered at all, is called unfortunate by Darwinists. Nevertheless some very elegant multidisciplinary research has confirmed the importance of song and dance to the unique development of human linguistic communication.

In the year before his death in 1882 he published *The Formation of Vegetable Mould* containing his observations on the behavior of earthworms. In this work he would return to the question of intelligence in lower animals that he had already broached in his major work. Late at night he would put on his heavy wool cape and go out into the garden with a candle to observe earthworms as they ventured forth well before the earliest bird. He would point to the instincts that can be observed even in higher animals, the senseless repetition of a behavior that is entirely inappropriate to the occasion, as the conformation of an autonomic, stimulus-response intelligence. "With animals, actions appearing due to intelligence may be performed through inherited habit without any intelligence," he would say, adding, "although aboriginally thus acquired." It was the unusual response, the unpredicted response, the response that is appropriate in an entirely foreign setting, that gave him pause. Nine out of ten worms would recoil from the light he would shine on them. One would not. It was the exception that led him to suppose, "Nor are their actions so unvarying or inevitable as are most true instincts."[19] Darwin's third unique contribution was that there was an evolution of intellectual capacity in the animal kingdom.

Having removed the creator from the trials of having to explain good and evil, he saw sentient beings evolving through natural selection such that pleasurable sensations served as their guide in order to avoid or at least reduce pain, hunger, thirst, and fear. This utopian stage of his intellectual development, if remembered at all, is called unfortunate by Darwinists. To find consciousness in lower forms of life would violate one of the founding ideals of science permitted to Descartes by the Catholic Church. The most important element embedded in Charles Darwin's theory, however, a principle that Charles Darwin continued to subscribe to all of his life, was Uniformitarianism. He lived by the canon of *"Natura non facit saltum."*

The image of Charles Darwin that is beloved today is not the tense young man who was being constantly reminded of his grandfather's genius; it is not the young man undergoing the horrific trial of initiation in the around-the-world trip on the *Beagle*; it is not the grim Darwin who was sick with anxiety, who allowed Thomas Huxley to stand up against Bishop Wilberforce; it was not the Darwin toiling in the coils of his mother's faith. The image of Charles Darwin that is beloved today is the man dressed in black with the St. Nicholas beard of white living in the moment with the help of his loving family, servants, and friends. This is the Buddha-like Darwin for whom life was a happy state of affairs. This is the tranquil image of Darwin beloved today. The creative ideas he was having at this time are ignored.

The history of science is riddled with fallacies, fallacies that become paradigms, paradigms that become régimes. Darwinism refers to a theory of evolution that became acceptable to upper-class Victorian culture, not to the man himself. John Farley points out that historians have done a very poor job of telling the story of nineteenth century biology; too many of them seem to suggest that the only significant work done in nineteenth century biology is to be found in the Darwinian notebooks. The history of science is also dominated by the perspective that the evolution of ideas is a series of brilliant ideas that formed more or less spontaneously in the capacious minds of a few individual men of genius belonging to our tribe without any reference to the peculiarities of time or place. This makes the achievement a miracle and the individual a saint. To gain a much better perspective on Victorian science we need to compare it with parallel developments on the Continent with the help of Farley. Only when this is done can we begin grasp how social setting shapes the ideas that we have.

CHAPTER 4
Science on the Continent

IN GENERAL THE CATHOLIC countries suffered less than the Protestant countries during the Little Ice Age and the Reformation. Catholic countries such as Italy and Spain had the best universities at that time. We don't know if there were any universities worth considering in Scandinavia, but we do know that playing the violin was banned because it was the instrument of the devil. With warmer weather the Ladder of Perfection was still propped up against the wall of the Vatican: all creatures stood in a hierarchy in the Chain of Being on the way to Providence. The snake was on the bottom rung of this ladder in the animal kingdom; below them came the plants and the minerals. The angels were just below God.

In France the academy was dominated by Catholic theology in the eighteenth century and by the scholastic formalities of its Jesuit missions. In the middle of the eighteenth century George-Louis Leclerc de Buffon, the naturalist, mathematician, and cosmologist, began publishing his great work *Histoire naturelle, générale et particulièr* in thirty-five volumes. It included everything that was known about the natural world; it was read by everyone with the slightest pretense of knowledge. He suggested that the solar system might have formed out of an impact of a comet with the sun. He thought that the Earth might have been older than 4004 BC, perhaps as old as 75,000 years. He proposed that a constantly recycling mass of organic molecules was released from the organism at death and re-formed again into a new organism. Form was imprinted on this indestructible mass by the *moule interieur* that

Buffon likened to the new gravitational force discovered by Newton. Change imposed by a vital energy such as this was a thesis upon which the Church would not relent. It was not part of the agreement worked out by Descartes. For outrages such as this, Buffon's books were burned and he was forced to recant. Buffon recanted, but in a section on birds in one of his books he included an illustration of the black and white magpie—a bird that mimics speech without any knowledge of meaning—standing in front of a church.

The French Revolution at the end of the eighteenth century would occur after the secularization of French culture had been well advanced by the Enlightenment, at least at the upper strata of culture. The Reign of Terror that came out of a crop failure in 1789 smashed the classical mold of governance. The French Revolution was quite unexpected by the Philosophes. The upper and educated classes were suddenly aligned against the outraged and starving poor looking for revenge, not enlightenment. Aristocrats were decapitated. Science went underground. Out of the rubble Napoleon created a new secular aristocracy, revived science, and used it as the basis and justification for creating a modern French state. It was a dictatorial conversion of Natural Science into Natural Law. Napoleon cleanly and violently separated the French academy from the Catholic Church. This destroyed the détente that had been achieved by Descartes in the seventeenth century. The relationship between religion and science in France became an angry divide.

Lamarck published his first suppositions about evolution in this environment, in 1801. His theory was influenced by Erasmus Darwin. Unlike the Darwin family theory of evolution his theory would co-adapt with political theory over time, against the church. Like Charles Darwin he believed in the inheritance of acquired characteristics, but in the different political environments Darwin's theory would migrate to the position of microscopic random changes while Lamarck's theory would migrate to the position of macroscopic changes through volition. It could be adapted to political reform.

With the Church out of the way, there was a heated argument between the catastrophic geologists, the Neptunists and the Vulcanists. Georges Leopold Chretien Frederic Dagobert Cuvier supposed that violent revolutions marked the different geological epochs and that life was completely wiped out by floods—hence the fossils of extinct species. He is best known for clearly establishing the reality of extinction. Life, he thought, had then been created anew by the Creator. Cuvier's old-Earth thesis did not require an evolution of species. Cuvier, the greatest zoologist of his time, saw animals as functionally integrated wholes. There was no change in the species between catastrophic events. The last of the epic overthrows was Noah's flood. Cuvier was a Neptunist. This theory supplied an answer for fossils, always a thorn under the saddle of biblical fundamentalism that allowed for one epoch and one epoch alone.

In contention with this theory was that of Jean Baptiste Armand Louis Leonce Elie de Beaumont. Elie de Beaumont appealed to somewhat less violent overthrows marking the epochs that were in abundant evidence in the geological record. He pointed to the evidence of volcanic eruptions of large scale. He was a Vulcanist. Life was not completely extinguished; therefore, there was no need to appeal to secondary acts of creation. Charles Lyell visited with Elie de Beaumont in 1833, and he tried to dissuade him of his catastrophism but was unsuccessful. The celebrated geologist and naturalist Leopold von Buch sided with Elie de Beaumont. In von Buch's *Description Physique des Isles Canaries* he clearly expressed the belief that varieties slowly become changed into permanent species, which were no longer capable of intercrossing.

Von Buch's thesis was enthusiastically elaborated by Etienne Geoffroy Saint-Hilaire. Geoffroy stated, "The external world is all-powerful in alteration of the form of organized bodies ... these [modifications] are inherited, and they influence all the rest of the organization of the animal, because if these modifications lead to injurious effects, the animals which exhibit them perish and are replaced by others of a somewhat different form, a form changed so as to be adapted to the new environment." Buffon and Voltaire applied this thesis to New World flora and fauna that they thought were degenerate species of European forms.

This created the setting for a famous debate between Geoffroy and Cuvier. The question at hand was did form or function determine the phenomena of life? Whereas the meeting in 1860 at the British Association at Oxford was for the purpose of discussing evolution versus Genesis, the debate sponsored by the French Academy of Science in 1830 concerned two different evolutionary theories. Despite their differences, they remained friends and they respected each other's research. In the future there would be a synthesis of the two ideas: organismal lineages change over time in response to changing environments, and their form constrains the functions that they can take. In the popular press at the time, however, Geoffroy was depicted as an ape wearing glasses showing that the English commons and French communards were equally offended by scientific intrusion into the issues of genesis. The Geoffroy/Cuvier debate was thought to be more important than the political revolution that was going on at the same time by the great German philosopher Johann Wolfgang von Goethe.

The German-speaking peoples at the time were represented by 234 independent countries, fifty-one free cities, and about 1,500 knightly manors. German evolutionary theory was rather more focused on the need for social unity than a need to throw out an old, decayed state, as was the case in France, or to protect an established, middle-aged state, as in England. Frederick II of Prussia was an enlightened king who followed the Philosophes in France with interest and the developments of science in general. The antagonism directed toward science from the Church would be largely restricted to southern, Catholic Austria. In the north, Goethe led the German enlightenment. Looking through their microscopes German biologists would see an embryological unity in life in harmony with the philosophy of *Wissenschaftideologie*—the great organic unity of all knowledge. The intimation of biological evolution was palpable. Goethe let the genie out of the bottle in 1784, on the night of March 27, in a missive to Herder:

> *I have found neither gold or silver, but something that unspeakably delights me—the human Os intermaxillare! I was comparing human and animal skulls with Loder, hit upon the right track, and behold—Eureka! Only, I beg of you, not a word—for this must be a great secret for the present. You ought to be very much delighted too, for it is like the keystone of anthropology—and it's there, no mistake! But how?*[1]

Goethe had experienced a candlelight epiphany in his discovery of a human intermaxillary bone integrated into the upper cheekbone. It was homologous to a bone hitherto known only in animals. Goethe had discovered the key that pointed to the evolution of the human being from an ape. This concept had existed heretofore in European tradition only as a metaphor in occult tradition: the ape representing man untransformed by alchemy. His biographer Emil Ludwig observes, "How did Goethe come to discover what had escaped the adepts? Because, as a dilettante, he examined the skull with an open mind, because his eye was unprejudiced—he was not looking only for what system and instructor had pointed him to. That eye was thinking while it gazed; and during its years of a roving apprenticeship to natural phenomena, it had perceived relations, transitions, gradations. And how, once more? Because there was a soul behind that eye which divined, from the gradual development of its own powers, from the slow difficult unwinding, coil on coil, of the mighty cable, that Nature obeyed a kindred law."[2]

In 1790 Goethe published the *Metamorphose der Pflanzen* four years before Erasmus Darwin published his magnum opus. As described by Linda Orr, Goethe proposed, "that all parts of a plant represent modification of a type-leaf and that all plants are the morphological developments of one type-plant."[3] This was called an *Urbild* or primordial structure. Goethe's poem *Metamorphose der Tiere* extends the principle of *Urbild* to the animal kingdom: "Every organism shapes itself after eternal laws. And the rarest form mysteriously preserves the primordial structure."[4] The *Urbild* principle was the polar opposite to the evolutionary principle of Darwinism, where contests between individual organisms ratify microscopic evolutionary accidents.

Goethe was a polymath of the stature of Erasmus Darwin. His accomplishments spanned the fields of poetry, drama, literature, theology, philosophy, humanism, and science. He magnum opus was *Faust*. It was possible that he was familiar with Marlowe's *The Tragical History of Doctor Faustus*, since he was well read on Shakespeare. Goethe's Faust seeks knowledge through Nostradamus, by manipulations of the sign of the Macrocosmos, and from channeling an earth spirit but in the end was not satisfied. He translated the Gospel of John, but is having trouble translating the word "*logos*" as in, "In the beginning was the Logos ..." A dog that had been following him is irritated by this and reveals himself to be Mephistopheles in the guise of a traveling scholar. Mephistopheles offers to give Faust a moment in which he would be free of all striving after knowledge. The wager is that if Faust begs for that moment to continue, Mephistopheles would win Faust's soul. Goethe never seems to experienced such a moment.

After Frederick II, Germany was ruled by mediocre kings and at the beginning of the nineteenth century came the threat of Napoleon. After Napoleon, the search for the ideal man and of natural law of the Enlightenment gave way to a search for social realism and a unified German state. In his monumental *Deutsche Grammatik*, Jacob Grimm (1819) helped to set the scene. Grimm's grammar was an affectionate survey of German cultural history. The great nineteenth century German historian Leopold von Ranke gave his discipline a formal academic standing with a statement of objective processes necessary to the business of doing history. Von Ranke saw that in each state some particular moral or intellectual principle predominates: a principle prescribed by an inherent necessity, expressed in determinate forms, and giving birth to a peculiar condition of society or character of civilization—a *Zeitgeist*.

Georg Wilhelm Friedrich Hegel would redefine the linear concept of history. First, he supposed, there was a historical imperative to history. There was a core of change that could not have happened any other way. Second, change was in the direction of progress. So far this was a restatement of Augustine's theory of history. Hegel then argued for cycles of change. Any given phase tends to be confronted by its opposite as the honest opposition. This, in turn, tends to be replaced by a phase that is a resolution of the two opposed ideals. This dialectical process is remembered by the formula: thesis, antithesis, and synthesis, with synthesis being the next thesis. This was the sort of thing that English historian Thomas Carlyle thought was beyond the scope of the Victorian mind.

One of the most stylish presentations of evolution on the continent came from the great French historian Jules Michelet. Michelet evolved with his times. His portrayal of the geographical basis of French history was superior to anything attempted by a historian until recent times. First he suggested an analogy between Napoleon's violent rise and fall and Buch's volcanic theory of geological change by heat and pressure. This was the intellectual revolution that overthrew the awesome establishment of Cuvier and the biblical deluge. The second revolution in Michelet's thinking occurred around 1830 when Buch and Elie de Beaumont were supplanted by a new vision authored by Charles Lyell, "where for the first time the Earth appears as a worker manufacturing herself, through constant and patient toil without any violent jolts."[5]

Michelet would see Lyell as the school of peace and a welcome relief from either Cuvier or Beaumont, as well as Napoleon and the school of war. With Lyell, of course, would come Darwin. *In La Montagne*, composed in 1864, Michelet would embrace the school of peace as a total science that would become part of an international movement. "What strengthens this geology of peaceful transformations is the fraternal support it has among naturalists, among the great masters of metamorphosis, or Geoffroy Saint-Hilaire, Goethe, Oken, Owen, and Darwin, who reveal how an animal, under the varied influences of its environment, and through the instinctive impulse leading it to choose what is best for it, how, I repeat, it has made and modified itself."[6] This, of course, is not very Darwinian, except in the general sense of its Uniformitarianism. In English science metamorphosis was something that only happened to rocks under pressure, and evolution did not proceed through intention.

The metamorphosis of Michelet, as Roland Barthes points out, is in the camp of Lamarck and the transformists. In order to get Lyell Michelet had actually collapsed Lamarck into Darwin. "Let's stop a few minutes," says he, "at the solemn passages where uncertain life seems still to linger, where nature questions herself, testing her will. Shall I be fish or mammal? Important question, troubled hesitation, long and varied combat."[7] This was the spirit of the Enlightenment injected into an English evolutionary principle. This was the evolution of the butterfly released out of the caterpillar, in contrast to the three-toed horse evolving out of a slightly smaller three-toed horse. Michelet inaugurated a French tradition for broad, integrated scholarship. The French academy would, by the end of the century, have Emile Derkheim investigating the machinations of society under the heading of sociology and Ferdinand de Saussure discovering the metalanguage of communication called semiology, the study of signs.

When Camille Flammarion published *L'Atmosphere* in 1872, it was the first popular presentation on the subject of the weather. Chapter seven in the second book covered shooting stars, bolides, aerolites, and stones falling from the sky. The short résumé of such events, as they were then known, was temperate and reasonable. By the year 1872 the religious wars had long passed, the French Revolution had finally expended its energy for social reform, the climate had warmed up, and Sir Charles Lyell had pronounced that nothing of consequence ever fell out of the sky. The largest meteorite then identified weighed 240 pounds. The origin of these was a mystery. Since meteorites were apparently of non-terrestrial origin, it was suggested that they were stones hurled out of volcanoes on the moon.

Ernst Haeckel popularized Darwin's theory of evolution in Germany. He earned a doctorate in physiology at the University of Jena. He visited Darwin, Lyell, and Huxley at Down House in Kent in 1866. Once again the translation of English evolution into a Continental culture resulted in some unexpected usages. Haeckel continued to lean toward Lamarck and posited the concept of heterochrony, which involves changes in the timing of embryonic development over the course of evolution. Haeckel was responsible for the idea that ontogeny is the short and rapid recapitulation of phylogeny: the development of the fetus from the egg is a recapitulation of the evolution of a species from simpler forerunners. If a land animal had ancestors that lived in water and used gills, then each embryo of that animal continues to develop gills as did its ancestors, even though the gills may be lost during a later stage of embryonic development. These ideas are a reflection of Goethe's *Urbild* and of a skill set in science that had gone missing in England—microscopy.

The great English genius Robert Hooke was one of the fathers of the microscope, and it was he that coined the word 'cell' to describe the basic unit of life. But after him little happened in that field in England. English biologists thought the discoveries being made with the microscope were trivial. Darwin would see what he could see with a hand lens. Darwin was agnostic on *the first great cause* endued with animality, on the microscopic life forms revealed by Leeunwenhoek and Hooke, and on embryological development and its relation to evolution. By contrast to Haeckel and German science we can see that Darwin proposed a gross anatomy of evolution in his magnum opus. Haeckel would also coin the word 'ecology' to mean the interacting balance of relationships between organisms and their environment. This was not principle that could go very far with Darwinism as its organizing principle that was largely devoid of an understanding of the recursive processes of an interacting universe.

Darwin is famous for predicting a pollinating moth with an unusually long tongue, necessary to suck from the one-and-a-half-foot nectaries of an orchid from Madagascar, *Angraecum sesquipedale*. In 1903 such a moth was discovered and named *Xanthopan morganii praedicta*. At the time when Darwin published *On the Origin of Species by Natural Selection*, no remains of human ancestors had yet been identified. Haeckel predicted that such remains would be found in the Dutch East Indies, he described what they would look like, he charged his students to find them, and one of them did. Eugene Dubois found the remains of Java Man.

Haeckel also promoted the idea of the white European male as the pinnacle of evolution. The English were certainly not averse to this Social Darwinian application of Haeckel's thesis, whereby primitive, colonialized societies needed guidance from the higher forms of civilization, hence the White Man's Burden. This theory was widely embraced by all Western European cultures and only became politically incorrect after the Second World War in the next century. German evolution saw a deeply-rooted *Urbild* common to life overlaid by cycles of change unfolding to a great, national *Wissenschaftideologie*; French evolution was a metamorphosis containing elaborate interconnections; and English evolution imposed microscopic change, without the benefit of the microscope, change that satisfied John Locke's empiricism and was well suited to the slow, labored pace of empire.

Theories of evolution have always been of great moment because of their connection to social tradition, but they contributed little to the development of modern biology. The great advances in biology in the nineteenth century occurred on the continent in the area of sex and cell theory, while the English rested on their laurels. The single most challenging intellectual issue to overcome involved the understanding of the seed. The practitioners of the oldest medicine, veterinarian medicine, have always known that the biological ideology followed by Solomon and Aristotle was false. The Epicureans, who appreciated woman as an equal, thought that the female contributed semen to the embryo. The Roman naturalist Lucretius also saw past the limitations of the patriarchal precept:

> *A female generation rises forth*
> *From seed paternal, and from mother's body*
> *Exist created males: since sex proceeds*
> *No more from singleness of seed than faces*
> *Or bodies or limbs of ours: for every birth*
> *Is from a twofold seed; and what's created*
> *Hath, of that parent which it is more like,*
> *More than its equal share; as thou canst mark,*
> *Whether the breed be male or female stock.*[8]

Although the Epicures and Lucretius stand out as an anomaly to convention, they do allow us to see is that there was a social issue at stake not an intellectual thesis. Paul enjoins the female to cling to the male for only through the male can she know God. Here we see the principle applied with its inevitable and fully rationalized spiritual conclusion. St. Thomas Aquinas supposed that woman contributed only passive matter to the offspring, while man contributed active form. Woman, he supposed, was the weaker vessel; she was emotion over reason. Charles Darwin followed this tradition without comment.

The modern story of the seed begins with William Harvey, the English doctor who solved the enigma of physiological function of the heart while being schooled by the anatomist Vesalius. Harvey dissected the uteri of does in King Charles the First's forests at different stages after coitus. Harvey did not find the mass of blood and seed that was expected, having been schooled on Aristotle. Harvey supposed that the matter that was created or formed by the vital element of the semen was an egg, by analogy to a chicken's egg, but without a shell. He supposed the egg was quickened by some insensible spirit from the male, like magnetism.

By the end of the seventeenth century the Cartesians had excluded all psychic, or occult, or vital spirits from their explanations. This led them to the doctrine of germ preexistence. Preformed embryos, it was supposed, must have been created by God in the original act of creation, and every germ has preexisted since that miraculous day. Ovism, as it was called, became the common understanding of the philosophers of the day. Eggs are large and are easy to see. Sperm would not be discovered until the invention of the microscope.

One of the first great microscopists, Antoni van Leeuwenhoek, was a jealous and secretive man. He spied into the Lilliputian world with his lenses, allowing no one to see but he, with one exception. In 1680 he discreetly attributed the discovery of the animalcules found in semen to his apprentice. Where and how the apprentice got this semen is not mentioned. In any case, the discovery of animalcules in human semen stirred the pot. At first these animicules were thought to be parasites of the testes. Five years later Leeuwenhoek changed his mind and decided that these sperm were the preformed human. This was the origin of the theory of spermism.

In 1678 Nicolas Hartsoeker also discovered animalcules swimming around in semen. He imagined that he could see a tiny little human homunculus gathered up in the head of the animalcule. He imagined that they would attach themselves to the egg by an umbilical vessel in the tail. He even drew a picture. The ovists, however, were still in the majority. To them the function of the sperm, if any, was to stimulate the egg. The spermists held their ground.

They quoted from Aristotle's *Compleat and Experienc'd Midwife,* his analogy between the male seed and the seeds of plants.

As early as the eighteenth century Joseph Koelreuter, professor of natural history at Karlsruhe, laid the groundwork for the theory of epigenesis. Epigenesis rejected any sort of preformation in the egg or the sperm: fecundation was supposed to be the joining of two fluids that react and produce a new being. This dynamic chemical view of spontantous fertilization became popular among some German naturalists even though it severed that vital connection with Adam. The battle between the ovists, the spermists, and the epigenesists would continue on into and through the nineteenth century until the mounting burden of absurdity would topple over with a nudge from an Augustinian monk. But fall was set up by Hartsoeker's famous homunculus that would eventually push the gimcrack theory of preformationism and the theory of epigenesis into well-deserved retirement.

One of the most important books on the history of biology in the twentieth century was *Gametes and Spores, Ideas about Sexual Reproduction, 1750–1914* by John Farley. In this book he describes how Schwann argued that the ovum described by Karl Ernst von Baer—the father of embryology—was probably a cell while the germinal vesicle reported by Jan Evangelista Purkinje—who created the first department of physiology at the University of Breslau in Prussia—was a nucleus. Between 1838 and 1842 Matthias Schleiden and Theodor Schwann developed cell theory: the bodies of plants and animals were made up of cells. Of equal importance was the pollen theory of Schleiden. Schleiden announced to a startled botanical community that from a grain of pollen a tube actually grew down into the embryo sac and implanted the embryo—not in the form of a preexisting but miniaturized body, but as a cell.

In 1858 Rudolf Virchow, called the father of pathology, enlarged upon this thesis, publishing his theory that all diseases were diseases of the cell. He also supposed that all living things came from living things in contradistinction to the concept of spontaneous generation that was a reasonable assumption before the microscope and the discovery of microbiological life. Cell theory took hold, giving us the catchy axiom that cells multiply by dividing. From this Louis Pasteur and Robert Koch would develop the germ theory of disease. With cell theory modern biology was born.

In 1865 Gregor Mendel violated a fundamental principle of the Judeo-Christian scientific tradition. "If the influence of the egg cell upon the pollen cell were only external," he wrote in obvious reference to Schleiden's theory, "if it fulfilled the role of nurse only, then the result of each artificial fertilization could be no other than that the developed hybrid should exactly resemble the pollen parent, or at any rate do so very closely."[9] Mendel went on to assume that there was an equal contribution of male and female factors to the seed. The equal contribution to the seed allowed for his Law of Segregation and Law of Independent Assortment. Here was an explanation for the varieties within species, including the genders themselves, which could be observed in nature. This was a possible mechanism for the evolution of species.

The Darwins, both grandfather and grandson, fall on the neither side of this watershed in the evolution of biology. They belong to biology before the microscope, field biology. The Darwins were part of the long line of justly famous English naturalists. The English stayed away from the general turmoil between the eggists, the spermists, the preformationists and the epigenesists. Mendel read his paper, "Experiments on Plant Hybridization," at the Natural History Society of Brünn in Moravia in 1865. Mendel read *The Origin of Species*, but he made no mention of it in his 1865 paper. What Mendel did say in reference to evolution was that the continuation of the sort of work he was doing was essential to the understanding of the history of organic evolution. The Augustinian monk sent a copy of his seminal paper to Darwin; it lay unopened among the old man's papers, but even the biologists who did read his papers were not able to understand them. His work would have to wait for the idea and the *Zeitgeist* to come into alignment.

While Mendel's theory was germinating, microscopists were using improved nuclear stains that enabled them to actually see the male and female chromosomes. It is to August Weismann, beginning in 1887, that we owe the thesis that the cellular germ line is isolated from the somatic cells of the body. With his increased clarity Weismann could see no mechanism whereby the alterations of body cells could be communicated to germ cells. The development of body cells and germ cells appeared to occur independently. He used this supposition to show the improbability of Lamarckism. Weismann was a convinced Darwinist, which meant that he agreed with Darwin's supposition that random variations could result in the evolution of species. On the other hand, his idea that germ cells were completely isolated from somatic cells negated Darwin's theory of pangenesis. This set the stage for the political dialectic between Darwinism and Lamarckism in the next century.

Mendel was discovered by three scientists at the turn of the century: Hugo de Vries in Holland, Carl Correns in Germany, and Erich Tschermak-Seysenegg in Austria. De Vries, who was deeply impressed by Darwin, would suggest a mutation theory of evolution. Correns was the student of Karl Nageli, with whom Mendel communicated but who failed to understand the importance of his work. Correns discovered cytoplasmic inheritance that extended Mendel's theory of nuclear inheritance. Tschermak was the grandson of the man who taught Mendel botany in Vienna. It is appropriate that we add Thomas Kuhn's corollary about major paradigm shifts in science: the older generation that knows what it knows has to die out first before a much change can occur. I will also add that the younger generation may be motivated to distinguish itself from the elders especially if the social environment itself is changing. It seems to apply here.

The English biologist William Bateson read about Mendel's work on the train between London and Cambridge in 1900. He immediately recognized the importance of it and went around proclaiming that Darwin was now of historical interest only. Bateson believed in rather larger jumps in evolution such as those shown by Mendel with no blending. As he saw it, Mendel's theory struck at the heart of Darwin's Uniformitarian selection. Darwin thought that the function of sexuality was to weed out such jumps in inheritance. The disappearance and reappearance of recessive traits in his peas in successive generations with the cooperation of the two sexes shown by Mendel suggested that such changes could possibly lead down the path to speciation. Mendel's equality of sexuality would lead to the creation of the scientific discipline of genetics and to DNA, the sine qua non of biological evolution in the twentieth century.

Scientists cannot be divorced from their social context. For thousands of years the role of the patriarchal society, both pagan and religious, had determined the primacy of the male sperm against the circumstantial evidence. As John Farley suggests, the role of sexual reproduction in the eighteenth century biology was dominated by the wealthy, often aristocratic amateur. We remember the lusty persona of Erasmus Darwin. "The biological theories to which the nineteenth century scientists subscribed—an almost sexless egg laying female and a reproductively insignificant energizing male—were as much a reflection of these middle-class virtues as they were the result of the biologists' scientific discoveries."[10] This was the sexual milieu of Charles Darwin. Beginning with the Socialist revolts in the middle of the century, however, the role of the woman in society was changing and culture beginning to be sexualized. Mendel's proposal, pushed ahead a generation, would come to coincide with and help to promote a revolution in the perception of the gender roles in society.

Behind the ivy covered walls however the apostasy of Mendel that emasculated preformationism has been a great embarrassment to science and is one of the least known and least celebrated intellectual achievements in the history of science, given its importance. When Mendel is remembered, what is commemorated is his presumption that heredity is particulate. Most modern scientists are shocked to discover that any intellectual community could even hold a view so transparently political as preformationism, let alone their greatest hero Charles Darwin.

Botany had long presented serious difficulties to the narrow proscriptions of Darwinian evolution. No amount of haggling and niggling could force plant life into the Judeo-Christian–friendly metaphor. Arthur Tansley, a plant ecologist and lecturer in botany at Oxford, led the attack on the old evolutionary taxonomists, the pride of English biology. The authors of the *Tansley Manifesto*, published in 1917, attacked the Darwinian teleology that attempted to explain the appearance of a structure on the grounds of its utility. The authors argued, "It has to be frankly recognized that the study of the detailed evolution of the plant world, which has acquired a factitious importance owing to the overwhelming effect on the imagination of botanists of the doctrine of descent, has no valid claim to the dominating position which, especially, in this country, it has so long held."[11]

Frederick Bower, who was one of the last of the old timers, responded to the attack as "the exhortations of preachers of other doctrines than those of natural science."[12] Once this initial attack upon the English tradition of field naturalists was made, the academic specialists would operate well below the level of general theory. They would offer nothing in the way of improvement; they would simply become disinterested in the whole abstract perspective. The knowledge base for a coherent theory of evolution simply did not exist yet. In the meantime Darwinism was being retailored to fit the new discovery that sex was not a bizarre sideshow intended to eliminate precipitous changes, it was the method of evolution. Darwinism was already occupied justifying a social ideology that would keep it the default theory of evolution for two generations.

Darwinism got its second breath with the assistance of a cousin of Charles Darwin, Sir Francis Galton, who coined the term "eugenics." Eugenics was the principle of the survival of the fittest applied to human populations. The eugenics movement concerned itself with purifying the race of the newly forming national states. In the seventeenth century a reactionary movement in Europe to both the Christian Church and the Enlightenment came into existence. This took the form of romanticizing pagan Celtic and Nordic culture that existed before it was overrun first by Roman culture and then by Roman Christian culture. By the Victorian era Viking revivalists would gather wearing horned helmets—which the Vikings never wore—practicing hocus-pocus rituals in an attempt to exorcize both foreign religions and industrial blight. When the conditions were right they made common cause with Eurocentric intellectuals that were at the peak of their hubris over the superiority of European racial stock. Maintaining national bloodlines was the big political issue.

The English had already been pruning their redundant population by transportation to the Australian colony. Social Darwinism was a useful justification, and it led to other social remedies such as mental defectives being removed from the breeding stock. Charles Darwin had attempted to explain the altruistic behavior of worker ants giving their lives for the anthill. They were doing it for the greater good he suggested. He called this family or kin selection. He did not enlarge the analogy to the human situation, although he would not lost much of his audience had he done so. Herbert Spencer discussed the apparent conflict between Christian ethics and Social Darwinism in his book *The Study of Sociology*. Egoism was right, in his view, if it was in the name of the greater welfare of the Christian state acting for the welfare of the people. In Germany Haeckel was promoting a familiar line: "The Caucasian, or Mediterranean man (*Homo Mediterraneus*), has from time immemorial been placed at the head of all the races of men, as the most highly developed and perfect." Richard Wagner would put the whole Nordic legend to music. Wagner, curiously, was the name of the assistant to Doctor Faustus. This movement would be labeled Aryanism, from the Sanskrit word meaning "noble".

This theme of racial superiority is nothing new of course, in spite of its sophisticated intellectual elaborations. We find the north/south dimension of national prejudice worldwide rooted in primitive tribal tradition, with its local flavor being shaped by geography. There is also an east/west version of this prejudice. In Europe it is clearly marked by the boundary of Slavonic language culture in the east and Indo-European language culture in the west. With the secularization of culture and the weakening of the role of the church, the old hatred between the Eastern Orthodox Church and the Roman Catholic Church was weakening. Few could find energy anymore for that ancient antipathy, the definition of the Trinity that had saints tearing each other's throats out for centuries in the early days of Christianity. Since East and West had been well practiced in the art of defining themselves in opposition to the other, the old dialectic would easily metamorphose a new secular form.

Russia, at the behest of Peter the Great in the seventeenth century, was westernizing itself. Russia was the poor man of Europe, with famine a constant threat throughout the country. The reform began by the wholesale importation of a German bureaucracy and professional class. There was some opposition on religious grounds. There was also a Slavophil tradition represented most heroically by Leo Tolstoy. He thought that all western innovations such as science, rationalism, Protestantism, humanism, and the like reduced the noblest of human spiritual and intellectual achievements to the comprehension level of the masses. During the European social upheavals of 1848, the Czar brutally suppressed any signs of revolution. Aristocratic families from the serf-driven agricultural estates were falling on hard times in the 1860s after the largely peaceful abolition of twenty-three million serfs in 1861. Some of the sons of these impoverished landowners were motivated by new ideas coming from the West. Among these was Prince Petr Kropotkin. Petr Kropotkin would become the Russian Darwin, but the fact he was a naturalist was about all they had in common. Their life stories and their theories of evolution give us a good summary of the cultural differences between eastern Europe and western Europe, and they anticipate the Russian Revolution and the Cold War to come before Communism appeared on the scene.

Kropotkin was especially interested in the French Philosophes that, at a distance, were probably seen as the actual motivation and intellectual basis of the French Revolution. He didn't have the means to enter St. Petersburg University, so he accepted an appointment as an attaché to the governor-general for East Siberia Irkutsk. In 1864 Kropotkin accepted charge of a geographical survey expedition, crossing North Manchuria from Transbaikalia to the Amur, and soon was attached to another expedition that proceeded up the Sungari River into the heart of Manchuria. Darwin left bucolic England and sailed around the world to scenery so new, so majestic, at 12,000 feet in the Andes, around the Horn to the volcanic strangeness of the Galapagos Islands, followed by the beauties of Tahiti; Kropotkin would earn his spurs in the depths of Central Asia. The forests and steppes of Central Asia are some of the most challenging environments on the planet. A short growing season and the greatest range of temperature extremes except for the great desert regions make this an awe-inspiring but species-poor environment. On the whole Kropotkin was looking at an ecology that models the normal environment for Pleistocene planet Earth.

In England Malthus warned that the greatest challenge faced by a rich culture was overpopulation, Darwin saw natural selection through competition bringing the maximum amount of happiness over the long haul, and Spencer was supplying the largely meaningless but catchy slogan, "survival of the fittest." Kropotkin stated that Darwin had not only turned biology into a science, and he had created the basis for a scientific study of human society. Kropotkin made his own evolutionary thesis regarding human society in *Mutual Aid, a Factor in Evolution*, in 1902. He didn't see competition between individuals over the spoils as required by Darwinism as much as he saw life struggling to survive the intemperate weather of the Russian steppes. Going to the heart of the issue, cooperation versus competition, he states:

> *Life in societies enables the feeblest insects, the feeblest birds, and the feeblest mammals to resist, or to protect themselves from, the most terrible birds and beasts of prey; it permits longevity; it enables the species to rear its progeny with the least waste of energy and to maintain its numbers albeit a very slow birth-rate; it enables the gregarious animals to migrate*

> *in search of new abodes. Therefore, while fully admitting that force, swiftness, protective colours, cunningness, and endurance to hunger and cold, which are mentioned by Darwin, are so many qualities making the individual, or the species, the fittest under certain circumstances, we maintain that under any circumstances sociability is the greatest advantage in the struggle for like. Those species which willingly or unwillingly abandon it are doomed to decay; while those animals which know best how to combine, have the greatest chances of survival and of further evolution, although they may be inferior to others in each of the faculties enumerated by Darwin, save the intellectual faculty.*[13]

Living in a country whose social structure was rotten from top to bottom and where the Church was still a powerful actor, Kropotkin became an anarchist. He equated all government with slavery: the many in the service of the few. The sociability that he saw in nature required no laws, no lawyers, and no law enforcement. This ideology was popular in the Hobbesian environment in Russia at the time. While living near London in the late nineteenth century, Kropotkin became friends with the most prominent English-speaking Socialists, including William Morris and George Bernard Shaw, but anarchism was already old hat by that time. Marxism was just starting to make inroads into Russia. Kropotkin would eventually be ignored by the mainstream of both sides in the coming ideological struggle between the East and the West, but from 1861 up to the Revolution in 1917, as described by Alexander Vucinich, "If there was one dominant feature in the work of Russian naturalists, it was an emphasis on plant and animal ecology. Very few naturalists deviated from the basic theories of environmentalism: the living world and its physical environment were viewed as interacting and constantly changing variables."[14] Petr Kropotkin was the most prominent voice for this point of view.

The first champion of Darwinism in Russia was Kliment Arkadievich Timiriazev. Timiriazev's wholehearted acceptance of Darwin's ideas allowed that there was no need of further elaboration and improvement. Darwin had done for biology, as he saw it, what Lyell had done for geology, Comte for sociology, and Marx for economics. While Western Europe was already beginning to sharpen its teeth on the dialectic between Darwinism/Capitalism versus Lamarckism/Marxism, Timiriazev thought that Darwin had largely agreed with Lamarck. For Timiriazev the real opposition was religion. Once again we see that when Darwinism arrived on the continent, it acquired a local flavor unique to each adopting culture. There was French Darwinism, German Darwinism, and now there would be Russian Darwinism.

When Mendelism arrived after the turn of the century, Timiriazev suddenly became alert. The invisible factors involved in inheritance coming from Germanic microscopic science, as he saw it, upset his sense of a rational, tangible naturalistic science. At first he related to the new German biology as a revival of the metaphysics of vitalism, but he came to respect Mendel for the way his work allowed for saltatory changes in inheritance that Darwin had handled rather badly: progeny that will show a supposedly lost trait again in later generations. While most were taking sides, Timiriazev thought that Mendelian and Lamarckian assumptions were not mutually exclusive; they were complementary contributions to the general principle of Darwinian evolution. Darwin himself had allowed environment a more important role in natural selection over the role of the sexual contest at courtship.

Timiriazev would make a pilgrimage to Down House five years before Darwin passed away. His description is a rare and charming vignette of the old man. Timiriazev describes the village of Downe relating to Darwin as it might to a church elder, he describes the butler who answered the door as a member of the family, and Francis Darwin, Emma and Charles's son, the major domo of the family:

> *Frequently the son led in Mrs. Darwin, an amiable old woman, without a shade of primness or a desire to show off her worldly manner and her way with guests, with a simple and easy grace that bespeaks a truly educated and cultured person. Her tone and her conversation did not reveal a trace of provincialism or strain in dealing with strangers … Unfortunately I was too anxious to see Darwin to pay much attention to her and only the moving and heartfelt things that the son wrote about her in the memoirs about his father made me aware of how much humanity is indebted to that modest unassuming woman who had performed her quiet feat of love: by daily and constant care she had allowed her husband, who had hardly known a day of full good health and had despaired of his life thirty years ago, to complete his Herculean work. A few minutes later, and quite unexpectedly, Darwin entered the room. I have already had occasion to describe my first impression of him. It must be said that the familiar portrait of him with a long grey beard was not yet known at the time. The only known portrait of him was one in the German edition of his "Origin of Species" (and in my book Charles Darwin). That portrait, dating back to the*

1850s shows him as a man of about forty, well-shaven and with trimmed side-whiskers and because the portrait showed him from the waist up one could not help seeing him in one's mind's eye as a shortish plump man looking rather like a businessman or perhaps a sportsman but certainly not a profound and great thinker. And now I was confronted with an impressive old man with a large grey beard, deep-sunken eyes, whose calm and gentle look made you forget about the scientist and think about the man. It couldn't help comparing him to an ancient sage or an Old Testament patriarch, a comparison which has often been quoted since.

Timiriazev was amazed at Darwin's lively and boisterous manner of sprinkling their discussion of science and life with jokes and pointed ironies. He was neither condescending nor ethnocentric. He never asked Timiriazev if it was true that it was very cold in Russia or that they had many bears. Learning that Timiriazev was a plant physiologist, Darwin candidly observed how shocked he must be to find the English so far behind in that area. He was particularly interested to discover that the Russian biologist was working on chlorophyll, observing that it was a most interesting organic substance. They went on to the experiments conducted at Rothamsted, an experimental station in agronomy, where there was a struggle for survival among the meadow flora due to the use of fertilizers.

On this hot, cloudy day in July they walked the short distance across the green lawn to Darwin's greenhouse. Timiriazev observed that it was small, the kind that any Russian landowner could afford to own for his hortensias and pelargonias. He was impressed that Darwin was conceiving experiments on obtaining artificial plant growth, a new area of science that, if not a necessary component of Darwinism, was its natural extension. Timiriazev was observing a man who had gone through the alienation of adolescence, the initiation of manhood, who had accepted the religion of his father, and had been able to return to the protective fold of his youth as an old man. We recall T. S. Elliot, who said, "We shall not cease from exploration and the end of all our exploring will be to arrive where we started and know the place for the first time."

Nikolay Ivanovich Vavilov opened a new century of Russian science, graduating from the Moscow Agricultural Institute. In 1913 and 1914 he traveled to Europe to study with William Bateson, who had taken the insights of Mendel to found the discipline of genetics. As the leader of a group of

Russian laboratory biologists, Vavilov studied the variability of species in terms of both external and internal influences. Vavilov's group acknowledged the role of habitat in morphological and physiological changes, Darwin's natural selection, but they thought that Darwin's random variation was scientifically inadequate because it allowed for an infinite and unpredictable variability in living forms. They claimed that both the deeper regularities in biological evolution and the actual role of external influences could be meaningfully detected only by recognizing the importance of inherent, or automatic, determinants of change. These Russian biologists were making a creative synthesis of the English tradition and the German tradition of evolution. It could not, however, be sustained.

Perhaps the most visionary voice for Russian evolutionary theory and environmental science was Vladimir Ivanovich Vernadsky. He was one of the first scientists to recognize that the oxygen, nitrogen, and carbon dioxide in the Earth's atmosphere result from biological processes. He published works in the 1920s arguing that life forms could reshape a planet as surely as any physical force. He supposed that just as life fundamentally transformed the geosphere, the emergence of human cognition fundamentally transformed the biosphere. This too could not be sustained. The utopian stage of Marxism was already drowning in the blood of political reality.

Marxists began to have second thoughts about Darwin, especially his relationship with Malthus and Spencer. Marxists began to lean in the direction of Lamarck, who they supposed favored evolution through volition rather than chance. This was more in accord with their own economic science, a part of the revolutionary process of change. As Marxism moved east, the contrast between science and religion became more obvious. I. V. Michurin carried the Lamarckian theory of godless biological transformation into the Soviet era, igniting the bitter ideological conflict between eastern Europe and western Europe that biologists ignored at their own peril. Following the Russian Revolution, deviation from a crudely politicized version of Lamarckism promoted by Trofim Lysenko, with the blessings of Stalin, was dangerous. Lysenko did not believe in prescriptive inheritance through units called genes. Vavilov was imprisoned on the charge of wrecking Soviet agriculture, and while in prison he died of the effects of starvation. The West ignored Vavilov, and Lysenko gave the West a weak opponent against which to direct its scorn.

It is amazing how rapidly Russia adopted Western science and advanced the cause. Kliment Timiriazev was born into a culture in the last gasps of medieval spiritualism, made a pilgrimage to the home of the last great English field biologist, incorporated the lessons of German laboratory science, and lived long enough to debate Einstein's Theory of Relativity in the Soviet

Academy. As soon as a strong central government was in charge after the Revolution, the freewheeling period of intellectual speculation was over. Following a pattern that would become familiar for the next century, the physical sciences would do best under autocratic control. Their revolutionary days had passed centuries ago with the rearrangement of the planets around the sun. The biological sciences involved with the definition of evolution and culture would feel the heat of political pressure, the social sciences would be an arm of the ministry of information.

Secularized versions of Genesis slowly began to remove epistemology from the withered grasp of religion in the eighteenth century and began to give it a systematic basis, but of the literally dozens of theories of evolution proposed in the nineteenth century all of them had to admit to difficulties of interpretation or of countervailing circumstantial evidence. Two theories would take precedence in the twentieth century, and it would be for purely political reasons. They would be at the heart of a life and death struggle between the two political antagonists, Capitalism and Marxism. Social Darwinism in the West sponsored the need for the newly emerging national states to maintain their racial identity in turbulent times; Socialistic Lamarckism in the East reflected the need to completely redefine social structure in terms of class struggle to achieve a new Pan-Slavic empire and the eventually universal communism. Whatever happened on one side of this East/West cultural boundary was considered ignorant and superstitious on the other side of the boundary.

In retrospect Darwinian evolution was the view of the evolution of species as seen by a taxonomist. Its actual dynamic or physiologic process was largely proscribed by the Judeo-Christian cultural tradition he belonged to. The outlines of the physiology of biological evolution and environmental science in general were beginning to appear in the perspectives of Buffon, Goethe, Vernadsky, and Vavilov, among others. This is a natural progression in scientific discovery—anatomy precedes physiology—but scientific speculation in Europe would come to an end with the Russian Revolution. The only possible venue for resolving the intellectual dialectic would come from a new actor on the stage of world power in the twentieth century, America.

CHAPTER 5

Evolution Comes to America

THE PURITANS WERE MOVED to settle their New Jerusalem in the New World in order to rid themselves of the sins of English culture, thereby to relieve themselves of the pox of civilization. From the icy shores and rocky soils of New England they came to realize some of the home countries virtues. They were eager to adopt some of the latest inventions, and recover some that they had left behind, for the betterment of life. The first American medical text was by Thomas Palmer. He was influenced heavily by the work of Nicholas Culpeper. Culpeper was a doughty English Puritan who had translated the Royal College of Physician's *Pharmacopoeia Londinensis* into English from Latin for common usage to the accompaniment of much hew and cry. Higher education in America begins its history with Harvard founded in 1636 in the Massachusetts colony. The name comes from its first major benefactor, John Harvard, a minister educated at Cambridge.

While Galileo was suffering under house arrest in the Old World, the first generation of American college students would learn Aristotelian cosmology and astronomy. Without much ado, however, the Copernican scheme of things was soon being taught at Harvard in 1659. American historians are prone to attribute this to early Americans being uniquely disposed to revolutionary advances in science, because of their adventuresome spirit. This was not necessarily the case. Americans were curious about practical solutions to everyday dilemmas, but were not disposed to experimental philosophies that introduced doubt about the revealed truth of their religion. The fact of the matter was that the Bible had precious little to say about astronomy and whether the sun was in the middle of the planetary system. The Bible read in the vernacular, without the oversight of an intellectual academy, offers little to choose between an Aristotelian and a Copernican cosmos. The trial of Galileo stemmed from the insult committed by him to an entrenched Italian academy. There was no such academy in America, making a tidy living writing astrological charts. In New England the new cosmology could slip into bed with fundamental Christianity without protest. The president of Yale from 1739 to 1766, Thomas Clap, designed and constructed one of the earliest orreries (a clockwork model of the solar system) in America. It was sun-centered. This did not, however, make him an adventuresome spirit. Clap also delivered sermons condemning the use of artificial light such as candles as dangerous to Christian morality.

Orreries were one of the first scientific teaching tools and a high achievement of the machinist's craft in the eighteenth century. They also unwittingly augmented the normal human tendency toward concrete thinking; they conveyed Aristotle's general mood of a geometrically perfect, immutable solar system, albeit with the sun now at its center. Orreries were among the early automatons of the Age of Reason, along with Jacques de Vaucanson's digesting, shitting duck.

For mercantilists who were the major philanthropists of colonial higher education, mathematics was clearly a practical science. Mathematics kept the books, mathematics calculated the taxes, mathematics improved the odds of completing an ocean journey, and mathematics computed the insurance that covered the losses that did occur. The possibilities and the limitations of natural science in the eighteenth century are clearly indicated by the subject matter taught from the Hollis chair at Harvard. They included:

> *Natural Philosophy, and a course of Experimental in which to be comprehended, Pneumaticks, Hydrostaticks, Mechanicks, Staticks, Opticks, etc. in the Elements of Geometry together with the doctrine of Proportions, the Principles of Algebra, Conic Sections, plain and Spherical Trigonometry with the general principles of Mensuration, Plain and Solids, in the Principles of Astronomy, and Geometry, viz. the Doctrine of the Spheres, the use of the Globe, the motions of the Heavenly Bodies according to the differint Hypotheses of Ptolemy, Tycho Brahe and Copernicus with the general Principles of Dialling, the division of the world into various kingdoms, with the use of the maps, and sea charts; and the arts of Navigation and Surveying.*[1]

Herein we see the basis for the future pecking order of American science that would place the mathematical disciplines or hard science first in praise and material reward. Without realizing what they were doing, these scientists were making the same intellectual compromise with their Puritan culture that Descartes had made with the Catholic Church. America would become the home of utilitarian science. At this stage, medicine was in the care of minister-physicians and was learned by apprenticeship. Psychology and sociology and all their offshoots were unheard of. All of science was under the oversight of religion. John Winthrop, who was the second to sit in the Hollis chair, would observe sunspots from Boston Common, and papers on the transits of Mercury and Venus were published in the Philosophical Transactions of the Royal Society with the help of his friend Benjamin Franklin. Both Increase Mather and his son Cotton Mather observed the pass of Halley's comet in 1682 with the Winthrop telescope. They preached extensively about that portentous event, intermingling both Newtonian and Puritan precepts.

Franklin's first tour of England took him to the Midlands in 1758 looking for English relatives and the Birmingham printer John Baskerville, whose typeface was much admired. It was during this trip that he met Erasmus Darwin for the first time. When Erasmus was elected to the Royal Society of London in 1761, Franklin was already a member, in fact, its most illustrious member. Later Franklin would join Darwin's Lunar Society. Franklin, branded as the evil genius of the American Revolution by George III, was America's only first-class scientist. His city, the Quaker city of Philadelphia, was first in scientific standing. Darwin commemorated him as the man who snatched the lightning from the heavens and their scepter from tyrants. He is commemorated in taxonomy by association with a particularly attractive flowering bush, Franklinia, found on the Altamaha River by the naturalist William Bartram. No one has found that plant in the wild since 1790 however. It exists only in a cultivated form. In the present day Franklin is mainly remembered by Americans for his Poor Richard rustifications and for being a signer of the Declaration of Independence.

A century before the Statue of Liberty, it was Thomas Paine who would light the torch of the French Enlightenment in the New World and attempt to drive a wedge between religion and science in America. He was a straight-talking Englishman who came to America at the suggestion of Benjamin Franklin. He was much celebrated for his stirring rhetoric about the times that tried men's souls. He evoked the Tricolor maiden of Liberty with a torn blouse, not the matriarchal, heavily robed Statue of Liberty of a later time. His pamphlet *Common Sense* was the most widely circulated revolutionary publication of its time. After the American Revolution he joined the French Revolution and became a member of the National Convention. When the reactionaries led by Robespierre took over, Paine's brand of intellectual was attacked, and he was thrown in jail. While in jail he wrote *The Age of Reason*. His argument in this work comes directly from the vernacular Bible that the Reformation had provided for him. He found the Bible a mass of contradictions, inconsistencies, and pure fabrications. In *The Age of Reason* Paine attacked the Bible on its own merits. He supposed that only the creation that we behold is real and orderly and that the true meaning of God may be read directly out of Nature. This was the natural religion of the Quakers, Erasmus Darwin, and Franklin. His knowledge of science did not go beyond the previous century with Galileo and Newton. He was unaware of more recent developments such as the reform of Genesis by Erasmus Darwin.

Paine was brought back to America by President Jefferson, but he found himself a social outcast for his anti-religious views. Natural religion might be tolerated, but it could not be espoused. As Americans were largely refugees from the Religious Wars in England, they were interested in practicing their religion without being persecuted. They were not much interested in the French or the French Enlightenment and were generally frightened by the French Revolution. They followed English models and the mood of conciliation between church and state, church and science that prevailed in English culture following the English Civil War. The great political accomplishment of American culture at that time was the union of the various Christian sects, the Puritans in New England, the Catholics in Maryland, the Anglicans in Virginia, and the Protestants in Georgia, under the one non-denominational God of the Quakers, in Philadelphia.

Many of the Founding Fathers were Freemasons and therefore of an anticlerical leaning, including George Washington and Benjamin Franklin. It was because of them that the Great Seal of the United States shows the Grand Pyramid at Cheops surmounted by the all-seeing eye of God. The Freemasons pretty much saw eye to eye with Deists such as Thomas Jefferson. The Puritans, well established in their New Jerusalem in New England, were comfortable with the anticlerical leanings of the Freemasons and the Deists. Therefore the outbursts of opposition from John Adams were bracing, not destructive. They all believed in the perfectibility of humanity in this life, they believed that they were laying out the principles for self-realization and/or salvation through education, and they believed in a culture that would eventually rule itself without the need for laws in the spirit of Isaiah.

That the union of states survived the enticements to become just another kingdom in its first hesitant steps was due in large part to President George Washington and his Freemasonish, frontier personality. This was one of those rare moments when the individual who was powerful enough to lead did so for the social group, not for his race, class, or religion. At almost no time down to the present in America has there ever been such a tolerant and universal vision in religion and politics. Acting like a midwife, the Quakers and the Freemasons would never again have so important a role in American culture. In the future the Freemasons would generally restrict themselves to craft rather than speculation; that is, they would be involved in good works, not history and the ultimate mysteries. The Age of Reason would eventually find its larger audience in the twentieth century, when the well-endowed, well-entrenched institution of science would launch an attack on the church and feel justified in the face of a large sect of religious fundamentalism in the country.

The opening up of North America to European discovery was a powerful impetus to reviving the question of creation, change, and evolution in the eighteenth century for those not totally engrossed with human revolutions. Two questions immediately occurred: whether the New World represented a second original creation or simply a dispersal of the one and only creation; and, second, whether the New World was a degraded version of the European creation or not. Buffon and Voltaire both thought that New World species, including the native humans, were degraded distant relatives of European life forms.

In America the range of opinion extended from New England's mathematical boat captains, who were content with Genesis, to America's premier philosopher-president and skilled naturalist from Virginia, Thomas Jefferson. Jefferson had a collection of the bones of animals that were no longer seen to exist. Among these were the bones of a rhinoceros-sized sloth that had a wicked set of claws. Cuvier identified the taxonomical relationship of this giant sloth or *Megatherium* and established that the animal no longer existed. He would assign the demise of this animal to the last of a long line of catastrophic events, Noah's Flood. Jefferson assured the Philosophical Society in Philadelphia, "In fine the bones exist: therefore, the animal has existed. The movements of nature are in a never-ending circle. The animal species, which has once been put into a train of motion, is still probably moving in that train. For if one link in nature's chain might be lost, another and another might be lost, till this whole system of things should evanish by piece-meal."[2] This resembles Lyell's argument against catastrophic geologic events: it would be too depressing to think so.

At this time the word "fossil" had its simple Latin meaning: something dug up out of the ground. The word "extinction" meant snuffing out a candle or, by metaphorical extension, the termination of a royal line. Jefferson presumed that a mammoth, whose bones had also been uncovered, would be discovered out west. One of the goals of the Lewis and Clark expedition was to discover the *Megatherium* and the great North American elephant. Concerning things falling out of the sky, Jefferson allegedly said about a reputed meteorite that struck Weston, Connecticut, in 1807, which two Yale professors verified, that he would find it easier to believe that two Yankee professors would lie, than that stones should fall from the sky.

Jefferson was a follower of Joseph Priestley's materialistic religion. This religion supposed that there were no angels, there was no soul, and there was no God as spirit. Although the devices of the divine workman were incomprehensible at the moment, there was an absolute faith that the Creator's ways were always simple, economical, and understandable to the meanest intelligence. In this brave new world Jefferson would strip his personal Bible of all its miraculous and catastrophic events, leaving only its ethical teachings. As we have seen, Uniformitarianism was not so much something that was discovered by Charles Lyell as it was a reflection of the Zeitgeist of the times, at least for those with the luxury to enjoy them. The habit of reading the Old Testament literally was still the norm for common folk.

It is hard to find anyone in America that didn't believe in a creative as well as a benevolent designer behind the works of nature. For John Adams and the Puritans the truth had been revealed in the Bible; for the Quakers and Jefferson it was revealed through nature. The truths that were self-evident in the Declaration of Independence were truths that could be derived from either tradition. Natural selection through chance, the central mechanism of Charles Darwin's evolutionary thesis in the next century, was not on the table at this time. Later, however, Jefferson would be remembered by taxonomy as a result of his association with the giant sloth *Megatherium jeffersonii*, in spite of the fact that *Megatherium* would be shown to be well and truly extinct along with *Mammuthus jeffersonii*.

Through the first half of the nineteenth century American geologists remained convinced of the possibilities for violence as depicted in the Old Testament. The American landscape seemed to offer ample evidences of catastrophic events in the past. Benjamin Silliman was the first professional geologist in New England. He studied at Edinburgh in 1805. He did a chemical analysis of the infamous meteorite that fell near Weston. He was also the first to fractionate petroleum by distillation. Silliman was impressed by the geological theories of Cuvier. The cosmogony in Genesis, he said, "considered in a purely scientific view, is extremely remarkable, inasmuch as the order which it assigns to the different epochs of creation is precisely the same as that which has been deduced from geological consideration."[3] The immutability of the species between catastrophic events also seemed to be well represented in the geological record. Edward Hitchcock, a Congregationalist pastor and professor of chemistry and natural history at Amherst, introduced the question of the actual length of the "days" of Genesis, and this created an opening for metaphor. Geology and religion could still occupy the same platform, but the beginnings of a rift were there.

A curious American public flocked to hear a new version of geology from Sir Charles Lyell in 1841. Lyell's book *Principles of Geology* had already appeared when he arrived in America, and Lyell, a religious man, had ended this book with the statement, "in whatever direction we pursue our researches, whether in time or space, we discover everywhere the clear proofs of a Creative Intelligence, and of His foresight, wisdom and power." Lyell was amazed at the widespread interest and utmost decorum of his audience and pleased with the equanimity with which his Uniformitarian geology was received. He had half expected Americans to be semi-barbaric or, at least, rigidly fundamentalist. While Old Testament fundamentalism was a better formula for those in the wilderness of the frontier, the New Testament and Uniformitarian geology were not unattractive to the Unitarian tendencies of the well-settled East. Not even the New England Puritans were as stiff-necked in the interpretation of the Bible as they once had been.

Robert Chambers' book *The Vestiges of the Natural History of Creation* was published in America in 1845 by the respectable firm of Wiley and Putnam. This book offered evidences of biological evolution. It was attacked equally from the pulpit and the laboratory. It was during the process of a critical review of the book, however, that the Harvard botanist Asa Gray began to have questions. Gray had studied the system of classification espoused by De Jussieu, Lamarck, and De Candolle that had already replaced the Linnaean system in Europe. Gray had brought North American taxonomy up to European standards. He, along with all biological taxonomists, had a natural predisposition to see the implications of an evolutionary scheme out of the book of Nature. The minor geographic variations in the plant kingdom had inspired him in the same way that the tradition of English natural historians had inspired Darwin. In 1846 Gray was not yet prepared to go to the lengths that Chambers had. "How can a baboon originate anything above a baboon, with only a baboon's mechanism?"[4] he wrote, but by 1854 he wrote a letter to the English botanist Hooker opening the question of evolution in the plant kingdom. Hooker showed Gray's letter to Darwin. In 1857 Darwin wrote to Gray. The letter is now famous in the history of the Darwinian myth, because in it he outlined his theory for the first time:

> *I am convinced that intentional and occasional selection has been the main agent in the production of our domestic races; but however this may be, its great power of modification has been indisputably shown in later times. Selection acts only by the accumulation of slight or greater variations, caused*

by external conditions, or by the mere fact that in generation the child is not absolutely similar to its parent. Man, by this power of accumulating variations, adapts living being to its wants ... I have found it hard constantly to bear in mind that the increase of every single species is checked during some part of its life, or during some shortly recurrent generation. Only a few of those annually born can live to propagate their kind. What a trifling difference must often determine which shall survive.[5]

It was this letter that was read at the Linnaean Society in England on July 1, 1858. This was the occasion upon which two theories of evolution were presented, those of Charles Darwin and those of Alfred Russel Wallace. Darwin concluded that Wallace's own theory had arrived at almost exactly the same general conclusions that he himself had on the origin of species. The purpose of this meeting was to establish the priority of and, therefore, the eternal fame of the first discoverer of the origin of the species with the eclipse of the other. It was Darwin, who was awarded the prize, and it was Wallace, ironically, who would coin the use of the word "extinction" to describe the disappearance of unsuccessful species in the struggle for survival.

The major opposition to Darwin in America in the academy would be Gray's colleague at Harvard, Jean Louis Rodolphe Agassiz. Agassiz was a Swiss-born naturalist and a welcome input from the European Continent to the American academy. He had introduced the idea of glaciations and Ice Age to the New World. He had taught a generation of naturalists to recognize the damage wrought upon the landscape by glaciers. Charles Lyell had heard Agassiz give his thesis in Scotland in 1841. At first Charles Lyell resisted the idea of massive glaciations because it sounded catastrophic. The evidence for glaciations—u-shaped glacial valleys, glacially polished and etched stones, terminal moraines of stones left at the foot of glaciers, and the like—had embarrassed Lyell most of his professional life. The argument was so convincing; it solved such a host of difficulties.

Lyell developed a theory to show glaciation could be accomplished gradually. Indeed, Lyell would theorize that gradual glaciation could explain many geological features of Scotland that James Hall had proposed were due to a catastrophic diluvial wave that had passed over the whole of the country. His thesis killed two birds with one stone. His argument was simple. As the Earth cools during an Ice Age, the glaciers increase when the cooler summer temperatures fail to melt what had been added during the winter. Glaciers gradually and incrementally build up over tens of thousands of years. The English physicist John Tyndall didn't think so. In a paper on *Alpine Sculpture* in 1864, he observed that it takes considerably more energy to change water into snow to create glaciers than it does to melt ice back into water. Glaciers couldn't be created by simply reducing the temperature of the environment, he argued. In fact, they will tend to melt away incrementally over time in a cooling environment. However, by this time nearly everyone was in love with Lyell's Uniformitarianism, and Tyndall's paradox has never been answered to the present day.

Agassiz was of the camp of the French catastrophists. He was a great admirer of George Cuvier, the great French paleontologist. Following Cuvier and from his own work on fossil fishes he argued "the impossibility of referring the first inhabitants of the Earth to a small number of branches, differentiated from one parent stock by the influence of the modifications of exterior conditions of existence."[6] From a passage in a lecture in 1863 he stated, "To me the fact that the embryological form of the highest Vertebrate recalls in its earlier stage the first representative of its type in geological times and its lowest representation of its type at the present day, speaks only of an ideal relation, existing, not in the things themselves, but in the mind that made them."[7] With this we see French catastrophism combined with the Germanic tradition of an underlying type or *Urbild* that was qualified from the microscopic studies of the cell and of embryology. When Agassiz died, Uniformitarianism lost its major opponent in America.

Joseph Le Conte reviewed the causes of the Ice Age through the eyes of Uniformitarianism in his *Elements of Geology* at the end of the century. Le Conte supposed that the facts could be well explained by a northern elevation of the land producing ice accumulation; the weight of the glaciers, in turn, would produce subsidence; subsidence produced a moderation of the temperature and a melting of the ice; and this last, by lightening the load, produced a elevation again. All of the speculations that he made followed the basic assumption that change occurred very slowly over a long period of time. One disturbing anomaly to this thesis was the one-mile-wide crater discovered in Arizona south of the Grand Canyon. It could be ignored.

The transition from the Ice Age to the Psychozoic era or age of the mind, as Le Conte labeled it, represented a shift from the reigns of brute force and animal ferocity. In the Psychozoic era the dangerous animals decreased in size and number, and the useful animals and plants were introduced or else preserved by man. The old regime of ferocious animals had been found in California dramatically preserved in the La Brea tar pits. What was not fully comprehensible to the scientists at that time was the cause of the massive lava-flows in the Northwest. The gradual processes of weathering cannot be finessed to explain the appearance of major lava flows. And there was the problem of the very clear record of a sudden, complete turnover in the plants and animals along the Pacific coast in very recent time that included the "evanishment" of a giant bison, a giant bear, a giant wolf, a horse, a camel, a saber-toothed tiger, and a mammoth species or two.

By this time Halley's Comet had become prima facie evidence for orderly, predictable Uniformitarian science. Halley's Comet would sweep by once more in 1910 with just enough clearance. Only the death of Mark Twain was commemorated by it. Before he passed he would comment:

> *Upon arrival in heaven do not speak to St. Peter until spoken to. It is not your place to begin. You can ask him for his autograph, there is no harm in that. But be careful, and don't remark that it is one of the marks of greatness. He has heard that before. Don't try to Kodak him. Hell is full of people who have made that mistake. And leave your dog outside. Heaven goes by favor. If it went by merit you would stay out and the dog would go in.*

The straight-talking sage from Missouri covered an abyss of despair with his humor, despair rooted in the hypocrisy that he saw between Christian ethics and the behavior of Christians, and there wasn't much in the way of respite in the mechanics of science. A archetypal American, Twain's adolescent level of spiritual attainment and intellectual training would betray his curiosity, leaving him bereft and without support. In the vacuum left by faith there was only his ironic posture toward culture. This would come to be the posture of many an American since, although without so fine a sense of humor.

The brutality and horror of the American Civil War brought on the same spiritual and intellectual malaise in the culture as occurred in England following the English Civil War two centuries earlier. Just as the English Civil War marked the end of the stunning creativity of the Elizabethan age, the American Civil War ended a period of turbulent creativity in America in the 1840s and 1850s. The shadow thrown by that war is so deep and so long that the prewar period is barely remembered today. Historian Henry Adams gave his Hobbesian view of the new, post-bellum engine of American progress in his classic autobiography, *The Education of Henry Adams*. It was a period of unleashed and uncontrolled Capitalism under the benign neglect of President Grant. Social Darwinism began to enter America after the Civil War, introduced by Charles Sumner. Social Darwinism was the philosophy of a victorious culture justifying its spoils.

A famous banquet was held in 1882 at Delmonico's Restaurant in New York City for America's leading robber barons. They jammed in to the restaurant to swill down bushels of oysters, hogsheads of champagne, and sides of roast beef, all without the aid of vomitoria. While they drank whiskey, smoked cigars, and fondled gold chains hanging down from greasy waistcoats that girdled their midsections they listened to encomiums of praise raised in the name of Herbert Spencer. What Adam Smith had done in the previous century—excused the suppression of wages paid to labor in the name of the economies of scale—survival of the fittest did for these self-satisfied Capitalists.

In Europe the traditional craft workers and others that were being thrown into poverty by the Industrial Revolution turned to Socialism. When they came to America, they became democrats. Andrew Carnegie was one of these. He was relentlessly ambitious. This would turn him into a democratic Capitalist. Andrew Carnegie would use Social Darwinism as the justification for his robber-baron behavior, yet he was also laissez-faire when it was seen to be to his benefit.

Being a God-fearing man at the last, Carnegie gave away wealth for the building of public libraries for the public good as the day of his judgment approached. Perhaps he was troubled by the ghost of Jacob Marley. This sort of charity was the Capitalist version of redemption, a sort of indulgence to appease a lingering guilt. In Carnegie's libraries would be found books by the Pragmatists, books on Social Darwinism, even books on biological evolution, and, of course, books on religion. Education sufficient to read the Bible in the vernacular was a theme of the Reformation; an educated public was the fundamental basis of the Enlightenment and a basic assumption of the American social contract. Now it was a kind of afterthought. Nevertheless, secularization was being methodically advanced throughout the nineteenth century by education. This was not exactly what Martin Luther had in mind when he translated the Bible into German while shivering in the grip of the Little Ice Age in Wartburg Castle.

Another famous library was the one built in 1887 in Thomas Edison's laboratory of scientific invention located in West Orange, New Jersey. In this invention factory Edison told his staff he expected a major discovery every six months and a minor one every ten days. His factory operated on what he called the American system. Each worker performed a single, simple, task repeated over and over. No special education was required. No variations were allowed. Descartes would let the genie of mechanical invention out of the bottle for the Age of Reason; Adam Smith would formally give the process a Christian financial basis; the mechanical genius of West Orange would provide it with an assembly-line application on the highest scale. Henry Ford would be among those who were impressed, and it would pave the way for the highly specialized publish-or-perish academy of the twentieth century that would rattle out better living through chemistry for a consumer society.

In the late stages of the nineteenth century the agnosticism of Thomas

Henry Huxley, which suited the purposes of English science while still living under the roof of High Church Anglicism, was discovered by Charles Sanders Peirce, Oliver Wendell Homes, and John Dewey. They decided to call it Pragmatism. Pragmatism was embraced by a nation that had torn itself apart and was looking for a new guiding philosophy, a philosophy that was not associated with the ideologies that produced religious intolerance or the freedoms that led to the Civil War. In the Zeitgeist of Pragmatism, Lyell and Darwin captured an American audience with a unified and coherent front and because their thesis required no apparent need for translation for the predominately English-speaking American public.

When English biology was catching up to Germanic laboratory tradition and the Tansley Manifesto was attacking Darwinism as crude and out of date in the home country, Darwinism would be adopted in America though not without a few, basic, unconscious translations in order to fit into the American culture. In England the Capitalist, free-market debate was a resurrection of the Magna Carta compromise that was a dispute between the noblity and the crown. The Capitalist debate involved a new power elite demanding lower tariffs at the border and a dissolution of the royal trading monopolies, after which the enriched economic class would kneel at the foot of power and proclaim, *The king is dead; long live the king!* Charles Darwin was a member of the intellectual elite; he was consistently upper class and high-church observant, at least in public. English gentlemen of this time thought that democracy was a pernicious form of government. Darwinism was the biological version of Victorian Capitalism that was hierarchical and non-democratic.

In America, of course, no political or intellectual thesis, no matter how promising, had a chance of general acceptance flying those colors no matter how pragmatic it may appear to be. In America the Declaration of Independence and the establishment of unalienable personal freedoms comes first and foremost, followed by the corollary government of the People, for the People, by the People. Capitalism in America is a reflection of the freedoms won from the autocratic class in England. Capitalism is conceived as the constitutional power of the individual to operate free of government restraints. The government's role is to get out of the way. Power, ultimately, comes from the People. Just as Darwinism had to be tailored to fit the cultures of France, Germany, and Russia, American culture would unconsciously apply a different definition to natural selection and the survival of the fittest.

The intellectual understanding of gender was conventional and clear-cut in America up to the twentieth century. The American naturalist James Weir would attempt a secularized version of the story of Adam's rib, saying, in 1895, "The Almighty, in creating the female sex, had taken the uterus and built up a woman around it."[8] With little adjustment, this was conformable to Darwin's view of sex. The American tradition of incest was not disturbed by the new biology although hybridization or out-breeding was occurring at an alarming rate from an English perspective. Darwin's public utterances never gave offense when it came to the act of creation or descent from apes.

The separation between church and science in America occurred through gradual accommodation as it occurred in England. It was not a violent rift such as occurred in France or Russia. Evolution had begun as a loud debate between the church and the academy, but by the end of the century the issue was as meek as a lamb in both England and America. The president of Cornell University, Andrew D. White, summed up *A History of the Warfare of Science with Theology in Christendom* with rather more conflict than actually occurred, supposing that Darwin's theory of evolution was generally opposed by religious authorities for instance. The gradual alienation in America between religion and science was more realistically summarized by Dirk Struik:

> *Organized religion was never fundamentally attacked; "natural" theology had time and opportunity to gradually adjust itself to scientific evidence. This adjustment, however, was not performed without considerable loss of prestige. Post-Civil War scientists rarely bothered to study the theological implications of their results. The era of the specialist had arrived; and the increasingly secularized Protestant churches for many years no longer took a strong polemical interest in the relations between religion and whatever branch of science was currently popular in the public eye. The scientist, on the other hand, often took little interest in the more general social and philosophical implications of his work. Science, losing contact with theology, also lost contact with philosophy as well as with human relationships, to the detriment of all.*[9]

The next major signpost on the road to scientism in America after Charles Darwin was imported from Russia. Ivan Pavlov was born in 1849 in Ryazan. He began his higher education at the Ryazan Ecclesiastical Seminary, but he dropped out and switched to the University of St. Petersburg to study natural history and become a physiologist. He received his doctorate in 1879. He investigated the gastric function of dogs in the 1890s. He did this by externalizing the salivary gland so he could measure its response to food under different conditions. Most intriguingly he noticed that dogs began to salivate before food was actually in their mouths. This led him to pursue what he called "psychic secretion." He described reflex responses that were conditioned by previous experiences that he called "conditional reflexes." Further work involved the study of different personality types to overwhelming stress or pain, elaborating upon the classical typology of phlegmatic, choleric, sanguine, and melancholic. What Darwin had done with the concept of natural selection, Pavlov would do with his concept of the conditioned reflex.

He earned a Nobel Prize in 1904 in Physiology and Medicine. Unlike many pre-revolutionary scientists, Pavlov was praised by the Soviet government and encouraged to continue with his work. Carl Jung extended his work, finding that introverted people were more sensitive to pain than extroverted people. The British psychiatrist William Sargant used conditioning to achieve memory implantation and brainwashing. John B. Watson in America developed a comparative psychology on the basis of Pavlov's experimental approach and called it Behaviorism. Behaviorists supposed that all behaviors or conditioned reflexes could be studied scientifically without recourse to physiological events, to constructs like the mind, or divine purpose expressing itself through the material world.

Calvin S. Hall stated in his chapter on the genetics of behavior in the *Handbook of Experimental Psychology* that although psychology had always taken a lively interest in heredity it had generally attempted to investigate the relationship between behavior and genetic constitution by methods of pre-Mendelian vintage. This would change in the laboratory of R. C. Tryon at Berkeley in 1927. Tryon began a long-term research project to breed two strains of rats based upon their ability to solve a T-maze for a reward. From a single set of parents bright rats were bred with bright rats, dull rats with dull rats until two strains of rats were created based upon their ability to solve the T-maze problem. They were even identifiable by their hair color. The dull rats had a darker hair color.

This was a centerpiece for Darwinian evolution using the Pavlovian laboratory setting to ask questions. Even though selection and inbreeding had been going on in more informal settings for countless millennia, this was an astonishing achievement for a new and different public. For a college-educated populace, the new science was the fair-haired boy whose future was filled with the light of truth and the hope of personal freedom from disease, material hardship, and fear.

At the same time, Darwinism was also beginning to come under fire. The Butler Act in Tennessee forbade the teaching in any school in the state, "any theory that denies the story of the Divine Creation of man as taught in the Bible, and to teach instead that man has descended from a lower order of animals." The American Civil Liberties Union had offered to defend anyone who would test this law. A local business man by the name of George Rappelyea convinced a group of fellow businessmen that the town of Dayton, population 1,800, would be put on the map if a "Monkey Trial" were to be held in their town. George convinced a friend of his, John T. Scopes, who was the high school football coach, to substitute in the science class in order to teach the profane principle. Clarence Darrow challenged the Tennessee law against the teaching of evolution in 1925 by defending Scopes. William Jennings Bryan stood for the prosecution. The die was cast, and one of the most famous legal trials in United States history was set in motion.

The northern tendency has been to associate Bryan or indeed any anti-evolutionist with Christian fundamentalism if not the Klu Klux Klan, but in fact Bryan was not a Christian fundamentalist. On the seventh day of the trial Darrow had Bryan enter the box for cross-examination in one of its more arresting moments. Darrow's purpose was to attack the believability of the Bible. Darrow asked Bryan about the time of creation and length of the days in Genesis. From the northern perspective, as captured in the play *Inherit the Wind*, the Bryan-like character is tricked into confessing that the actual days were probably not actual twenty-four-hour days. In reality it seems that Bryan knew exactly where Darrow was going and pre-empted and undermined the attack by admitting the point. Jerry Falwell has observed that Bryan lost the respect of fundamentalists when he subscribed to the idea of periods of time for creation other than twenty-four-hour days. Then as now the only thing that mattered was which side was one on. There was no middle ground. One side of the argument has tarred Bryan; the other side has feathered him.

In any case, the Scopes Trial was not about whether Tennessee had the right to pass such a law—the state was not about to let some northern carpetbaggers impose their values in their own domain; the public trial was simply about whether Scopes had actually taught a forbidden course. It was decided that he had. The jury deliberated for six minutes and pronounced him guilty. He was fined one hundred dollars. The North declared a moral victory with H. L. Mencken leading the attack with liberal doses of malice and vitriol, as he was wont to do. He called Bryan a "buffoon" whose speeches were filled with "theologic bilge." In fact, Bryan showed evidence for a greater knowledge of science than the defense lawyers or even the one expert that the judge allowed to give evidence. Dr. Maynard Metcalf was the single Darwinian expert allowed to give evidence. He testified by giving a definition of evolution:

> *Evolution, I think, means the change—in the final analysis I think it means the change of an organism from one character into a different character, and by character I mean its structure, or its behavior, or its function or its method of development from the egg or anything else—the change of an organism*

from one characteristic which characterizes it into a different condition, characterized by a different set of characteristics, either structural or functional, could be properly called, I think, evolution—to be the evolution of that organism; but the term in general means the whole series of such changes which have taken place during hundreds of millions of years which have produced from lowly beginnings—the nature of which is not by any means fully understood—to organisms of much more complex character, whose structure and function we are still studying, because we haven't begun to learn what we need to know about them.

H. L. Mencken described this testimony as, "One of the clearest, most succinct and withal most eloquent presentations of the case for the evolutionists that I have ever heard ... The doctor was never at a loss for a word, and his ideas flowed freely and smoothly." The judge ruled that evidence such as this was irrelevant and allowed no more expert witnesses to testify.

The only even-handed and reasonable presentation during the whole trial was made by Dudley Malone in the defense of Scopes. Malone, a practicing divorce lawyer, began by complimenting Bryan, his old boss at the State Department: "Probably no man in the United States has done more to establish certain standards of conduct in the ... world of politics," but, Malone chided, "Bryan is not the only one who believes in God." Malone argued that now was not the time to fear truth. "The children of this generation are pretty wise. If we teach them the truth as best we understand it, they might make a better world of this that we have been able to make of it ... For God's sake, let the children have their minds kept open." In his booming baritone Malone moved the audience, and he was given the longest and loudest applause of the entire trial. Outside the trial, however, the rest of America had other fish to fry. Malone is not remembered.

The biology text that Scopes used was Hunter's *Civic Biology*. This book mentions biological evolution only briefly at the end of the text, but there was a marked interest in promoting eugenics. Eugenics was based on the theory that mating with the genetically damaged or genetically inferior degraded the species. The theory of eugenics came from the theory of Social Darwinism and was specifically formulated by Charles Darwin's cousin Francis Galton. Eugenics was big dogma in the intellectual academies of Europe. In England eugenicists were using deaf-mutes, imbeciles, drug addicts, prostitutes, vagrants, the blind, the insane, and epileptics as evidence of defective germ plasm. Wearing glasses was viewed with suspicion. One estimate had nearly one-quarter of the English population as genetically defective. In the earlier days of empire, many of these would have been shipped out to Australia.

America was already leading the way in the scientific application of eugenics. In 1904 an African Pygmy named Ota Benga was captured in the Belgian Congo and brought to the Bronx Zoo and kept like a monkey as an example of an inferior type of human. Franz Boas, one of the first anthropologists, was one of the very few intellectuals who opposed the general spirit of eugenics. During the Scopes Trial the principle of eugenics was still widely acceptable among the educated and by many of the religious as well. Darwinism could not be taught in public schools, at least in Tennessee, but Social Darwinism could be used as an excuse for racism and the promotion of eugenics. This was well suited to the South's deep rooted and overt racism against African Americans and it confirmed the imperial approach to racism of the more sophisticated North.

Bryan was not only opposed to the teaching of evolution, he was also opposed anything, such as eugenics, that impugned the Christian equation of personal responsibility and the possibility of redemption. He quoted from Darrow's ground-breaking, diminished-responsibility defense of the murderers Leopold and Loeb as evidence for the evil consequences of the modernism represented by Darwinism: "Is any blame attached because somebody took Nietzsche's [evolutionary] philosophy seriously and fashioned his life upon it? ... It is hardly fair to hang a 19-year-old boy for the philosophy that was taught him at the university." Darrow quarreled with the morality of this defense.

America eliminated immigration from the Eastern Hemisphere by the Quota Act of 1921 and from everywhere else with the exception of northwestern Europe by the Immigration Act of 1924 to protect itself from bad blood. This was the first formal act of immigration limitation by America. No longer were the huddled masses welcome. Franz Boas, of Jewish, eastern European descent, would no longer see cousins arriving on America's shores to start a new life. Like America, Germany would focus the spotlight of blame for bad blood on foreigners. When Germany was on the road to taking the eugenics principle to its horrific conclusion in 1940, Hitler would use California's 1909 sterilization law for genetically defective people as the model for a German sterilization law.

The Scopes Trial occurred the same year that Franz Kafka's *The Trial* was published, but that was a trial of an entirely different sort. Mainstream America was still psychologically in the nineteenth century compared to Europe. America was acting out its psychodrama with precious little second sight. Eugenics is not part of the self-serving histories, plays, or movies that have followed the Scopes trial, nor the lucrative reenactment of the trial that now occurs in Dayton every year like a religious Christmastime pastoral. No matter which side one was on the Scopes trial a battle between good and evil, it was not an existential search for sanity as was Kafka's trial.

American scientists were now coming out of the middle class. A conversion from religious tradition to scientism was a common occurrence. The psychological expression of self-righteous outrage, common to the recently converted, would frequently manifest itself. The inability to teach the principle of Darwinism at the elementary levels of education only made the conversion dearer. The loss of the Scopes Trial would keep the college-educated on the defensive. They would see themselves as a precious minority surrounded by the ignorant masses. For the solid majority, intellectuals or "eggheads" would become an object of ridicule. Intellectuals would pursue their business isolated from public view. American science as a secular profession was now beginning to make its presence felt and was learning its lessons from Europe.

In Europe at the close of the twentieth century the close association between the chromosomes observed by the microscopists and the discreet nature of the Mendelian or inheritable factors suggested that inheritance indeed came in big chunks since there were only seven pairs of chromosomes in Mendel's prime research subject, the edible pea. In addition, the fact that recessive Mendelian factors could disappear for a generation only to reappear was seen as confirmation of his thesis that evolution occurred in big jumps, not by small, accumulated changes as supposed by Darwin. Yet there also seemed to be more inheritable or Mendelian factors than there were actual chromosomes. The supposition that Mendelian factors were equivalent to chromosomes was resolved in the American laboratory of Thomas Hunt Morgan.

Morgan had observed in the large salivary glands of the fruit fly a phenomenon called crossing over. Crossing over occurs when parts of a chromosome break off and are exchanged with its chromosome mate. This gave Morgan a method for mapping the chromosomes and determining the smallest unit of inheritable information. He called this functional unit of inheritance a gene. Genes appeared to be strung along the chromosome like beads on a thread. With this information in hand Morgan turned to sex. Sex was clearly distinguished in the chromosomes: females had a matched pair of chromosomes where sexual traits were clearly involved; males had an unmatched pair. With the publication of *Heredity and Sex* in 1913 Morgan stated that the sole determination for sex resided in the disposition of the X- and Y-chromosomes and their unique genes. Morgan would become so enamored with the functionality of his gene-infested chromosomes that he would state, "The cytoplasm can be ignored genetically." This would set in stone a concept of gender that had begun with Weismann. Sex cells were supposed to be isolated from body cells, now chromosomes were isolated within the cells cytoplasm like an egg yolk from its white. XX versus XY in isolated sex cells would become biological dogma for the next seventy years.

Shared sex was the missing link that would complete the conversion of Victorian biology into dynamic, twentieth-century American culture. Competition over sexual access would become the cause célèbre by the end of the twentieth century for the intellectual, urban culture in general. Darwinism justified the sexual liberation begun by feminists in the nineteenth century. Darwinists would now travel the hinterlands, converted to the principle of shared sexuality, beating their drums and brandishing their spears with primate enthusiasm. A fundamental assumption of Charles Darwin's understanding of the biological world—the male carried the seed—had been overhauled, but it would still be called Darwinism. In retrospect, however, this new version of sexuality seemed to leave the old Ladder of Perfection in place, albeit in a secularized format: the husband on the top rung, below him the little wife, and the rest of creation on down to the bottom.

One of Morgan's coworkers was Hermann Joe Muller. The scientific odyssey of Muller is a revealing portrait of the intellectual landscape during the 1930s and 1940s. While in Morgan's lab, Muller recorded the frequency of the natural rate of mutations in the fruit fly. He then discovered that if he irradiated the spermatozoa of the flies with X-rays he could increase the mutation rate by 150 percent. The pugnacious Muller brought out the anti-Semitism in Morgan as well as the end of his own marriage. In a general haze of paranoia, a divorce, and a failed suicide attempt Muller launched a picaresque tour of European academia.

Muller was a fervent Socialist and a vocal believer in eugenics. He arrived in Berlin just after the rise of Hitler to power. Horrified, he watched his genetic principles being used against his own culture as the laboratories of Oscar Vogt were destroyed for not removing the Jews under his charge. Moving further east to Leningrad, he arrived at the laboratory of Nikolay Vavilov just before Trofim Lysenko would begin his purge of Western genetics in the Russian academy with the blessings of Stalin. He returned to the West with stops in Spain in the midst of a Civil War, where there were no supplies, and Edinburgh, where it was freezing cold and dark. While Europe finished its twentieth-century bloodbath, the best of all possible worlds would be America. He ended up back in America at Indiana University. In the following year he won the Nobel Prize for his discovery of artificial mutation.

Muller's X-ray experiments focused the debate about the order of magnitude one would expect mutations to be. The X-ray–induced mutation suggested a microscopic level of damage, damage at the molecular level. There were still doubts about how such small mutations could bring about such radically new and complex structures as eyes and wings. Most biologists were convinced of the existence of macro mutations. This uncertainty was allayed by the first of the mathematical geneticists, Sewall Wright, R. A. Fisher, and J. B. S. Haldane. Considering the time spans involved in biological evolution, they argued that the relatively low rates of micro or point mutation together with only mild selective advantages would be sufficient to allow the gradual accumulation of new, favorable attributes. They would use the mathematical tools of massed-population statistics pioneered by Francis Galton. Biologists began to understand the technological analysis of the mathematical biologists in the 1930s. This would mark the beginning of proof by mathematics in biology.

Most biologists at the time believed that proteins were the basis of inheritance. Proteins form the structure of cells and catalyze chemical reactions as enzymes. It was the latter function that attracted the most attention, because without the assistance of enzymes the chemistry of life does not work. For proteins to be the chemical of inheritance it was assumed that they copied themselves, but there was another complex chemical contained in the nucleus of cells called deoxyribonucleic acid that was piquing the interest of a few. DNA was first isolated by the Swiss physician Friedrich Miescher in 1869. Phoebus Levene identified the parts of the DNA molecule. Levene was born in Russia and educated at the Imperial Military Medical Academy. In 1893 he immigrated to the United States because of the anti-Semitic pogroms. He was the head of the biochemical laboratory at the Rockefeller Institute of Medical Research when he deconstructed DNA, but his hypothetical reconstruction of the molecule from those parts was incorrect.

DNA was generally thought to be a "stupid molecule" serving only as structural support for proteins in the nucleus. In 1928 Fredrick Griffith showed that traits from a smooth form of bacteria that was dead could be transferred to a rough form of the bacteria and adopted. A major breakthrough came in 1943 when Avery, MacLeod, and McCarty showed that DNA was the transforming substance that changed a rough bacterium to a smooth form. The race began in earnest to discover the chemical key of life. The competition was fierce, indeed nasty.

Attracted to Indiana University because of Muller's presence was the precocious nineteen-year-old James Watson. Unable to work with Muller, he shifted his attention to Salvador Luria's research group that was looking at how viruses infected bacteria. Watson would begin his own, more abbreviated scientific tour in search of vindication for his belief that DNA, not protein, was the chemical basis for the gene that encoded life. Denmark was the only place on the Continent that Watson would visit. Europe was very much an intellectual and a spiritual wasteland after the Second World War. The world wars unlined the much-reduced role that religion now had in culture. Unlike the religious wars of the Little Ice Age, these were wars between modern, secularized nations for dominance over raw materials, especially oil. Religion's only roll was to supply hope for those whose job it was to go into the trenches and die. With nothing much happening in Denmark Watson would go to England, which was on life support and rapidly socializing.

In 1951 at Cambridge Watson chanced to meet Francis Crick. Watson and Crick had similar ideals about DNA being those governing atoms for encoding the instructions for building organisms and to be the molecule of genetic memory. They had both read Erwin Schrödinger's influential essay "What is Life?" They were particularly attracted to the passage where Schrödinger states, "In Darwin's theory, you just have to substitute 'mutations' for his 'slight accidental variations' (just as quantum theory substitutes 'quantum jump' for 'continuous transfer of energy')." And later when he says, "We conclude that the dislocation of just a few atoms within the group of 'governing atoms' of the germ cell suffices to bring about a well-defined change in the large-scale hereditary characteristics of the organism."

In England a technology developed at the end the nineteenth and the beginning of the twentieth centuries was being used to visualize large molecules. Rosalind Franklin had already used X-ray diffraction crystallography to show a double-helical structure for DNA. Using the parts discovered by Phoebus Levene, Watson and Crick proposed the correct molecular structure for DNA. By proposing a molecule that could separate into two strings and recombine at conception, they had met the goal Mendel had pointed to nearly one hundred years earlier, the equal exchange of inheritable material between the sexes. This they deemed to be the key of life. Their first paper was published in 1953, and for it they would win a Nobel Prize. It was one of the most important biological discoveries in the twentieth century.

Summing up the tradition of Weismann, Mendel, Morgan, and Muller, in his text *Molecular Biology of the Gene* Watson states, "The geneticist's view of a gene is thus: a discrete chromosomal region which (1) is responsible for a specific cellular product and (2) consists of a linear collection of potentially mutable units (mutable sites), each of which can exist in several alternative forms and between which crossing over can occur."[10] Over the years there were all sorts of more profound genetic mutations, substitutions, and augmentations discovered, but it was small-point mutations that fit the Darwinian rule of very small, gradually accumulated changes and resonated with the physics of chemistry.

Supplementing this proposed genetic method of inheritance was a Central Dogma, as defined by Crick. There was one gene for each type of enzymatic protein that was made, and information was passed from the nucleus of the cell to the factory located in the cytoplasm. This chemistry is not reversible; there is no feedback. Also under the heading of this Central Dogma is the principle that DNA is the only self-replicable molecule in all of life, thereby enshrining its role as the essence of life. Our growing acquaintance with radiation and the damage it produces now had a chemically defined target and an evolutionary implication. If the English had had a cell theory in the nineteenth century it would have looked like this: monarchical, top down, assembly line, and absolutist—God Save the Queen!

Through American eyes the Victorian way of science represented by Cambridge and Francis Crick appeared to be nothing but the objective truth. And so a benign bonding of the practical methods of modern German biology with English ideology occurred in America. The statistical mathematics of Galton and molecular genetics of Watson and Crick became the sorcerer's apprentice to Darwinism in America. Americans viewed Continental science through the jaundiced eyes of English culture if we attended to anything beyond our borders at all. This is the reality of our exceptionalism.

The aura of intellectual purity that surrounded molecular genetics radiated from the Nobel biologist Jacques Monod. Monod proclaimed in *Chance and Necessity,* 1972, that the kingdom of ideas had been invaded "by the one according to which objective knowledge is the only authentic truth."[9] The book of nature was now open to read. The presumptions of molecular genetics would naturally carry over to engineering, and the Central Dogma would be applied to what is called the Darwinian programming of computer chips: a set of instructions (an algorithm) called a fitness program that don't 'low no promiscuity, no parasitism, no incest. There are no politicians making shady deals, no bankers inventing Ponzi schemes, no religious fanatics, no redundancy, no welfare, no gypsies, and no music allowed round here, just pure ideology.

In 1976 the explainer of science to the general public for Oxford University, Richard Dawkins, published *The Selfish Gene.* In this book Dawkins gave a literary version of Descartes' human simulacrum, Vaucanson's digesting, shitting duck, Wolfgang von Kempelen's chess automaton, and Jaquet-Droz' writing automaton. Rather than relying on electricity to bring Frankenstein to life, Dawkins divined motivation out of the genome that runs the body like a puppet. The sole aim of this gene robot is to insert it parts into a female in order to replicate itself and advance its cause.

Early on in the book he attempts to deal with one of the thornier challenges entailed with supporting the theory. Dawkins asked, "What is the good of sex? This is an extremely difficult question for a Darwinist to answer."[3] Darwin made a similar statement in 1861: "We do not know why nature should thus strive after the intercrossing of distinct individuals. We do not in the least know the final cause of sexuality."[4] But Darwin, as we have noted, had sexuality separated from natural selection. When Mendel caused natural selection to be collapsed into sexual selection, the extravagant behavior of sexual males became troublesome. A good bit of this behavior seemed to be counterproductive.

An attempted solution to the puzzle would come from the Israeli scientist Amotz Zahavi. Zahavi is the father of the handicap principle, which goes like this: Animals and humans prosper in spite of their risky and most extravagant behavior, such as the behavior that leads to the peacock flaring its feathers, or behavior that would seem to be sacrificial or charitable because these behaviors advertise how well endowed, how fit, how fearless they are. Females that select males with the most exaggeratedly developed character traits can be sure that they have selected from the best genotypes of the male population, according to Zahavi. This certainly sounds like the basis for a biological Ponzi scheme, but we continue.

The handicap principle was reviewed by Richard Dawkins in the first edition of *The Selfish Gene*. He called it maddeningly contrary. He wrote it off, but an Oxford biologist named Alan Grafen came up with a mathematical model that would allow the handicap theory to work according to the rules of Darwinism. Therefore, in the second edition of *The Selfish Gene,* Dawkins recants: "if Grafen is right—and I think he is—it might even necessitate a radical change in our entire outlook on the evolution of behavior."[2] What Dawkins doesn't mention is that the handicap principle, or conspicuous consumption as Thorstein Veblen described it in human culture, resonates with the motivating principle of American capitalism. Get rich quick; everything else follows. So much for the peacock, but what about species where the male and the female look exactly alike and there is no conspicuous expenditure to attract the female that might be construed as a model for Russian communism? In the black and white, good versus evil world occupied by Darwinism versus Lamarckism the defense of one side of the dialectic exposes it to attack from another direction.

Another dilemma concerning sex that Dawkins has to answer is why does his gene robot ever share power with another gender in the first place? Sex, it is supposed, is better because it produces variety, but in Darwinism we are not allowed the sense of precognition, we cannot expect anticipation of good results in the future from present behavior. Change comes by accidents that turn out to be beneficial in the moment. As to how and why Dawkins' gene robot ever gave up cloning or parthenogenesis to share in the process of reproduction in the first place, he gives a tortured argument for the advantage offered by sex, but then he concludes, "This comes perilously close to being a circular argument, since the existence of sexuality is a precondition for the whole chain of reasoning which leads to the gene being regarded as the unit of selection. I believe there are ways of escaping from the circularity, but this book is not the place to pursue the question."[6] In the absence of another mathematical confabulation, Dawkins can't quite get the body off the table. Sex does exist, so his audience loves it. We are redeemed by our self-indulgences. God and Marx are both dead. Dawkins' book becomes an overnight classic.

The *New Yorker* heralds *The Selfish Gene* as a splendid example of how difficult scientific ideas can be explained by someone who understands them and is willing to take the trouble. Dawkins travels the hustings with his Darwin mask on, accompanied by a trained fool with no formal education in science. The fool asks questions that the audience is not clever enough to ask and is always appropriately amazed by Dawkins' answers, thereby helping the shape the response of the audience that might not know exactly when to clap. P. T. Barnum would have been proud. In a *New York Times* book review of July 8, 2001, molecular biologist Lee M. Silver critiques one of the many clones of Dawkins's book and suggests that like the exodus of Israel from Pharaoh's Egypt, the story of evolution must be told anew to each generation. He suggests that the reader rely on earlier texts. He is speaking of Richard Dawkins's book, not Charles Darwin's books and certainly not Erasmus Darwin's poems on evolution.

By the end of the twentieth century the bloom was off the rose. As much as we admire the discovery of DNA, we cannot help but notice that the most attractive aspect of the principle proposed by Watson and Crick is that it is politically correct. It is a theory that is perfectly suited to fit the Darwinian theory of evolution, but is it intellectually viable? Does it explain the exponentially expanding database of science? We will now proceed to the basic assumption used by Darwin to build his theory, Uniformitarian geology. The assumption that has to be met before we can proceed.

CHAPTER 6

Modern Geology

THE MOST SERIOUS DISTORTION imposed by Uniformitarianism geology on our understanding of life on Earth involves the question of the normal state of the planet in our present age and how we got here. This is where the issue of truth and consequences really hits home. This is where cosmology, in the guise of science or religion, has made some very wrong turns. Charles Darwin himself seems to have provoked the distinction between the Pleistocene epoch and our own Holocene epoch roughly dated around 10,000 years ago. Charles Lyell was prodded by the young Charles Darwin while on his *Beagle* trip around the world. Darwin carried with him a copy of Lyell's geology text, which contained his Uniformitarian thesis. Darwin posted Lyell a letter describing a remarkable burial site he had discovered in the basin of the La Plata River, at the Bahia Blanca in South America. Here lay the countless remains of extinct animal species buried intact, ligaments and skin were still identifiable. "All these animals coexisted during an epoch which, geologically speaking is so recent, that it may be considered as only just gone by," he says.

Then Darwin meditates on the case of extinct Siberian animals preserved in the ice and of several theories of sudden revolutions of climate, and of overwhelming catastrophes that were invented to account for their entombment. But he backs off from a discussion of causes of these entombments to discuss to the more technical problem of whether the quantity of food was sufficient to support those ancient rhinoceroses on the steppes of Central Siberia.

Obviously is was. Obviously the environment has changed radically since that time.

We know of no response to Darwin's post, but we do know that Lyell drew a line in the sand at this point in time. His solution to the anomaly was that the environment changed when the Pleistocene came to an end. The Holocene epoch or modern era was born. The correlation between the ocean levels dropping and extinctions is common. Mammoths were frozen by the herd in Siberia with their last meals still intact. Massive kills on a regional basis occurred; more general extinctions of whole species would mount up in a relatively short time. Burial sites of extinct animals are found the world over, and they were the centerpieces of theories that ranged from conventional biblical explanations to secular theories such as those of Cuvier and Beaumount at the beginning of the nineteenth century.

James Croll of Scotland communicated with Lyell in 1864 to discuss his theory that ice ages were the result of variations in the Earth's orbit. He supposed that ice ages appeared to have occurred on 22,000-year cycles lasting approximately 10,000 years each. The last one would have ended, by his calculations, 80,000 years ago. It was Croll that suggested that decreases in winter sunlight combined with the reflective properties of snow favored snow accumulation. This was the solution for glacier formation by Uniformitarian means that avoided other depressing possibilities. Other work showed that Croll's dates showed a very poor match with geological evidence, so all of his speculations, except the glacier formation idea, were dropped by the end of the nineteenth century. However, he did get the ball rolling.

The classic review *Climate Through the Ages* by C. E. P. Brooks, 1926, is a seamless presentation of paleoclimatology from the Uniformitarian perspective with all gaping uncertainties and rational incongruities smoothed out. Mountain ranges are portrayed as gradually elevated, smooth domes that were worn into irregular contours by the ordinary, timeless processes of erosion. Massive glaciations found at various times through the eons were related to mountain building. Several possibilities were given for temperatures slowly varying over time: the slow cooling of the oceans, the erosion of the mountains, the occurrence of a moderate increase in volcanic activity, or favorable astronomical conditions such as the distance of the Earth from the sun combined with the cycle of the obliquity of the ecliptic of the Earth as it faces the sun. The latter thesis was developed by the Serbian civil engineer Milutin Milankovitch in 1920 to explain the interglacial periods of the Pleistocene epoch.

A more sophisticated version of Croll's theory, the Milankovitch model proposes that there is a variation in the inclination of the Earth's axis with a period of 41,000 years and the wobble of the axis at a period of 21,700 years. This precession of the equinoxes caused a change in the phase of the seasons, relative to the location in the orbit. The third cycle that made up the Milankovitch model had a 100,000-year period, although according to Milankovitch's calculations, the 100,000 cycle was a minor factor. Although a better approximation to geological evidence than Croll's theory, it still fell short. Enter F. E. Zeuner.

Zeuner employed more details than ever: the rise and fall of the landscape, the changes in ocean currents, and so forth. It had become a Jesuitical argument about how many angels could fit on the head of a pin. The details of the argument were imminently rational. It was the initial assumption that was pure speculation. The astrological arguments were all very regular by necessity, since they only included orbital or axis cycles. Ice Age events, however, are not as regular as these orbital mechanics would demand. In C. E. P. Brooks' revised edition of *Climate Through the Ages* in 1949, he states that Zeuner's theory cannot account for the Ice Age as a whole, only for details within it. In other words, if you take Zeuner's pattern and slide it around on top of the actual data, it will occasionally match up. Brooks gives a summary of a century of Uniformitarian studies in paleoclimatology:

> *Most of the earlier adventurers in this subject had a single cause—oceanic circulation, carbon dioxide, eccentricity of the orbit, etc.—and saw in it complete explanation of the whole of climatic history. This spirit is not yet entirely dead, but of recent years a broader outlook has become manifest ... There are at least nine and sixty ways of constructing a theory of climate change, and there is probably some truth in quite a number of them. The greatest extremes of climate are not to be attributed to the abnormal development of any one factor, but to the co-operation of a number of different factors acting in the same direction.*[1]

A very important tool for dating recent events became around a decade later. Radiometric dating using a naturally occurring unstable radioisotope, carbon-14, was developed by Willard Libby, for which he was awarded a Nobel Prize in 1960. It was developed making Uniformitarian assumptions. It assumed that the amount stable carbon dioxide, carbon-13 and carbon-12, in the atmosphere stays the same through time. This would not be true, however, when volcanic eruptions release stable carbon dioxide into the atmosphere. Therefore, a catastrophic event like a major volcanic eruption or a major impact event that caused increased volcanic activity would dilute the amount of unstable radiocarbon in the atmosphere. If such a catastrophic thing were to happen, it would appear as though the event had occurred over a long period of time rather than abruptly. Even human additions to the carbon dioxide in the atmosphere with the industrial revolution have caused a distortion in this dating method. Libby's assumption of constant carbon dioxide through time is not a conspiracy to obscure the reality of catastrophic events; it is not even a philosophical posture. The scientific method is by nature Uniformitarian. It starts with the simplest possible formulation. The jump from the laboratory to nature always comes with a bias toward oversimplification. The stakes were raised and the mindset profoundly disturbed with the release of nuclear energy.

When the first atomic bomb was exploded at a place called Trinity an unconscious erosion of the assumptions behind Uniformitarianism began to occur. Suddenly the powers of nature were revealed at close hand. The need to constrain the possibility for violence to something that we the chosen people could maintain became paramount. Robert Oppenheimer worked on the Manhattan Project as an advisor on the development of the bomb. Unlike most of the physicists and engineers who worked on the project, Oppenheimer had wider interests, including a deep interest in Hindu religion. Oppenheimer's ambivalence to the type of work he was doing made him a Faust-like character in some people's eyes. To others in that competitive and paranoid environment, he was simply an unstable, untrustworthy colleague.

When the first test was conducted at Los Alamos resulting in the first nuclear explosion, Oppenheimer said, concisely, "It worked." In his mind, however, came a phrase from the *Bhagavad Gita*: "If the radiance of a thousand suns were to burst into the sky, that would be like splendor of the mighty one ..." and "Now I am become death, the destroyer of worlds." Did this herald the end of another cycle, a Great Yuga, as described in Vedic tradition? Suddenly there was closure with a much older wisdom that had not erased catastrophe as a reality of the human condition on planet Earth.

The conventional academic whose scale for massive violence was measured in tons of TNT or the hundred-year flood was thrilled by the release of nuclear energy, but was not prepared for a deeper understanding of the consequences of a nuclear war. The American public was completely naive. At first we thought that if we were hide under our desks or tables we would be safe from a nuclear war. After the fact, after the genie was out of the bottle, intellectuals began the serious business of visualizing what the result of a general nuclear exchange would be, if the enemy were to acquire nuclear weapons. It would be called the nuclear winter. It would bring an end to everything we understand as life. Catastrophe was officially now at our doorstep with a push of a button, catastrophe of biblical proportions.

Another hidden persuader for the shift in Uniformitarian consciousness was a literary form that would come of age after the Second World War—science fiction. Science fiction was a domain where imagination could range free outside the constraints of the Marquis of Queensbury rules of institutional science. A critical mass of new writers came together, including Isaac Asimov, Damon Knight, Frederik Pohl, Robert A. Heinlein, and Arthur C. Clarke. In the freewheeling domain of science fiction, not only are planets destroyed by other rogue planets or super weapons, whole galaxies smash into other galaxies. Visitors from other galaxies would visit in their advanced flying saucers to rapture us away or destroy us. Not only did this set the seed for a secular religion to augment or supplant traditional religion, still waiting for a Second Coming, it was shaping the adolescent minds of those who would become the new generation of hard scientists. From the perspective of the Futurians, Uniformitarian science was old fashion.

In 1950 Immanuel Velikovsky's book *Worlds in Collision* was published by the respected publishing firm of MacMillan. Velikovsky was not a Futurian; he was the last in the line of traditional catastrophists going back to medieval times and before. Velikovsky had been encouraged and recommended by no less than Albert Einstein. Velikovsky supposed that the catastrophic events described in the Old Testament were distorted by a mythical tradition and by the introduction of gods but that they were based upon actual catastrophic occurrences. He also supposed that natural processes of nature were the cause, not an outraged thunder god who might or might not be appeased by his chosen people.

Velikovsky found confirmation for such catastrophic events in ancient documents from pre-Columbian America, China, India, Iran, Babylon, Iceland, Finland, Greece, and Rome. He ascribed the catastrophic events to close passes with Venus rather than impact events. Velikovsky supposed that Venus was thrown off from the planet Jupiter around 3,500 years ago with near collisions with both Earth and Mars until it settled into its present orbit. Suspiciously, Venus's backward rotation is locked to Earth, always showing the same night side with the inferior conjunction between the two planets indicating a gravitational influence between the two. He got this idea from some observers at the time, when Typhon was a month or so away, who thought the comet Typhon was Venus out of its normal position in a fight with Mars.

In the preface to his book Velikovsky states, "Harmony or stability in the celestial and terrestrial spheres is the point of departure of the present-day concept of the world as expressed in the celestial mechanics of Newton and the theory of evolution of Darwin. If these two men of science are sacrosanct, this book is a heresy."[2] Professor Harlow Shapley, director of Harvard College Observatory, says bluntly: "If Dr. Velikovsky is right, the rest of us are crazy."[3] Because Velikovsky got the backing of Einstein it was published by the eminently respectable publishing house MacMillan, but the academic establishment would see to it that MacMillan was forced to sell the rights to the book to the popular press, where it became a best seller. This success confirmed its unscientific premise in the eyes of the academy.

The sacred names of Newton and Darwin would not be critiqued by a Russian Jew who did his studies in the New York Public Library, even if he did have the backing of Albert Einstein. And there was also the fact that Einstein was under investigation by the F.B.I. for his support of the Peace Movement. It was Einstein's supposition that matter was just another form of energy that explained the source of energy coming from the Sun and led to the release of nuclear energy on Earth. More than most he understood the implications of this discovery. Although the American academy was liberal in many ways compared to the mainstream culture, it was liberal in a nineteenth-century sense of liberal versus conservative. Engineering departments pressed ahead creatively where applied science resulted in new mechanical inventions. Medical departments could be creative in the search for new drugs, but Cold War scientific America did not favor tinkering with cosmology or epistemology; there would be no new Galileos, Newtons, Freuds, or even Einsteins in this environment.

A generation of American scientists would embarrass themselves marching to the podium in order to attack Velikovsky. I attended a Velikovsky roast at Berkeley in the early 1960s. The nasty infighting that is the norm within the academy suddenly vanished at this public event, and I listened to forced conclusions and statistical fabrications as Science presented a unified front in the face of this attack from an outsider. In a 1963 review of the subject, Norman D. Newell criticized the new catastrophism of Velikovsky as based on "little or no evidence."[4] Of course the evidence he was talking about had been systematically buried or misplaced for a century.

Although I found the Velikovsky thesis attractive for its power to explain and integrate things that conventional science was not able to do, I soon forgot the whole thing. The regular business of modern life was and is a constant brainwash to the Uniformitarian point of view. We face a constant stream of Uniformitarian explanations for things like the extinctions of various mega fauna around 10,000 years ago. One of the more familiar and popular attempts to explain world-wide instances of massive burials from the frozen mammoths in Siberia to the burial at Bahia Blanca described by the young Charles Darwin was the thesis by Paul Martin at the University of Arizona: over-hunting by stone-age natives. This fit with the thesis of our killer ape ancestry that was being developed by Robert Ardrey, Desmond Morris, and Conrad Lorenz. It was all part of our Cold War psychology at the time.

Another attempt to explain planet-wide extinctions without invoking meteors or comets was that a virus or a fungus attacked those species. This was the argument used by Norman Newell to explain the wholesale disappearance of the dinosaurs. In Newell's supposition pathogenic fungi caused the end of the Age of the Dinosaurs. Bernard Buigues and Ross MacPhee, the curator of mammals at the American Museum of Natural History, were hoping to test the theory of a killer plague that brought down the frozen mammoths found on the Taimyr Peninsula in Siberia. This general thesis violates what we know about disease-host relationships and vectors. Newell himself stated that there was no evidence for his theory happening at any time. This was also the response by John Alroy to the diseased mammoth theory. He made the point that there is no virus that can kill all the mammoths but spare rodents and rabbits.

Alroy came to the support of the Martin argument with a powerful computer model. In his simulation one hundred virtual humans are placed near present-day Glacier National Park. In his computer run the one hundred virtual humans virtually destroy forty-one animal species over a period of 1,200 years. Of course the human population would have had to expand considerably to accomplish such a thing, and there is precious little evidence for it. The physical evidence for real humans over this period is very rare and for Darwin's La Plata site, none at all. In general we need only watch what these Uniformitarians say about each other's fantasies. They do us the favor of deconstructing each other's theories.

Far away in Australia, far from the centers of scientific tradition at Oxford and Cambridge, Harvard and Yale, a new catastrophism was beginning to take shape in 1966. Prof. S.T. Butler, professor of theoretical physics at Sydney University, makes a statement in an interview with the *Sydney Telegraph* describing a close pass with an asteroid called Icarus that was in store for the planet just two years hence. Butler speculated that if Icarus were to hit us, it would have the explosive power of 1,000 hydrogen bombs.

M.I.T. professor Paul Sandorff decided to assign a hypothetical problem for his systems engineering class: how to avoid an impact event from such an asteroid. This evolves into a book called the *Icarus Project*. The conclusion reached in this book is that the consequences of a collision with Icarus would be felt the world over. In dissipating the energy released, 100 million tons of the Earth's crust would be thrust into the atmosphere and would pollute the Earth's environment for years to come. A crater fifteen miles in diameter and three to five miles deep would mark the impact point, while shock waves, pressure changes, and thermal disturbances would cause earthquakes, hurricanes, and heat waves of incalculable magnitude. Should Icarus plunge into the ocean a thousand miles east of Bermuda for example, the resulting tidal wave, propagating at 400 to 500 miles per hour, would wash away the resort islands, swamp most of Florida, and lash Boston with a 200-foot-high wall of water.

It was now remembered that Karl Reinmuth in Germany had discovered an asteroid in 1932 that crossed Earth's orbit. He called the asteroid Apollo. It was subsequently discovered that Apollo is part of a group of asteroids. Over the years more Apollo objects have been discovered. Many of these Apollo objects have the possibility of colliding with Earth every 100,000 to 200,000 years or so. In 1973 the first systematic program for their detection was started by E. M. Shoemaker and E. F. Helin at the California Institute of Technology. The most glittering moment for the C. I. T. group will come when Carolyn Shoemaker discovers the comet that spectacularly beaks up and collides with Jupiter in 1996. It produces impact sites bigger than the planet Earth.

While a group of scientists interested in impact science was beginning to form, the academy in general is unaffected. The scientific overview of paleoclimatology, *Climate change* [small *c* intended], edited by John Gribbin, 1978, is still firmly in the Uniformitarian camp. There were no considerations of the possibility of meteoric impacts and the catastrophic effects on the weather even though the book opened with an article by D. H. Tarling where he states, "If ice sheets disappear rapidly, as is indicated by the geological record, then major ice surges from Antarctica may be possible and almost instantaneous with resultant rapid rise in sea level in days rather than over several thousands of years."[5] Although this is certainly a break with the rule of Uniformitarianism, it was an accident without a cause. Since there is nothing on the list of acceptable causes that could bring this about it was simply ignored.

Another thing that catches the eye in this book was the observation of Hermann Flohn. He suggested we should pay special attention to the Atlantic section, "where the climatic anomalies during the last glaciation appear to have been significantly larger than in other sections."[6] For some reason, unexplained, North America and northern Europe were heavily glaciated, but there were no significant glaciers in Siberia. Permafrost yes, glaciers no. But that was as far as this discussion went. As Peter Medawar points out about the scientific mind, scientists tend not to pursue questions until they can see the rudiments of an answer in their minds. Historian of science Thomas Kuhn has pointed out that in times of "normal science" scientists will tolerate a large number of anomalies in order not to rock the boat. William Corliss in Glen Arm, Maryland has made a living at Science Frontiers collecting volumes and volumes of scientific anomalies: data that is suspended in limbo because the basic operating system cannot accept it.

Most of the rest of Gribbin's book was largely business as usual. The only contributor to make a biblical reference is a Frenchman, Jean-Claude Duplessy. The French, although not favored by the gods of war in the nineteenth or twentieth century, have never completely capitulated to the English Law of Uniformitarianism or English science in general. Duplessy stated, concerning a similar, apparently catastrophic ice surge at Valders in Michigan, "that [it] could be the explanation for the deluge stories common to many Eurasian, Australian, and American traditions."[7] From an Anglo-American perspective, of course, anything from a Frenchman may be taken with a grain of salt.

At the same time Gribbin was organizing his conference on climate change Luis Alvarez, a Nobel physicist at Berkeley, his geologist son Walter Alvarez, and two space scientists, Frank Asaro and Helen Michel, were examining a thin, red-brown line of clay rich in iridium in the ancient geological record. Luis Alvarez had not been indoctrinated by an education in geology that had taught generations of geologists what not to think. On the other hand, he had a geologically skilled interlocutor, his son, with whom he could thrash out evidence, evidence that certainly seemed to suggest a major catastrophe. Far from the restraining influences of Oxford and Cambridge, with the self-confidence imbued by Nobel confirmation, the Alvarez team would make a startling discovery. A meteor strike was the likely cause of the extinction of the dinosaurs 65 million years ago. No land animal over twenty-five pounds had survived this impact. The impact site would be identified with the Chicxulub crater on the Yucatán peninsula in Mexico in 1988. It is the Alvarez meteor that officially blows up the foundation of nineteenth-century science including Darwinian evolution.

Nobel laureates often make fools of themselves when they attempt to make contributions outside of their narrow area of specialty. Alvarez was an exception. Alvarez came up with solutions to scientific problems the way Edison rattled off inventions. In my communications with Alvarez, however, it was obvious that he had no doubts about the business of science in general. He was the polar opposite to Robert Oppenheimer, with whom he worked at the Lawrence Radiation Laboratory in 1936. His several, widely varied contributions were as a technician bringing the tools of physics to bear on other problem areas. He burned a hole in the tapestry of speculations that had taken over a century and hundreds of thousands of individuals to weave, but he never varied from the conventional operating system of scientific tradition. He made no response when I suggested that his Nobel Prize was merely a gold retirement watch compared to his startling discovery of a cosmic impact, which, I suggested, was one of the most important scientific discoveries of the twentieth century. It certainly never occurred to him, as it occurred to Immanuel Velikosky, that Darwinism would be in need of overhaul if his theory was correct. Alvarez was able twist the arm of a recalcitrant academy, however, when it came to his discovery.

Dewey McLean began publishing articles in the 1970s on volcanic activity and the greenhouse effects created by carbon dioxide, followed by reproductive failures of animals, and then by extinctions. In a communication to me, McLean described a conference with Alvarez in Ottawa, Canada, in 1981 on the K-T extinction boundary 65 million years ago. McLean attributed Alvarez's iridium layer to a huge volcanic event at that time in India called the Deccan Traps. He said, "Luis Alvarez became angry with me for equivocating his impact theory. NASA had already chosen it to be the basis for the Spacewatch Project. Alvarez threatened my career if I opposed him publicly." The media followed Alvarez around, and McLean disappeared from sight. This is not unusual behavior in high-powered, Nobel Prize–level science, or in the public coverage of such quarrels.

In some ways this is a rebirth of the Neptunists versus the Vulcanists except the Neptunists are now called the impactors or catastrophists. The famous and friendly debate between Geoffroy and Cuvier in 1830, however, is a thing of the past. It is no longer a polite, philosophical discussion; it is a question of professional life and death and therefore of fame and fortune. This method of science precludes C. E. P. Brooks' hope for a broader synthesis through cooperation. The truth-versus-fiction dialectic, so common in the culture at every level, leads to conflict rather than comity. The winner-take-all style of problem solving has created many a scientific blunder over the decades as narrowly conceived rationalizations result in unexpected consequences.

Even before the Alvarez impact thesis was officially announced, D. A. Russell suggested that a major supernova event occurring at the same time could have created an iridium layer. This would have resulted in a major radiation event for biological organisms as well. More recently, researchers at University College London have suggested that a barrage of meteors was involved. Sankar Chattergee has now identified an impact site off the west coast of India at the time of the K-T boundary considerably larger than the Chicxulub, an impact that appears to have caused the Deccan Traps. He has called the site the Shiva complex after the Hindu god of destruction and renewal. The dramatic greenhouse effect suggested by McLean is back in vogue as part of the equation for the massive extinction, but not in his name of course.

The K-T boundary at 65 million years ago was the result of a major and complex extinction event that went on for tens of thousands of years. At the very least it involved multiple impact events combined with megatsunamis, poisoning of the atmosphere, and global wildfires that caused extensive mortality. Marine volcanic involvements in connection with those events would have caused anoxic death in the oceans and multigenerational stress on all ecosystems. It would be one of a handful of major extinction events subsequently found in the life history of planet Earth. Ordinarily life survives and can even be stimulated by catastrophes of lesser magnitude.

In 1979, same year that the iridium anomaly was announced by the Alvarez team, Hollywood released the movie *Meteor* in which Earth is saved from a five-mile-across meteoroid by the combined strength of US and Soviet nuclear forces. Project Icarus inspired this film. If fiction had helped to open the question of catastrophic cosmic events, this movie helped to close the subject down in the public mind. The movie was a creative and financial disaster. The whole political spectrum in America was shifting to the right at that time and Ronald Reagan would soon to be elected president. Attention would shift back in the direction of Conservative Protestant sects who believed in an imminent Second Coming following the Tribulations of the 1960s and the 1970s. Some of those groups, including some highly placed military officers, believed that a nuclear exchange with the Soviet Union would be that return and that the chosen few would be "raptured" up to heaven.

A good-natured obfuscation of the new catastrophists came from Nigel Calder in his book, *The Comet is Coming*, 1980. Calder is a highly regarded explainer of Victorian science to us common folk. Most of his book is a jolly good send-up of the superstitious fears of people who were anticipating another pass by Halley's comet with trepidation. Near the end of the book he introduces the subject of the Alvarez meteor. After showing pictures of foraminifers before the Alvarez meteor and smaller species that followed he states: "to pretend that the story of life on Earth is just a succession of adaptations to catastrophic comet shocks would be at least as foolish as to ignore to inevitability of intermittent events of that kind."[8] Of course he might have shown *Tyrannosaurus rex* before and a lizard after for a much more impressive contrast.

Science would have its pudding and eat it too. The establishment would accept the reality of cosmic impacts without changing the fundamental operating system of Victorian science. At the end of the book Calder concluded that events like the Alvarez meteor are expected only once every 250,000 years or so and that we will build a shield for Earth of nuclear weapons to defend us by the time the next meteor or comet arrives. By the time Halley comes around again in 2061, he predicts, "precautions against comets will seem as natural as keeping down mice."[9] And this has been the opinion of the educated public since. As far as evolution is concerned, if Charles Lyell and Charles Darwin could live with the contradictions to their theories in the interest of civility and peace of mind, how can we do any less?

Some heterodox thinkers such as the Cambridge astronomer Fred Hoyle were not content with that sort of obfuscation and began to put other pieces of the puzzle together. The Milankovitch thesis, even with the Zeuner upgrade, had fallen on hard times in the 1960s and 1970s because of its inability to explain very much. Having already dispensed with Milankovitch model as well as another of his own invention, Hoyle proposed that major meteor strikes were the cause of the great glaciations that had occurred in the last million years or so in a book entitled *Ice*. Having dispensed with Uniformitarianism completely, Hoyle was informed enough to see that the Darwinian thesis for evolution was also out of date. He claimed that the random emergence of life had the same likelihood that "a tornado sweeping through a junk-yard might assemble a Boeing 747 from the material therein." He wrote a piece of science fiction called *The Black Cloud* in which it turns out that the most intelligent life in the universe takes the form of black clouds, who are surprised that intelligent life can form on planets.

Victor Clube at Oxford and Bill Napier at the Royal Observatory at Edinburgh have shown evidence for a periodic terrestrial bombardment of Earth by a cloud of debris as it progresses with its solar system in a revolution around the galaxy. Passage through the Gould Belt three to six million years ago, in their estimation, would have resulted in glaciations, geomagnetic reversals, and extinction events, all of which have left evidence. In their book *The Cosmic Winter* the British astronomers show that Earth has been pelted throughout the Holocene period by Earth-crossing interplanetary objects that are part of the Taurid stream composed of thousands of bodies including asteroids, mountain- and island-sized boulders, smaller meteoroids, Encke's Comet, and other assorted fragments of celestial refuse. The yearly passage through different planetary sectors of space that creates the Leonids and other periodic meteor showers is an example of this sort of thing on a much smaller scale.

Released from the blinders of Uniformitarianism, geologists were now finding impact sites all over the place. Up to one hundred fifty major impact sites have now been identified. It begins to be accepted that impacts from space junk are a regular part of the early life of the planet. Ironically it has been the slower, relentless processes of erosion by wind, water, cold, and heat that have erased evidence of most impact events, but all we have to do is look at the moon to see the true story. Clube and Napier think that impacts continue to be a part of life on Earth. Indeed it happens on a regular basis when our planetary system enters a busier area of space during its grand peregrinations around our sector of the galaxy. A dating system based on numbers of impacts was suggested: the greater the concentration of impact sites, the greater the age of the impacted body.

Working quietly in the largely musty archives of the climatology department, the greatest climatologist of our time, H. H. Lamb, began to sow the seeds of doubt concerning the received wisdom about the irrelevance of climate to human history. Unlike most meteorologists, he had taken a degree in geology at Cambridge and he worked with botanists and historians. He had published a paper in 1970 showing the influence of volcanic activity on climate going back to 1500. In 1982 Lamb published *Climate, History and the Modern World*. In this work he overcame the widespread reluctance to study a wide range of natural causes of change. He follows in the footsteps of Ellsworth Huntington (*Civilization and Climate*, 1915) and S. F. Markham (*Climate and the Energy of Nations*, 1942) in the point of view that climate is a significant factor in shaping human civilization. Jared Diamond would be the next to sacrifice himself on altar of Uniformitarianism in 1997. All of these books are helpful in describing affects, but they are all still Uniformitarian when it comes to the profound nature of the causes. And we continue to tell the story of history in terms of its heroes and villains.

The Milankovitch thesis had fallen on hard times in the 1960s and 1970s because of its inability to explain very much. But when something approximating a 100,000-year cycle showed up in oxygen-isotope sediment studies, people like John Gribbin remembered that one of the three cycles that made up the Milankovitch model had a 100,000-year period. The snag, according to Milankovitch's own calculations, was that the 100,000-year cycle shouldn't be a dominant factor. How then does this minor influence come to create such a big effect as an Ice Age? According to Gribbin, "The most likely explanation was that when the Earth's orbit is more elongated, the precession cycle has a greater influence on climate."[10] If it all seems a bit too abstruse, they reassure us that there really is no need to understand all the subtleties of the mathematical argument to appreciate that they work.

In their easy-to-understand book, *Ice Age*, Gribbin and his wife Mary make it sound like real progress has been made. The authors recognized that the Holocene epoch was an invention to mark the emergence of human civilization and had really nothing to do with meteorological reality as preserved in geology. The Holocene epoch was no different from many warming periods recorded during the Pleistocene epoch or Ice Age. They describe the Holocene as an invention out of hubris. They politely ignore the role played in its invention by Sir Charles Lyell and Charles Darwin.

The authors also seem to be coming to terms with reality when they describe the very rapid switch from Ice Age to the warm period we are living in, even using the word "catastrophe," but their true colors are revealed by the major theme of the book. Along with a majority of the academy, they are flying the flag of the Milankovitch model again. Everything can be explained by this Uniformitarian thesis. In a temperature graph covering the past 65 million years they show smooth, regular, sinusoidal waves of change rather like the sixty-cycle current coming out of our electrical sockets: no jagged upward spikes such as might have been created by an impact event. Their theory is a sheep in wolf's clothing. The review on the flyleaf of the Gribbins book is by Charles Munger, vice-chairman of Berkshire Hathaway Corporation, right-hand man to Warren Buffet. He states, "Best work of science exposition and history that I've read in many years."[11] The common identity here is the guiding principle of investment in true value advocated by Buffet, that which will never change over time. The excitements and the panics of the ignorant mob in marketplace are ignored in favor of what is thought to be eternal.

Every religion, every science, and every philosophy has its end-of-all-days scenario, its nuclear winter, its terrible plague, its fatal collapse of the stock market that ends life as we know it. We like to believe these things are under our control or will actually work in our favor. The only reason we have any memory of the Typhon impact in Christian culture is because the charismatic church embraces the hope that the next such event will come with the promise of resurrection. The general climate of opinion among those who are doing well or hope to do well still leans toward Sir Charles Lyell's thesis that geological catastrophes have never happened based upon the premise that it would be too depressing to think so. No one has to prove the Uniformitarian thesis to this marketplace. Their end-of-all-days is the collapse of Capitalism. For the curious few we will continue.

The Russians established Vostok Station in 1957 near the Antarctic pole. In a joint venture with the French they drilled a 12,000-foot ice core in 1996 that shows temperature, carbon dioxide, and methane levels preserved in the icecap for the last 420,000 years. This covers four previous interglacial events: 320,000 years ago, 230,000 years ago, 120,000 years ago, and around 12,000 years ago give or take a thousand years. Interglacials such as our so-called Holocene epoch beginning roughly around 13,000 years ago, the extraordinary elaboration of human civilization not withstanding, show nothing particularly new geologically or meteorologically speaking. Our warmer period does not seem to be much different from twenty-two other short warming periods that have left their mark on the 2 million years of the Pleistocene epoch. These warmer interglacial periods have constituted five percent of the total time. The difference in yearly average between the Pleistocene and the warm periods has been about five degrees Celsius.

It is the greenhouse effect that keeps the planet as warm as it is. One of the principle greenhouse gasses, carbon dioxide, is chemically precipitated out of the atmosphere by the formation of carbonated rocks, limestone. It is the carbon dioxide released by volcanic activity that makes up for this loss. Magma heats up limestone and releases carbon dioxide. It is now conventional to assume that the Pleistocene epoch is the result of lower than normal volcanic activity. As far as glaciers are concerned the poles of the planet see as little snowfall as the harshest deserts see rain. At the highest latitudes is it so cold that glaciers only get additions when weather conditions blow in moisture from warmer surrounding seas. The glacier located on Kilimanjaro only exists during the short warm periods. The Wisconsin period extends from 110,000 years ago to around 13,000 years ago. It is conventionally called the last ice age, but it is no longer possible to think of an ice age as a period when huge ice sheets slowly accumulate and mysteriously creep out from the poles to cover some parts of the northern latitudes, but not others.

First we notice that the ice age intervals don't come regularly at 100,000 years as required by the Milankovitch model; they come every 100,000 years plus or minus ten thousand years or more. This is a reflection of the regular cycle of the planetary system through areas of the galaxy where there is a higher probability of an impact with space junk, but exactly when such an impact will occur is variable and unpredictable. This suits the theory of Clube and Napier. When such an event does occur the cool, dry, dusty planet is suddenly changed overnight by an upward spike in temperature and moisture. There follows a spectacular but short-lived warming and greening of the blue planet. The climate slowly cycles back down over several thousands of years to its norm.

As far as the heat budget of the planet is concerned, virtually all our attention is focused on the solar-greenhouse effect. We pay very little attention to marine heating. It is only with increased marine volcanic activity, however, that we have a solution to Tyndall's Paradox, the paradox of fire and ice: the enormous amount of heat that is needed to convert water to vapor to create glaciers miles thick. In fact, the planet shifts from solar to marine heating for a while because of the reflective cloud cover that is created. The moisture precipitates as snow from the cloud cover, especially over northern landmasses. Massive glaciation follows immediately after a major impact event. Glaciation causes the ocean levels to be lowered between 300 and 400 feet. In warmer areas there are pluvial rains. This completes the equation of fire and ice first proposed by Tyndall.

The clearest record of a warm period following a heat spike outside of our own was the last one, around 120,000 years ago. Jeri Kukla has analyzed loess deposit records in central Europe covering the last such warming period. These records show a sequence from forest, to grassland, and finally back to tundra. Each period was separated by a loess period, indicating a severe cold/dry snap. The climate warmed up again, but never to the earlier level. The oscillations in the climate were due to the rapid melting of the glaciers: the glacial water cooled the oceans causing a temporary cooling of the planet; this stopped the melting of the glaciers; the oceans were now allowed to warm up again by the still-active volcanic vents, but not quite so much; and the cycle began again. On average the climate was getting cooler and drier until it was back to the norm for the Pleistocene. The warm period lasted for about 10,000 years all together.

Although big events occur on a rough 100,000-year cycle, we are finding that life was not entirely cool and tranquil in between. One of the signal events of the Wisconsin period was the explosion of the super volcano, Toba, on the Indonesian island of Sumatra. Bill Rose, Craig Chesner, et al., estimate that the amount of material ejected by this volcano around 70,000 years ago was around 2,500 times the amount of material ejected by Mount St. Helens in 1980. Although the Toba explosion was not an event marked by a significant number of extinctions, the "volcanic winter" created by the cloud cover lasted for several years and caused a planet-wide die-off.

There is evidence from mitochondrial DNA that the human species experienced a genetic bottleneck at this period. This led to a reduction of genetic diversity, which is a polite way of saying "inbreeding." Stanley Ambrose speculates that the human population was reduced from around a million to only a few tens of thousands of individuals. Following this tragic event greenhouse warming was restored and life not only improved, but there was likely a Garden of Eden period for the survivors. It is obvious that this is not the first time that such catastrophe, and such a bottleneck, and such an Eden, have happened. We are coming to realize that this is part of the normal process of biological evolution.

The sequence of volcanic winter followed by global warming would seem to be part of the natural feedback system that is the core of the Gaia hypothesis promoted by James Lovelock. The Gaia hypothesis sees the biosphere and the physical components of the earth, atmosphere, cryosphere, hydrosphere, and lithosphere integrated into a complex interacting system. This self-healing system keeps the planet in a homeostatic balance. If things such as volcanoes, meteors, and comets are a part of life on the planet, they too fall under the regime of the Gaia hypothesis if homeostasis is to be achieved and life is to continue. The shift from solar greenhouse heating to marine heating with an impact event would be another example. Lovelock and the Greens that follow his thesis are firmly Uniformitarian, however, and do not get into the issue of volcanoes and impacts with space junk. In order to have any political leverage at all they hide under the skirt of Mother Nature with the rest of us and refuse to even acknowledge the existence of Father Nature.

The next event to stand out during the Wisconsin period shows up in the radiocarbon record. There was a terrific spike around 41,000 years ago that threw off radiocarbon dates by 8,000 years without any evidence of geological upheaval. This is the signature of a silent killer. Firestone et al. in *The Cycle of Cosmic Catastrophes* identify this as a supernova event. They point to the star Geminga in the constellation Gemini as the source of the cosmic ray barrage that caused the radiocarbon anomaly. Geminga is a neutron star not far from us in the Milky Way. Neutron stars are the remains of collapsed Red Giant stars. The cosmic rays produced by such a supernova create genetic damage and they destroy the food supply, introduce toxic chemicals and heavy metals in the air and water, and seed the cloud cover to produce rain and snow.

Supernovas are implicated in some of the largest known extinction events. The massive extinction that occurred in Australia at this time suggests that the island continent was subject to the direct exposure to a blast of radiation. Evidence shows a sudden change in diet occurred in the surviving species. DNA scientists believe that there was only one human blood type, type O, prior to this period. Following this event there were four. This is the sort of genetic damage one might expect from radiation exposure. *Homo sapiens* had already spread out of Africa into Europe and Asia before either this event or the super-volcano in Indonesia.

Firestone et al. identify the supernova as the catalyst for a series of catastrophes to come. A destructive secondary shock wave hit the planet at 34,000 years. It was during this period that Neanderthal were replaced by Cro-Magnon in Europe. Was this really another bottleneck where inbreeding occurred? Cave painting in Europe would begin not long after that event. Was cave painting, at least in part, a celebration of an Edenic rebirth of life after the shock wave by our forefathers and mothers? Did cave painting also reflect an elevated dialog concerning the meaning of life that goes on to the present day? In any case, there was more to come. According to the Firestone group a debris cloud from the supernova, including the breakup of a huge comet in local space, began to rake the planets and their moons of our solar system around 16,000 years ago. The clearest records for this event come from ice cores taken from the Greenland glacier. Timothy Culler and Richard Muller at the University of California, Berkeley have analyzed lunar spherules on the moon produced by meteorite impacts and have found that the impact rate is higher that at any time in the last three billion years.

A considerable amount of evidence collected by the Firestone group shows a major impact event, including an iridium spike occurred around 13,000 years ago. The Firestone group found a charred, carbon-rich layer of soil at fifty Clovis-age sites across North America. The carbon layer reflects a terrific firestorm. They have hypothesized that this was a result of an air burst impact event. Clovis sites refer to the earliest Native American culture identified by their characteristic stone spear points. After this layer there are no more camels, horses, giant short-faced bears, or numerous other species. There are also no more Clovis points. The survivors resurfaced as a different culture.

The Firestone group follows the conventional habit of lining up deglaciation with the heat spike. I have proposed that the heat spike was the cause of massive glaciation. A problem here is that it is very hard to date glaciers that no longer exist. Radiocarbon dating the biological remains in the terminal moraines of glaciers is very inaccurate and there are considerable anomalies in the radiocarbon graph points at this time. Henrik Rother of the Australian Nuclear Science and Technology Organization has recently used surface exposure dating on Mongolian glaciers. It is believed that this technique will improve the rather provisional dates obtained by radiocarbon. Along with David Fink and Frank Lehmkul, he has presented data that show that 13,000 years ago, glaciers in the Khangay Mountains were advancing when glaciers in Europe, dated by radiocarbon, are supposed to have been in full retreat. The Mongolian glacial retreats occurred a thousand or more years after the sudden rise in temperature. While Rother et al. don't know what to make of this I believe the Mongolian date accurately lines up glacial advance with temperature spikes recorded in ice cores at the poles.

I speculate that Iceland was a central factor in the 13,000-year impact event. Iceland is officially considered to be a hotspot of very recent origin. Iceland happens to lie on the volcanically active mid-Atlantic ridge. Multiple impacts coming across the North Pole at a low angle impacting mainly in Canada certainly torqued the thin crust of the planet over its molten core, opening up the mid-Atlantic ridge. It is reasonable to assume that a considerable amount of the energy of impact was transferred to wrenching and displacing the Earth's crust in relation to its molten core. Much of this energy would be released slowly over centuries and even thousands of years in the form of increased volcanic activity, but initially the Iceland hotspot came alive on a massive scale. Huge lava flows have come from Iceland fairly recently. A profoundly reactivated Iceland would explain Hermann Flohn's observation of the odd pattern of glaciation during the last glacial period. Siberia was ice-free. Western North America was largely ice-free. Only Canada, eastern North America, western Europe, and central Europe were buried under snow and ice. A volcanically active Iceland could have been the bull's eye for a giant, cyclonic, moisture-producing weather pattern.

Following swiftly on the heels of the impact, huge amounts of water were transferred from the ocean onto the land in the form of snow or freezing rain. The glaciers so formed were miles thick. The weight of these glaciers caused deformations in the crust and more volcanic activity. As soon as the atmosphere cleared sufficiently, the combination of warmed ocean and restored greenhouse heating began melting the glacial deposits, creating glacial lakes in North America and along the Eurasian ice sheet. According to oceanographers William Ryan and Walter Pitman in *Noah's Flood* a lateral drainage pattern was created due to the temporary bulge created by the weight of the ice sheet. Instead of draining south down the Mississippi, the Dniester, the Dnieper, or the Volga these lakes drained laterally east into the North Atlantic, or west into the North Sea.

After the heat spike caused by the impact event and the sudden creation of massive glaciers there was a sudden drop in temperature in the higher latitudes of the northern hemisphere. This is the Younger Dryas period that is tentatively dated at about 1,300 years in length, although it could have been considerably shorter than that. The period is named for the arctic wildflower Dryas octopetala that is characteristic of boreal ecology. No climate change of this size, extent, or rapidity has occurred since. I believe that this was caused by the drainage of the glacial lakes that was turned like a hose of cold water on the mid Atlantic ridge and its hot spot, Iceland. The ocean was cooled down.

The Firestone group has found another iridium spike 9,000 years ago. This heat spike reached an even higher level. The Younger Dryas period was suddenly over, and destruction returned at a furious level. I believe this may have been a comet impact that created Hudson Bay, broke up the ice cap, and splattered ice bombs over what is now the lower forty-eight states. The Firestone group think this event was earlier. It doesn't matter much. Terrible things happened. As to the cause of extinctions, let us count the ways: cosmic radiation first and foremost; direct hits from flying objects from space or from the secondary effects of impacts; fires, floods, mudslides, or from being frozen in situ with the last mouthful of food still in place; poisoned air and waters; famine; disease; and, finally, the last few survivors of a species being killed for food by other starving animals, including humans.

The wooly mammoth appears to have become extinct in stages: in Europe with the first iridium peak, in North America in between the two peaks, and in central Europe and Asia after the second iridium peak. The remains of a small, isolated population of mammoths that survived until 1700 BC have been found on Wrangle Island in the Arctic Sea. Doubtless this was because they were not in the area of profound glacial destruction around Iceland and because they were free from secondary predation while in a reduced and weakened state.

A temperature graph developed from pollen cores in Canada by William W. Kellogg shows the complete cycle that follows a catastrophic impact event. It is mirrored by the cycle of wet/warm to dry/cold captured in the ice cores taken from the newly formed glacier on Mount Kilimanjaro. They both show a warming peak around 7,500 years ago, or 5500 BC. At this time Lake Chad was ten times its present size. What we now call the Sahara Desert was at that time a savannah teeming with life. Ryan and Pitman discovered a particularly momentous event connected with this meltdown. The rising ocean levels breached the Oxus, which had been a cattle crossing separating the Mediterranean Sea and the Black Sea depression that held a relatively small, freshwater lake at the time. The survivors of European Ice Age culture, those who were our Indo-Europeans ancestors, were located in the lower Danube River valley and around the Black Sea depression. The Mediterranean backed up by the oceans of the world suddenly began to fill the Black Sea depression with a horrendous flood of salt water: and the dead, "like the spawn of fishes, they fill the sea." The authors suppose that this was the flood that was immortalized in *Epic of Gilgamesh*. Immediately following the Black Sea flood came the first substantial human urban settlement, Çatal Hüyük, on the Konya Plain, Turkey. Not long after this Sumer would put down its roots within the Fertile Crescent. Because Sumer had a longer life and left us things such as the *Epic of Gilgamesh* it has come to be called the Cradle of Civilization. The Kellogg pollen core graph shows a climatic dip of two degrees centigrade (yearly average) into dry/cold period only a century or two after the Black Sea flood.

Lonnie Thompson, a glaciologist at Ohio State, has documented the evidence for the next climate downturn from glacier cores around the world at 5,200 years ago or 3200 BC. This is confirmed in the Kellogg record. Thompson supposes that a change in solar output was the cause of the climate downturn. This is also the favorite explanation for the Little Ice Age, but at least for the latter event weather anomalies have not been found to be quite synchronous worldwide. If the Sun's output were the cause one would expect a more uniform cooling event. Oceanic cooling, on the other hand, progresses more rapidly in some areas than other areas while mixing goes on over decades and centuries.

This climatic downturn triggered the coalescence of tribes living along the Nile River forming the Egyptian civilization in 3150 BC. It was a defense against invaders pouring out of the desiccating Sahara region. Another sign of the times was the construction of Sun/Moon calendar temples such as the famous Stonehenge in England and another in Western Canada that is almost totally unknown. Gordon Freeman started by decoding the Canadian stone circle, as described in his book *Canada's Stonehenge*, and used his successful techniques on England's ancient monument as well. Freeman was the Chairman of Physical and Theoretical Chemistry at the University of Alberta. He studies the kinetics of nonhomogeneous processes, in short, the process of creation. The degree of intellectual capacity on display and the force of social motivation reflected in the calendar temples astonishing. When we correct for the inflation of social knowledge, including its self imposed halo of hubris; when we subtract all the artificial aids to performance that we depend upon; we are only just able to understand what these Stone or Bronze Age peoples were doing. The only similar social event in recent time we can appeal to by way of analogy is the formation of the National Aeronautics and Space Administration that allowed the United States to put a man on the Moon. The motivation? The Soviet Union and the fear of nuclear extinction.

Following the Typhon impact around 1500 BC came the *Atrahasis Epic* that was an updated version of the Gilgamesh Flood story. In this Sumerian tradition, high civilization was burdened by overpopulation and plagues. In punishment for the miscarriage of civilization, Enlil decided to destroy creation and begin over. Enki told the hero Atrahasis to tear down his house and build a boat. Following the destruction, the dead "like dragonflies they have filled the river." It was this edition of the story that was carried on in a modified form in Jewish culture as Noah's ark. The desiccation of North Africa would be completed the impact event of Typhon. After this the Sahara would be the land of the camel.

Much of our favorite catastrophe literature stems from the Typhon impact. Some stories hold memories of older traditions. There are a few traditions, however, that appear to be a memory of four worlds separated by three descents into the underworld. The underworld may be a memory of the impacts of 13,000 years ago, 9,000 years ago, 3,500 years ago. It is astonishing to think that human tradition could go back 13,000 years, considering how many invasions, migrations, reformations, and so forth have occurred muddy up the water, especially in Eurasian culture, but if such a possibility exists, and I believe it does, it is in the relatively untrammeled Americas where we will find it. The cleanest, simplest instance of such a story comes from the Hopi, who live an austere, demanding life in the Four Corners area of the North American southwest.

In the Hopi story of the Four Worlds, the first destruction came about not as a result of evil, but as a result of the people forgetting the plan of creation. The First World was destroyed by fire. The chosen people who had followed the plan were told to take refuge with the Ant People. When they emerged into the Second World they found it almost as beautiful as the first, with the significant difference that the animals no longer trusted them. (There were far fewer animals.) Once again the people forgot the plan of the Creator. Once again the creator informed the chosen people of his intentions, and they retired to the underground to live with the Ant People. (Ants were the original innovators for food storage, agriculture, and husbandry long before *Homo sapiens*.) The Second World was destroyed by ice. After the emergence into the Third World, the people advanced so rapidly that they created big cities, countries, and a whole civilization. In the past thirty years we have found the remains of the Norte Chico civilization, just north of Lima Peru, whose radiocarbon dates place it a millennium after Sumer, but contemporaneous with the formation of Egyptian civilization in the world-wide climatic downturn. When the chosen people emerged into the Third World, the further they proceeded down the Road of Life and the more they developed, the harder it was. This time the world was destroyed by great floods, and the chosen people hid in the hollow stems of bamboo. The bamboo refers to the creator deity Kokopelli who presides over childbirth (he carries unborn children to their mother). He is also the deity of agriculture, and music. He is always recognized by the "bamboo" flute that literally grows out of his head: he doesn't simply play music, he is music.

Charles Darwin stated that he who did not believe in Charles Lyell's Uniformitarian geology could not believe in him. Although he cites Sir William Thompson as stating that Earth was subjected to more rapid and violent changes in its early history, Darwin would not budge from his allegiance to Lyell after his moments of doubt aboard the H.M.S. *Beagle*. Under the heading, "The Forms of Life Changing Almost Simultaneously Throughout the World" in *The Origin of Species*, Darwin wrote: "Scarcely any paleontological discovery is more striking than the fact that the forms of life change almost simultaneously throughout the world." So how does he argue against this apparent contradiction to the rule? According to his theory, "the process of modification must be slow, and will generally affect only a few species at the same time; for the variability of each species is independent of that of all others."[12] He hopes that in time the missing pages of geology will be found and we will be able iron out the discrepancy. Immanuel Velikovsky and Fred Hoyle saw the damage that catastrophic events in geology and climatology did to Darwin's theory, but other catastrophic scientists like Clube, Napier, Alvarez, or the Firestone group make no such connection. Either it is beyond their area of expertise or they know that no good comes to them who challenge the Great One. I will end this chapter with the sad, cautionary tale of Stephen J. Gould.

Gould was the most illustrious evolutionary biologist and American explainer of science in the second half of the twentieth century. I attended a talk that he gave at Stanford University. It was as if a rock star were in the house. The excitement was palpable. I found his speech to be rambling and strangely abstract. He was clearly preaching to the choir, and it showed. In my communications with him, I made point after point revealing the deficiencies of Darwinism. Some of his responses were quite bizarre. When I suggested that Charles Darwin had merely shaped his grandfather's theory to fit Victorian culture, he replied that he preferred the poetry of Coleridge and Wordsworth to Erasmus's poetry.

Gould had already postulated a punctuation theory of evolution in 1972 with Niles Eldredge before the announcement of the Alvarez meteor became public. Their theory stated that Darwin's theory of gradual evolution had not been and could not be confirmed and that while gradual change if not stasis dominated most evolutionary time, there were with occasional evolutionary jumps. Gould and Eldredge attempted to make Darwinism more coherent with the actual geological record, since Darwin's missing pages were not forthcoming. Richard Dawkins croaked nevermore. Dawkins managed to dig up a quote from Charles Darwin that suggested that the Great One moderated his central thesis on occasion by allowing for slightly different speeds of evolution while still holding to the basic formula, *"Natura non facit saltum."* Dawkins attacked Gould, suggesting that he massively over-hyped his own work and had a grossly exaggerated opinion of the worth of his ideas. Gould's sin was the sin of pride. Anyone with the nerve to edit the religion of Darwinism will be found guilty of that sin.

Although one might think that a punctuation theory would have been well positioned for the discovery of the Alvarez meteor, Gould resisted this thesis for a quarter of a century. The Alvarez meteor was certainly punctuation, but rather too big a punctuation. Eventually Gould would accept the reality of the Alvarez meteor and its role in a major extinction event while endlessly reassuring his public by proclaiming his continued love for Darwinism. Nevertheless he would slandered by innuendo for what smelled like a contaminating whiff of Lamarckism and, therefore, of Marxism.

The heated dispute between Dawkins and Gould was sound and fury signifying very little in the big picture of advancing knowledge. While Dawkins was heroically defending a failed thesis, Gould was attempting a Procrustean solution on a theory long past saving. At the end of his life Stephen Gould did say something that was the most relevant of all. He said that we are not likely to see any significant change in the theory of Darwinism from the modern academy in the foreseeable future.

As far as the discipline of impact science is concerned, it is hermetically isolated from the rest of the academy so that it won't contaminate a century's worth of carefully developed Uniformitarian myths on everything from the evolution of the species, the reason for the founding of human civilization and its subsequent history, right down to the contemporary discussion on global warming. We have the stories that we like and we are sticking to them. While geology has failed to support Darwinism, the studies in biology itself have laid the foundation for a complete revolution in the last fifty years. The revolutionary changes in this area have not only completed the retirement of Darwin's theory of evolution to the curiosity cabinet, they have made the Central Dogma of Watson and Crick a dangerous anachronism. We will now proceed to that subject.

CHAPTER 7

THE FITNESS OF DARWINISM

FUTURE SCIENTISTS GET A brief introduction to biology in high school and perhaps a little bit in undergraduate school. It is usually out of date, but even if it isn't, it only takes a few years in the career of a scientist before all they have time for is keeping up with their own narrow specialty. The history of science is hardly ever a part of the curricula and even if it were such histories are generally self-serving, heroic, and thoroughly prejudiced. Very few scientists understand the cozy social relationship between Capitalism, Uniformitarianism, Darwinism, and the Anglican Church. The full implication of the damage done by the Alvarez meteor on Darwinism was minimal, but it was unraveling in its own domain as well.

Casual observation of the kingdom of the birds might well suggest a Darwinian scenario. We see male birds defending territories and competing for access to females with the winners spreading their wings and their genes at the expense of losers, we suppose. When naturalists actually began to enter the field to do long-term, close observation of behavior, they revealed circumstances to be rather less than obvious. In the 1937 *Transactions of the Linnaean Society of New York*, Margaret Morse Nice gave us an intimate life history of the song sparrow. Nice observed that the male of the species seemed to dominate the female. She described this as pouncing behavior. The male pounces on the female. So far so good, but then she observed, "I have no evidence that the female pays the slightest attention to the appearance, character, or singing ability of her mate, nor even to the number of legs he possesses. And it is not that her judgment is prejudiced by the attractions of a superior territory, for she is equally uncritical in this matter also. Old females try to come back to their former homes; otherwise their 'choice' of mates appears to be perfectly haphazard."[1]

There is no sense that a selection for superior qualities is being made in many cases, only that a male is necessary. The female doesn't give a fig for the genetic endowments of the male reflected by his song or his feathers; the female is more interested in a familiar environment. In a recent book by Tim Birkhead entitled *Promiscuity*, we find out that this is part of a general theme with variations right across the animal kingdom. It is common for females to copulate with more than one male, for instance, suggesting that it might be more important to become pregnant than to become Mrs. Jones. It also found that sons may return to their mother's affections as husbands. Good God! Apparently some of the animal kingdom has not read the Ten Commandments of Darwinian biology and is on the road to hell or extinction.

The Central Dogma of the twentieth-century Darwinism stemming from the discovery of Watson and Crick is that it is the gradual accumulation of point mutations of the DNA that produces new species. Newly mutated genes, it is supposed, are expressed if they win out in the contest with the old genes--hence, the survival of the fittest. Rank and file biologists routinely describe animals with relatively little genetic mutation in their genome as being on the verge of extinction, yet speciation by this method is still purely theoretical. It has never been observed in life or created in the laboratory.

The most notable experiments to test the theory were performed on the rats of Bikini Atoll and upon humans and all the surrounding flora and fauna at Nagasaki and Hiroshima. All that has been recorded has been an increase in genetic diseases. People living downwind of the American aboveground nuclear test sites showed higher levels of cancer, not new superior genotypes. The people living in Kazakhstan, the site of the Russian aboveground nuclear sites, show twice the normal rates of genetic mutation and this is passed along to their children, but no new species have been found, nothing useful has been created. A similar story is found at Chernobyl. All of the radiation or chemical mutation experiments ever done in the laboratory—and there have been a lot—have only produced damaged organisms. It is supposed that such events happen very, very rarely and take a very, very long time to resolve, but even with the rate of mutation speeded up by several orders of magnitude, we still haven't seen the first instance of speciation.

There is a very heavy emotional bias that supports this apparently untestable long shot. First, it fits Darwin's supposition that change occurs on a virtually undetectable level. This was what high Victorian culture wanted in its theory of evolution, and was to this standard that Watson and Crick concocted their theory. Second, this theory has much currency because it can be used to cover a profoundly tabooed subject. The human genome, so far as we know, is the most damaged genome in creation. Humans have over 6,000 known genetic disorders. Dogs have an estimated 400 genetic disorders, cats 150 disorders. Obviously we would prefer to see such damage as a good thing. We proclaim that this is the basis for variety, which is a good thing. Third, as we begin to run out of fossil fuels we will probably turn to nuclear energy to sustain high civilization for another century or two. We prefer to put nuclear energy in a positive light. Already we are motivated to find absolution from its release in the first place. Edward Teller suggested that perhaps a little more radiation would be good for us. He was doing nothing more than extending the Central Dogma evolution to its logical conclusion. It is obvious that a great deal hinges on our theory of evolution, perhaps even our own survival.

I began my scientific career in undergraduate school in the early 1960s under the cloud of paranoia created by the Cold War. I was not a particularly good student up to that point, but when I discovered research I came alive with interest. I decided to apply Pavlov's principle of the conditioned response to the habit of smoking. Did the design on the pack of cigarettes trigger the lighting up of cigarette in the dedicated smoker? I never did finish the project since the semester came to an end, but the teacher told me something that I have always remembered. Seeing that I might actually go into the business of scientific research and seeing my propensity for asking troubling questions, she told me that scientists only propose a hypothesis. Only culture in its infinite and mysterious ways can decide what the laws of nature are. This, I would later discover, was what Cardinal Bellarmino told Galileo. In those days the Pope was the final decider. These days it is something decided by a process that is something of a mystery, but money is certainly involved.

I began my formal scientific career as a graduate student in experimental psychology at Berkeley in 1963. I joined the research group of David Krech, Mark Rosenzweig, and Edward Bennett, who were using the maze-bright and maze-dull rats selectively inbred by R. C. Tryon in 1927. Tyron's work was the centerpiece for natural selection in the laboratory. This work had been extended in 1958 by Cooper and Zubek, who put the maze-bright and the maze-dull rats in different environments. Some were put in solitary confinement; others were put in an enriched environment full of toys. When raised in impoverished environments both strains of rats made many errors. When raised in enriched environments both strains made fewer errors. Apparently behavior could be affected by changes in the environment. This was a bit too close to Lamarckian science for comfort.

In another interesting experiment with these animals Rosenthal and Fode gave naïve experimenters rats that they called either maze-bright or maze-dull animals and asked them to run the animals in a T-maze and give them scores. What they actually did was give the experimenters a random selection of maze-bright and maze-dull animals. It turned out that the scores of the animals in the T-maze reflected what they had been called by the researchers rather than what they actually were. Not only could the environment play a role on a rat's performance, so could the experimenter's bias. The experimenter was part of the environment that affected the animal's performance. We can't assume that the original experiment was so biased, however, since there was only one strain of rats to begin with. It was the scientific tradition that was inflexible, not the rats.

David Krech, Mark Rosenzweig, and Edward Bennett were looking for the biochemical differences in the brains of the two strains as well as looking at the possibilities of changing brain chemistry by altering the environment. I joined the ongoing research project of Hal Markowitz, an older graduate student who had invented a fully automated maze that would facilitate the process of behavioral testing. Lab techs and graduate students often had had to wait tediously long periods of time for the rats to run the maze and make their choices. A shock was introduced into the automated procedure in order to shorten the time taken leading up to the choice point. As soon as that trial was over they could turn around and go back down the alley and make another choice, and so on. The choices were automatically recorded.

When we used this automated maze to run the standard test procedure, something quite unexpected happened. The maze-dull rats suddenly become maze-bright rats, and the maze-bright rats suddenly become maze-dull rats. Whatever the presumed change through natural selection was that made some rats bright and some rats dull could be reversed in a trice using a different test procedure. When we took this result upstairs to Dr. Krech, he contemptuously called it a "snowball effect." It was never explained what a snowball effect was, but after a period of consideration it did begin to dawn on us that a multimillion-dollar research project was suddenly thrown under a shadow of doubt. This discovery could not be published. No further work was done in this direction. It remained an undeveloped avenue of inquiry. Hal, who wished to continue in the area of experimental psychology, prudently decided to switch schools. Such are the realities of big-time science.

In 1982 a textbook, *Physiological Psychology*, was published by Rosenzweig and Leiman. The work of Tryon was nowhere mentioned in the book. The words "maze-bright" and "maze-dull" are never mentioned. The work done by Rosenzweig et al. on the maze-bright, maze-dull rats is mentioned, but only in terms of looking at the changes in the brains due to enriched environments compared to impoverished environments. The embarrassing discovery in

1964 had vanished, along with thirty-six years of research that preceded it. What was left was the discovery that enriched environments produce enlarged brains and healthy and intelligent animals; animals raised in impoverished environments became unhealthy, nasty brutes. Surprise! Surprise! There is one sentence in Rosenzweig's book that seemed to refer to our research: "We now know that the anatomy and chemistry of the nervous system are far more changeable by experience than was realized ten or 15 years ago."[3] In the larger picture the fall of a Darwinian pillar went unnoticed and unheard.

After the fiasco of the maze-bright, maze-dull rats I decided to transfer to the neurology department. I joined the research team led by Walter Freeman. Freeman had an engineering degree from M.I.T. along with a medical degree from Johns Hopkins. Having only a B.A., I had to go back and take courses in physics, chemistry, and thermodynamics as well as physiology and anatomy and a seminar on DNA given by Sydney Brenner. While I was doing that, Freeman went to Europe to study with Ilya Progogine. Progogine was making fundamental contributions to our understanding of thermodynamic systems that would promote pioneering research in self-organizing systems. Freeman was studying intentional behavior in cats and using chaos theory to explain brain function, since the old telephone switchboard analogies had fallen on hard times. I was now situated on the growing edge of consciousness in the academy. In fact, I was now a level of intellectual advancement that was above and beyond the pail for the Nobel Prize, beyond what the academic establishment could handle. Freeman was finding conscious attention reflected in the brains of his test animals, cats.

In 1976 Donald R. Griffin proposed an evolution of consciousness in his book *The Question of Animal Awareness*. Griffin was the ethologist who studied bats and unlocked the secret of echolocation that they use to fly through space in the dark. In this book he attacked the mental roadblock ritually passed down from Descartes to John Watson that only the language-using animal can be conscious. E. O. Wilson wrote about Griffin's proposal, "The very suggestion of a cognitive ethology might have been considered dangerous or even foolish by anyone other than an experimental biologist of Professor Griffin's stature. We will owe him a debt for breaking the taboo."[6] In my communication with Griffin, he indicated that there were plenty of academic "rednecks" around, but a few people were listening to his ideas.

Charles Darwin had actually begun the deconstruction of the Descartian view late in his life by drawing a distinction between instinct, meaning "to incite," and intelligence, meaning "to understand." He stated that there was no important difference between the intellect of man and the higher animals, although he continued to give a special place for consciousness, meaning "to know." Certainly there should be a unique descriptive for consciousness, which has such a range of symbolic language available to it, but these subtleties are not available to the common Darwinist.

Among other things Griffin gives a linguistic deconstruction of the theoretical debate. He lists twenty terms used in the clockwork orange world of Behaviorism and lists them on a gradient. The first two words that are taboo to use for the mental experiences animals have are "consciousness" and "freewill." The most acceptable words to use are "pattern recognition" and "neural template." This sort of linguistic deconstruction raises the hackles of scientists who really don't like to be shown how they are being used by their own language.

I attended a seminar that Griffin gave to a neurology group at Stanford University after his book was published. Neurology students know a lot about neuroanatomy and a few things that happen when you put electrodes into brains. They know next to nothing about geology, climatology, psychology, philosophy, or evolution. Most neurologists at that time were simple behaviorists. Behaviorism fits the dogma of the Judeo-Christian-Scientific tradition godfathered by René Descartes and Charles Darwin circa 1861. Descartes' hypothesis was that it is only humans, with their blessing of intelligence and language, that can act with intention. The rest of life's material productions are stimulus-response machines. The response to Griffin's lecture was ignorant, monotonous, and without curiosity. The general tenor of the questions asked of Griffin was did he have proof for his statements.

There is precious little about a global concept like evolution that can be proved or disproved in a controlled laboratory experiment. A considerable amount of the most crucial bits of information are observations of rare and unexpected events that violate habit or training. A theory of biological evolution should be a picture put together piece by piece with a good knowledge of the past history wrapped up with first hand field experience. Evolution should be the frosting on the cake of intellectual discovery, not a taxonomic tradition that is adhered to. The picture that makes the most connections will be the one that will do the best at helping us understand the past and predict the future. Darwinism has survived without direct proof for all of its existence because it was politically correct and because it didn't do serious violence to the information that was available, at least in the nineteenth century.

After those fruitless challenges to Griffin's work went on for a while I raised my hand. The moderator pointed to me. Without identifying myself to Dr. Griffin as a person with whom he had communicated by letter, I started by saying that I fully agreed with his general thesis. Then I asked a question. Wouldn't the attribution of consciousness to the animal kingdom open a serious can of worms for the larger culture? The moderator quickly gaveled the session to a close, thanked Dr. Griffin, and they retired for a postprandial glass of wine. I was completely ignored. I looked around the auditorium. No one was looking at me. I was invisible. Our use of animals in research and in the food factories of high civilization is a profoundly troubling subject. I materialized as the superego for a group that was uncertain of its moral standing and unwilling to deal with it.

One of the most rigorous and unflinching critiques of Darwinism to ever actually reach print was *The Great Evolution Mystery* by Gordon Rattray Taylor published in 1983. Taylor was an intelligence officer in the Psychological Warfare Division of Supreme Headquarters Allied Expeditionary Force and then chief science advisor for BBC. He was trained in the area of breaking codes during the Second World War and, in his second career he translated scientific jargon into common language for television. Taylor clearly underlined the contradictions and the glittering generalities of Darwinism. Quoting such luminaries as William Bateson, Ernst Mayr, George Gaylord Simpson, and Theodosious Dobzhansky, he shows that the origin of species, the engine of evolution, is still a deep mystery. The great value of Taylor's critique is that it restrained itself to the same territory that Darwinism limits itself to. He doesn't get involved in such sticky issues as the incest taboo; he doesn't open the question of the origin of sex; he doesn't look outside of English culture for other perspectives; and the Alvarez meteor, which would strike a mortal blow to the foundation of Darwinism, had only just appeared on the intellectual horizon. This makes his work the ideal first step for those interested in kicking the habit in easy stages.

As Taylor points out, the very definition of species is still open to discussion. The origin of varieties is about all that can be agreed upon. If there are organisms that can steadily modify in various directions, why are there lines stable enough and distinct enough to be called species at all? Why, if the willow can show intermediate species of every conceivable kind, does the gingko tree exist in only one species? These definitions have been arrived at by the static considerations of the taxonomist, not the dynamics of evolution. If evolution is a story of organisms seizing a niche by superior adaptation at the expense of the competition, why do we find forty genera comprising 160 species of the water plant *Podostemaceae* occupying the same narrow environment of running water? J. C. Willis describes this water plant as having an amazing variety of coexisting forms.

Taylor shows that the dominant position of taxonomy promoted by mathematical geneticists, which is to say Darwinists, has led the biological sciences into an overemphasis as to type (anatomy) at the expense of physiology: how it works. The father of physiology formally characterized the difference between the two. In *An Introduction to the Study of Experimental Medicine*, 1865, Claude Bernard states, "In physiology, we must never make average descriptions of experiments, because the true relations of phenomena disappear in the average; when dealing with complex and variable experiments, we must study their various circumstances, and then present our most perfect experiment as a type, which, however, still stands for true facts."[4]

The anatomist and the physiologist have completely different mindsets. The classic demonstration of the healthy and productive relationship between anatomy and physiology was that between Vesalius and Harvey. Vesalius laid out the first accurate modern human anatomy. Looking over his shoulder was Harvey, an English physician. Without Vesalius's achievement in anatomy, Harvey would have had no place to begin. As Harvey supposed: nature is no where accustomed more openly to display her secret mysteries than in cases where she shows traces of her workings apart from the beaten path. Like genius itself, physiology is based upon an original idea, not on the statistical average of types.

Taylor also reveals how the fixation on point mutation of genetic material as the mode of evolution precludes the possibility of ever seeing speciation in the wild. In a standard college textbook for the biological sciences, Jerrum L. Brown in *The Evolution of Behavior* states that when it comes to changing behavior, genes causing increased efficiency are likely to be selected for. This is the standard Behavioristic explanation. Taylor uses the instance the collared dove to make his point. The collared dove was adapted to the rocky slopes of the Middle East, but invaded Europe not long ago, spreading into Alpine valleys and lowlands alike and displacing the native, well-adapted wood pigeon. The fact that animals can choose to better their environment pulls the rug out from under the Behavioristic nostrums of Darwinian evolution. The collared dove could not wait around for the crap game of the big dogma to take its laborious course. It came about by intention and a general creative adaptation by a group. We may well ask at this point, could it be that genes are the cart that follows the horse?

Taylor sums up Darwinian evolution as "not so much a theory, as a subsection of some theory as yet unformulated. Its greatest weakness is the fact that it cannot be disproved. For every circumstance it is possible to imagine some justification. If a species survives, we are told that it adapted. If it fails to survive, that it failed to adapt. If it displays some unusual feature, we are told that it is of adaptive advantage—but whether it really is or not can never be proved."[5] Even though neither Darwinism nor Lamarckism has ever been proven in the laboratory or the field, Taylor states that it is the mark of arrogance of Darwinians that they have managed to throw the onus of proof on to the Lamarckians. As we have already seen, this was a matter of politics, and the Darwinians and the Lamarckians have nuclear weapons aimed at each other's major cities. There was precious little interest in Taylor's book. When the subject is evolution, preaching to the choir is what is desired. Which side are you on?

In some ways there is always an evolution of public consciousness going on. We could call it fashion. A revolution in public acceptance of apish ancestry in America was laid in the area of popular fiction. If you happened to be walking by a newsstand in 1912 you might have stopped to look at the October issue of the *All-Story* magazine. The cover was sure to catch your eye. It showed a man clad in animal skins sitting astride a rampaging lion, his knife raised for the kill. Another man, probably the lion's intended dinner, looks up from the ground in horror. The title of this pulp fiction story was *Tarzan of the Apes - A Romance of the Jungle* by Edgar Rice Burroughs. Burroughs was born in 1875 in Chicago. He spent his early life rotating between the few remaining outposts of the wild-and-wooly West and work in his father's battery company, a clerical job for Sears, Roebuck and Company, and as a salesman for pencil sharpeners. While in a state of near poverty as a salesman he began to write tall tales. The rest is history.

The fascination of males for the romance of the ape-man seems to have risen out of the change in the social environment of America as she lost her frontiers, her El Dorado, her Arcadia, her New Jerusalem. In Burroughs' pulp fiction, the son of a British lord and lady was marooned on the west coast of Africa. John Clayton was adopted by an ape tribe and became Tarzan. While the academy was still carrying the burden of Victorian prejudice about apes and African culture, popular culture was falling in love with the feral life style. From this first novel sprang two dozen more, over forty movies, hundreds of comic books, radio shows, television programs, Tarzan toys, Tarzan gasoline, Tarzan underwear, Tarzan ice cream, and Tarzan running shoes. The list is virtually endless. Edgar Rice Burroughs became one of the twentieth century's most popular authors; the publishing record of this urban myth challenged the Bible.

The influence of pulp fiction in literate cultures was nothing new. The earliest productions of pulp fiction motivated young Spanish men to seek out paradise in the New World where there existed, it was writ, a black, Amazon queen on the island of California who accepted all comers to her bed. Puritans came to New England, Bible in hand, looking for the New Jerusalem. Young southern men marched with courage and dignity into the face of withering canon, musket fire, and sure death during the Civil War with visions of chivalry dancing in their heads put there by Sir Walter Scott. The tales of Grimm accompanied by the music of Parzival sponsored the formation of the National Socialist German State out of the wreckage of the Weimar Republic. It is through such fictitious inventions that integrated group mind that mass movements of one sort or another can occur. Everyone is acting independently; everyone is acting together. In American culture the Tarzan male in different costumes and generally more violent with every passing decade would become a mainstay of culture. The tide of public interest in the ape-man opened the gates to a general interest in the biological descent from the apes fathered by Darwin. In the second half of the twentieth century, human evolution from the apes suddenly bursts onto the scene as an object of intense academic study.

Up to around 1960 there was virtually nothing known about the primates in the wild and almost no attempts to extract principles from primate behavior for the application to human behavior. This would change with *African Genesis* and *The Territorial Imperative* by Robert Ardrey. He supposed that we learned to stand erect in the first place as a necessity of the hunting life. Desmond Morris in *The Naked Ape* supposed that with strong pressure on them to increase their prey-killing prowess, our ancestors became more upright to become faster and better runners. As to the question of speech, Morris supposed that the hunting ape had to increase his power to communicate in order to cooperate with his fellows. With new weapons to hand, he also had to develop powerful signals that would inhibit attack within the social group. Ardrey also supposed that the hunting life demanded division of labor just as the male lion flushes the game for the lioness to kill, and that division of labor demanded communication between interdependent partners. Ardrey supposed that even in Miocene and Pliocene days of pre-human experience our hunting ancestors laid the foundations of human language as they shifted to the ways of the carnivore during the extended droughts of those times.

The evidence for this line of argument had come from a discovery made by Raymond Dart, professor of anatomy at University of Witwatersrand back in 1924. He discovered an upright-walking, five-year old *Homo* called *Australopithecus africanus*. The age of this find had been placed at around 3,000,000 years ago. This finding was ignored for several decades, but the time was now ripe. People begin to pay attention to the supposition by Dart about the character of the upright *Homo*. Raymond Dart thought that he had found an *Australopithecus* tool kit that he connected with our ancestor's carnivorous and cannibalistic way of life although there seemed to be little correlation between these imagined killer apes and any of our close living relatives. The few self-reliant, stone age peoples, known from the firsthand observations of Colin Turnbull of the Pygmies or Elizabeth Marshall Thomas of the Bushmen, failed to meet the needs of Dart, Morris, and Ardrey. Pygmies and the Bushmen sing. The killer apes of Morris and Ardrey never sang. So they turned to the undeniably nasty baboon to be the archetype for early man. The baboons of Morris and Ardrey also became thinly icons for the dismally savage century wherein the high moral purpose of civilization was bludgeoned in pursuit of the alimentary requirements, or pride, or oil, or simply out of fear itself.

Conrad Lorenz supposed that human behavior was far from being determined by reason or cultural tradition in *On Aggression*, and that it is still strongly influenced by violent instincts. He used the analogy of rats. Variations on this theme also included the nasty reptilian brain championed by Carl Sagan in *The Dragons of Eden*. The snake was once again the culprit. It had a nice ring to it, a familiar ring. By the argument of a genetic reptilian propensity for violence, civilization would be let off the hook. The snake made us do it. In the devious labyrinth of the human mind, the subconscious can have a private violent, erotic, baboonish wish-fulfillment fantasy; in the public mind the social order is redeemed for its role in suppressing these primitive impulses of the unwashed individual. This was a return to the original perspective of Darwinism that Christian Victorian culture was our defense against the primitive: Dr. Jackel versus Mr. Hyde.

Additional social pressure had also been building in the culture by loss of territory to the female over the past century. The revolution for women's equality came directly out of the secularization of culture and the Enlightenment that sponsored justification for the ancient system of gender hierarchy. Mendel's discovery of sexual equality evoked a natural law for a total revolution in gender politics. While the killer ape absolutely dominated in the jungle of the mind, the civilized male was losing his unquestioned prerogatives at home, in the marketplace, and eventually even in the academia as well. At first the entrance of females into higher education was greeted with hoots and hollers. After awhile, however, it became one of the eagerly anticipated and occasionally creative enticements for the faculty as the wise old silverbacks mentored the young female students, who were eager and vivacious. But there were some unexpected turns with the advent of feminist scholarship. Not all female scholars were content with the patriarchal tradition of evolution.

In the attempt to identify the culprit for the abominations of civilization we heard the testosterone hypothesis of Maria Gimbutas. Gimbutas was a skilled archeologist from eastern Europe who made substantial contributions to our understanding of Stone Age European culture. In a skillful judo move, Gimbutas took the frontal attack of the naked ape, stepped aside, and directed it to its clumsy conclusion. But in her thesis the roles were simply reversed. The male was singled out as the culprit and contrasted to the peace-loving, estrogen-besotted female.

Elaine Morgan's Feminist view of human evolution in *The Descent of Woman* was representative of important contributions made by women in science. Where Darwinists saw a male-dominated social society, Morgan saw a matriarchy. Where Darwinists saw blood and guts, Morgan saw cooperation.

Where Darwinists had our ancestors coming down out of the trees, Morgan had our ancestors coming out of the water. Morgan pointed to hairlessness, bipedalism, and the ability to hold the breath as marks of a period of marine existence, developing a theory first proposed by the marine biologist Alister Hardy. Morgan decided that the topknot of hair that grows out of the human head had the selective function of giving youngsters a handhold in the water, where their mothers had fled in fear of big cats.

The odd distribution of human hair is largely overlooked by most evolutionists, and Morgan's explanation is rather far-fetched. Morgan's little triumph in finding utility in hairlessness cannot mitigate the fact that she basically goes along with Victorian principles when it is convenient. In fact, no basic reconsideration of Uniformitarian geology or climatology has ever come out of Feminist scholarship, and there is no intelligent discussion on the trials of birthing by human females or the role of selection that may be connected with that. Feminists simply erase Genesis. This is an issue that Feminist scholars could be in the forefront of deconstructing.

Morgan's model for early man was the society of the chimpanzee. Amity reigns among the chimps, according to her thesis. They are hedonic, flexible in social relationships, and curious. Morgan recklessly assumed that baboon society is incapable of learning because of the general deference to the alpha male. Baboon society can be described as agonistic and hierarchal. All look to the dominant alpha male before deciding what they may do. She cited the inability to teach baboons as compared to chimps as a critical difference. Baboon culture was an evolutionary dead end as she saw it. In Morgan's view we are a baboonish society.

Eugène Marais, well ahead of his time, described the decided absence of instinctive or phyletic memories in baboons in *The Soul of the Ape* at the beginning of the twentieth century. If anything, it may have been the absence of phyletic traditions that was crucial for the survival of baboons in their demanding environment. Frans de Waal would follow the baboons around and discover that an alpha-male baboon would intervene on behalf of lower-ranking males in the interest of group civility. This was the same animal that had competed ruthlessly, or at least loudly, before he had ascended to alpha status. This sort of brotherhood among mobsters does sound like human society under some conditions. It turns out that one of the problems we face in our interactions with baboons is that they view white Europeans as very low-status baboons and see no point in learning from us. Black Africans are able to train baboons, however, using baboon techniques.

The romantic view of chimps has its ups and downs. In response to a perceived threat by a predator, such as a stuffed leopard introduced by Adriaan Kortland, chimps yelled and barked with every member charging about in a different direction. There was much jumping about and brandishing of sticks. These communal or individual charges were interspersed by periods of seeking and giving of reassurance by the holding of hands by neighbors. As the aggression-fear cycle waned over the course of an hour, it was replaced by curiosity. With inquisitiveness the stuffed leopard was investigated: poked, smelled, head detached from the body, body dragged into the bush. Here was a model for amity much desired by some humans. On the other hand, Jane Goodall would discover that the even-tempered chimpanzee could and would actually organize warlike attacks on neighboring groups, could and would become cannibal under the right circumstances. Selective pressures are becoming very intense for chimps. Chimpanzees are losing territory to farming Africans at a great pace. This certainly does not promote a pacific temperament.

As with most ideologies there is a kernel of undeniable truth that the honest opposition can only answer by ignoring, or by dashing around, yelling, and brandishing sticks. The manifest difficulties that are involved in establishing any sort of proof for Darwinism suggested to molecular biologist Gunther Stent that there is a limit to the refinement possibilities of analysis. In *The Paradoxes of Progress*, 1978, he supposed that beyond a certain point the inherent disorder of quantum mechanics will only leave us with a haze of uncertainty. Darwinism continues to be inflated by mass action, computer modeling. John Maynard Smith calls this fact-free science. The philosopher of scientific method Carl Popper points out that proof, or certainty, exists only in mathematics and in logic, where it is trivial in the sense that the proven conclusions are already hidden in the premises. Popper supposes that the difference between science and metaphysics is not by the proof of the pudding, but by the possibility of disproof. Darwinism is just as impossible to disprove, as it is to prove. Popper considers Darwinism to be a metaphysical research program.

The Cold War was beginning to run out the string at the end of the 1970s. A complete reaction to any form of evolution, let alone social revolution, in America was marked by the election of Ronald Reagan for President in 1980. Mikhail Gorbachev, seeking to revive a brain dead Soviet Union introduced *glasnost* (freedom of speech) and *perestroika* (restructuring) to save the State. This led to a very unexpected result. After fighting shoulder to shoulder with Protestant culture against godless Communism, Pope John Paul II let the principle of the lesser of two evils influence him. The Pope allowed that Western science was not so bad after. In 1988, the Pope lent himself to bridging the gap between religion and science. The pontiff proposed that science could purify religion from error and superstition, while religion could purify science from idolatry, false absolutes, and materialism.

In 1992, John Paul publicly endorsed Galileo's philosophy, noting how intelligibility, attested to by the marvelous discoveries of science and technology, leads us, in the last analysis, to that transcendent and primordial thought imprinted on all things. The Pope addresses the Pontifical Academy of Sciences on October 22, 1996, arguing that between ancestral apes and modern human beings, there was an ontological discontinuity—a point at which God injected a human soul into an animal lineage. The ontological discontinuity he spoke of was the seventeenth-century construction of René Descartes, under the tutelage of the French monk Marin Mersenne. It is the disembodied consciousness stemming from language given by God to Adam: In the beginning was the word. No more do we have the creator, in a loincloth, spitting into the iron-rich soil of the Middle East and forming species the way a potter throws pots.

John Paul had done what Urban VIII could not do as the leader of the most powerful European social institution of his time. Urban VIII could not accept the Copernican solar system because of resistance from his intellectual academy. John Paul had not only rectified that unfortunate circumstance, he endorsed the English version of evolution. Like Thomas Aquinas, who brought Aristotle into the fold of the Medieval Church, John Paul made peace with the Uniformitarian ideologies of the Anglican world. The patriarchal perspective of Darwinism and Christianity is still basically the same; there are no mothers in Darwinism. The interpretation of the violence in the Old Testament as metaphor by modern academia is well suited to the high church. Darwinism still imposes the incest rule of Judeo-Christian tradition on the whole of life as a law of nature. In the other direction it frowns on hybridization. Darwinism secularizes personal redemption, the core of Christianity, into personal selection, but that should leave the door open for the church as the next step. So what's not to like?

What was fundamentally different from the days of Urban VIII was the shift in power. The shoe was now on the other foot. John Paul argued from a position of relative weakness in western European society versus the secularized culture. The old triumvirate of power—the church, the crown, and their Jewish bankers—was trumped in the nineteenth century by Capitalism. Science is a lavishly supported advisor to industry and government. Just as there is a revolving door between politics and business, there is a revolving door between science, business, and politics. Religion, isolated from seats of power, is the poor relative. While science builds a new campus for biotechnology every other day, religion has to have garage sales to raise money to repair its leaky roof.

From his position of power as a favored explainer for science, Carl Sagan gave a predictably arrogant response to the suggestion of a détente with religion. The idea that God is an oversized white male with a flowing beard who sits in the sky and tallies the fall of every sparrow is ludicrous, he said. If by God one means the set of physical laws that govern the universe, he continued, then clearly there was such a God, but such a God was emotionally unsatisfying. It did not make much sense to pray to the law of gravity, he concluded. In short, there was no need for science to make the slightest accommodation to the church. Of course his ability to voice such an opinion in public can change overnight with an unexpected bolt out of the blue.

When science makes a monumental gaffe, it is swept under the rug as the cost of doing business. We recall Galton and his eugenics movement in the nineteenth century. Eugenics, based upon Social Darwinism, became a widespread intellectual movement among the advanced civilizations in the nineteenth and early twentieth centuries. It culminated in the German extermination camps in the Second World War. Suddenly the purification of national blood went out of fashion. Now we hear from the Galton Laboratory that research results now show there is a greater degree of genetic variability within the traditional ethnic races than between them? Oh well. The Second World War probably would have gone on in much the same way even without the intellectual justification.

With the fall of the Berlin Wall, Western science triumphantly celebrated its victory over Lamarckism. Almost immediately the original dialectic had to be resurrected: Darwinism versus Creationism. It was a historical re-enactment of the argument formalized by the rector of Bishopwearmouth, William Paley, at the end of the eighteenth century as a reaction to Erasmus Darwin. Rather like Victorian culture after the French Terror, the whole culture was shifting to the political right after the tribulations of the 1960s and 1970s. People stopped asking embarrassing questions and questioning authority. While Darwinism marinated in self-gratification it was given a patsy for an opponent.

Rank and file Darwinists who might stumble upon exceptions to the Central Dogma were kept in line by the threat of humiliation and the occasional object lesson. Colin Patterson, for instance, senior paleontologist at the British Museum of Natural History, is only remembered today for a candid comment he made in 1981 at a talk he gave at the American Museum of Natural History in New York. His agnostic statement, "Can you tell me anything about evolution, any one thing that is true?" was recorded by a Creationist in the audience, and it continues to make the rounds of Creationist circles to the present day. The new dialectic revolves slowly, slowly: on one side is the dark-haired Jesus Christ in a white toga; on the other side is the white-haired, black-caped Charles Darwin. No matter where you push on it, it corrects itself and keeps on spinning.

In high society Ralph W. F. Hardy takes the Central Dogma of molecular genetics and uses it as the model for the ideal culture in 1999. As the president of the National Agricultural Biotechnology council and former director of life sciences at DuPont, he stands in front of a Senate committee concerned with engineered transgenic plants that are entering the American food supply. In order to convey the Central Dogma in terms that they can understand, he draws this metaphor: DNA (top management molecules) directs RNA formation (middle management molecules) that in turn directs protein formation (worker molecules). We are clearly a middle-aged culture feeling the burden of tradition and the necessities for survival; it is very hard to imagine a youthful America or our Founding Fathers embracing such an overblown, imperialist ideal.

A serious blow to the hopes of an ethically pure, utilitarian, cellular CEO came with the Human Genome Project at the end of the century, when it was discovered that only about two percent of the human genome seemed to be composed of genes that code for proteins for the building of the body or producing enzymes. The rest included other such strange entities as pseudogenes, retropseudogenes, satellites, minisatellites, microsatellites, transposons, and retrotransposons, collectively known as junk DNA. As Matt Ridley put it in *Genome*, junk DNA is genetic material that uses energy but does not appear to contribute to the health and welfare of the body. As Ridley describes it, our genomes badly need worming. This was a new and unexpected revelation about the Selfish Gene. Fortunately the Monica Lewinsky scandal was attracting attention in another direction as this information leaked out.

Not only did the genome contain more junk than sense, as sense is commonly understood, there turned out to be far too few normal genes to account for the proteins that we did use. Barry Commoner, senior scientist at the Center for the Biology of Natural Systems at Queens College, made this striking observation of one of the largest and most highly publicized projects of our time in *Harper's Magazine,* in 2002:

> *There are far too few human genes to account for the complexity of our inherited traits or for the vast inherited differences between plants, say, and people. By any reasonable measure, the finding signaled the downfall of the Central Dogma; it also destroyed the scientific foundation of genetic engineering and the validity of the biotechnology industry's widely advertised claim that its methods of genetically modifying food crops are "specific, precise, and predictable" and therefore safe. In short, the most dramatic achievement to date of the $3 billion Human Genome Project is the refutation of its own scientific rationale.*

Celebrating the fiftieth anniversary of the first paper published by Watson and Crick, Natalie Angier harvested widely from the fields of discontent in the academy. Her summary for the *New York Times* concluded that the molecule that had exemplified youthful bravado, vast promise, and vaster self-regard has become another aging, pot-bellied baby boomer. This is similar to the conclusion reached by Ernst Mayr at Harvard decades earlier. Mayr was critical of mathematical approaches to Darwinian evolution, calling them "bean bag genetics." In many of his writings, Mayr rejected the reductionism in evolutionary biology that came with the discovery of DNA, arguing that evolutionary pressures acted on the whole organism, not on single genes, and that genes can have different effects depending on the other genes present. Also in 2002, Phillip Engle published a book, *Far From Equilibrium*, that summarized the devastation that had been done to the deterministic science of Newton and Darwin by people like Prigogine and dozens of others. There was very little interest in these matters. There would, however, be a great deal of interest in what would happen in September of 2005.

Another big showdown for old-time science would come with Scopes trial II. The dialectical contest of the original Scopes trial would be restaged in the trial of Kitzmiller versus the Dover School District. Generally speaking, you were either a Darwinist or you were a Creationist; there were no other options. The Creationist side showed a greater range of opinion from biblical literalism to scientist. An important voice in the Intelligent Design movement is Michael J. Behe, a professor of molecular biology at Lehigh University, Bethlehem, Pennsylvania.

Behe has gone back to the basic intelligent design argument of William Paley, but instead of using the eye as his example of an irreducible complexity, Behe uses the flagellum of a bacteria. The flagellum is a tail-like appendage made up of many proteins that the single-celled organism uses to motivate itself. According to Behe the flagellum foreshadows the work of a divine intelligence. Behe supposes that not until all these parts are in place does it become a functional organ. This is a direct attack on the Darwinian prejudice that only functional genes, tissues, or organs are allowed to survive and draw a salary and the only sort of evolution that can possibly occur is a bit-by-bit, mutational sort of change. Behe's fellow biologists at Lehigh have gotten together to write up a petition saying that he has the right to say what he wants, but that as far as they are concerned there is no room in Bethlehem for him.

It is interesting to note that in many ways it is actually harder for Darwinists to make their point than it was eighty years ago. The supposed proofs of evolution that Darrow's experts would have brought forth in the Scopes trial, had they been allowed, are now considered invalid evidence for

Darwinian evolution even by Darwinists. The concept of the linear evolution of the extinct *Eohippus* into the modern horse has taken a beating as more pages of the fossil history have been recovered. *Eohippus* was part of the standard proof for evolution of species and was included in Hunter's *Civic Biology*, which caused such a ruckus at the Scopes trial. An exhibit of horses as an example of orthogenesis (straight-line development) had been set up at the American Museum of Natural History, photographed, and much reproduced in elementary text books, where it is still being reprinted today. With more fossil evidence the neat evolutionary progression turned out to splay into a phylogenetic net. Horses had grown taller, but had also become shorter again with the passage of time. Perhaps horses were more like dogs, which brought up the question whether they had evolved at all.

On the other side of the table small group of professional Creationists are now proving to be better armed in the area of scientific knowledge, using the disorder in the academy as their key. They see that science is changing, but Darwinism is not. This gives them increased leverage. Creationists such as James I. Nienhuis, in an edition of *The Barnes Review*, observed that the so-called survival of the fittest is actually just a result of certain characteristics of some races exchanging genetic material to allow greater survivability for those creatures in certain ecological niches. In short, evolution is nothing more than we have seen in the dog since he got off the ark. Nienhuis believes in the genetic micro-adaptation of species to changes in the environment that an intelligent designer had designed.

For a few Creationists, the only substantial difference between themselves and Darwinism is chance versus design. The mechanical crapshoot of Darwinism also alienates Pope Benedict XVI and Tenzin Gyatso, the fourteenth Dalai Lama, who are bound to represent a caring, integrated universe. If watched carefully, Darwinists reciting the mantra of random selection, are mostly looking for deterministic, linear regularities of nature: the causes and the effects, the basis for truth versus evil moralistic thinking. In most cases there is no philosophical processing involved, it is received wisdom that is being applied.

The most interesting part of the trial from my perspective was not the attack by the ACLU on the woefully uneducated and superstitious opinions of some of Dover's citizens, it was the defense mounted by the Catholic public interest law firm, The Thomas More Law Center, hired to represent the Creationists. We recall that Thomas More was one of Henry VIII's closest and most well-rewarded nobles until Henry broke from the Church of Rome. When Sir Thomas refused to back him, Henry had him beheaded. The king then proceeded to lay the foundations for the Anglican Church. So now, over five hundred years later it is Sir Thomas and the Catholic lawyers in defense of the Low Church on one side, Sir Charles and science standing in for the Anglican Church on the other.

Among the lawyers from the Thomas More Law Center were Robert Muise and Patrick Gillen. Gillen was involved with questioning the Darwinian expert witness Robert Pennock. Pennock is well known for designing a computer program called Avita that is supposed to demonstrate natural selection. After drawing out Pennock about the program, Gillen surmised that if one looked at the resulting electronic organisms that had evolved out of his Darwinian program, "he would actually be correct if he inferred that there was an intelligent designer behind it." Pennock huffed and puffed, diverting attention with a disclaimer that neither he nor Darwin were really interested in the electronic creatures created by the program. Whether Pennock or Darwin or the Lord God himself for that matter cared about their creations was not an answer to the question of course. The point successfully made by Gillen was that in playing the role of God as a scientist, Pennock actually made a case for Creationism, not Darwinism.

Muise's approach was simply to unveil the confusion and double-talk that was going on behind the ivy walls of the academy, confusion of which the public is largely unaware. Over and over Muise forced the expert witnesses for science into subversions, circular arguments, and recanting their own positions or those of their peers, and finally they were forced to admit that everything about science is just a theory. Muise led the cross examination of Professor Kenneth Miller, who wrote the high school biology text that Creationists were attempting to replace.

Kenneth Miller is a biology teacher at Brown University and is a practicing Catholic. The Catholic lawyer defending Intelligent Design was cross-examining a Catholic biologist who had written the textbook that "reeked of Darwinism." If one is fascinated by Talmudic conjurations, it doesn't get any better than this. Muise began by inquiring about a passage in the third edition of *Biology: The Living Science* by Miller and Levine that says, "It is important to keep this concept in mind. **Evolution is random and undirected** [sic]." Curiously the statement about random evolution is only found in one of eleven editions of the biology text in question. Under cross-examination, Miller ascribed the insertion to his co-author Joe Levine, who, he said, got it from a book by Stephen J. Gould. Next Muise asked Miller his opinion on the matter directly, "Is evolution random and undirected," Miller responded:

> *I don't think that that is an appropriate scientific question. First of all, evolution most definitely is not random. There are elements of evolutionary change that are unpredictable, but the*

principal force driving evolution, which is natural selection is most definitely a non-random force, and then the second part of your question, undirected, that requires a conclusion about meaning and purpose that I think is beyond the realm of science. So my answer for different reasons to both parts of your question is no.

It was the Texas School Board that caused the emboldened statement to be removed, the board the controls much of the education for the country, but we move on.

In another innovative bit of questioning it was determined that if space aliens came to the planet to manipulate evolution on Earth it would not seem to be random to those aliens, but it would certainly seem to be so to those upon which the experiment is conducted. In other words, things that appear to be random might simply be due to ignorance or delusional thinking. Muise then continued by quoting Stephen J. Gould, Richard Dawkins, and George Gaylord Simpson to establish that the superstars of science believe that Darwinian evolution is random and undirected, and to force the issue that this is a cornerstone of Darwinism. Miller insisted that the statement that evolution is random and undirected is not something that can be tested by the normal methods of science and must be accepted on faith. Wow! Stop the presses!

The press that covered this trial, however, barely knew what was going on. There was a considerable amount of yawning and nodding off. While the lawyers for both sides failed to live up to the pyrotechnical standards set by Bryan and Darrow in 1925, at least *Harper's Magazine* covered the trial with some pizzazz. They covered it as they would a gun show or a hotdog-eating contest. What could possibly be learned about science in Dover, Pennsylvania after all? They send a great-great-grandson of Charles Darwin to cover on the trial. Matthew Chapman admitted to moving to America to free himself of the oppressive burden of expectations that fell to his lot, being a Darwinian descendant. He was now in Hollywood doing movies. He began his coverage of the trial with a comic turn, proclaiming that only he knew the truth, and the truth was that Dover was only thirty miles from Three Mile Island, and that the nuclear leak in the 1970s must be the explanation for the weird behavior to be seen in the locals. He said he had no evidence for this belief, and his lack of evidence was a matter of pride, in a parody of Creationism as well as himself.

The fact of the matter is that this black sheep of the Darwin family still brings his English upper-class standards and sensibilities with him. He was shocked by the level of willful ignorance he found in the New World. In

America Chapman is a big fish in a small intellectual pond. Yet at the same time we hear shocking "rumours" from across the waters that both biologist Richard Dawkins and chief spokesman for the Anglican Church, Right Reverend, Right Honorable, Bishop of Oxford, Lord Harries of Pentregarth, had to chastise Prime Minister Tony Blair for supporting the teaching of Intelligent Design by the Emmanuel Schools Foundation. This was thought to be strictly an American disease. First there was Mad Cow, now there is Holy Cow!

Judge George Jones, a George W. Bush appointee, apparently copied the scientific opinions that he used for the judgment almost verbatim from a paper by the ACLU that had its own collection of errors. The judge certainly didn't know what was going on although highly praised for his erudition in the liberal press. The trial was material for the likes of an Anatole France, or a Bertolt Brecht, or a Eùgene Ionesco. *Harper's* was on the right path, but only covered half the story thereby keeping its partisan credentials clean. In general it was just a sad, sad commentary on American culture.

While the political left and the expert witnesses from the academy celebrated the Dover decision, one of the plaintiffs and a former Dover science teacher, Bryan Rehm, surveyed the wreckage at the local level and concluded, "There is no way to have a winner here. The community has already lost, period, by becoming so divided." He agreed with parents on the other side that the fuss over evolution had obscured more pressing educational issues like school financing, low parent involvement, and classes that still trained students for factory jobs as local plants are closing down. These are exactly the conditions that have revived the need for Creationists and elected George W. Bush as president of the United States. Creationism and Wal-Mart offer the only hope when Democracy has failed. People who are not materially, intellectually, or spiritually well off need more than platitudes about the Founding Fathers or Victorian theories of science that are justification for empire.

To celebrate the two-hundredth anniversary of the birth of Charles Darwin, I take a trip to the Galapagos Islands. Darwinian Eco-tourism is a rich source of foreign capital for Ecuador, which owns the islands. I am excited to see the Galapagos mockingbirds as they come over to observe us, the Darwin's finches as they labor over cactus flowers, and the giant tortoises. I swim with the sea turtles and the penguins. The Pisco sours and the gourmet meals at the captain's table on the tour vessel are excellent. My fellow travelers are retired, middle-class professionals. They have only a passing acquaintance with biology. The strange, varied environment is as enchanting to them as it is to me, but they have virtually no knowledge of biological evolution, and they have no penetrating questions of our government guides. They have one simple motivation for going to the Galapagos as opposed to Egypt or China. They are interested to see the evidence that the species evolve, thereby proving Darwinism, thereby confirming a lifetime of non-church attendance. The Galapagos Islands have become the secular version of Lourdes to the true believer.

On the top of a rise on Bartoleme Island one day, while surveying the fantastic volcanic landscape, I ask the official state guide, in her national park tee shirt, if she can explain how such weak fliers as the Darwin finches could have arrived on the islands thousands of years before stronger fliers such as the mockingbirds. Chance, she says. Year after year, decade after decade, century after century, millennium after millennium, it was all a matter of chance. I wonder to myself how many monkey years does it take to type Shakespeare. Then I suppose that she is really just thinking of her business, an online cosmetics dealership that is her one chance to free herself from the low-paying and tedious chore of telling overweight, dull-witted, aged gringos to stay on the path and not feed the natives. By reflection I can see that the trinity of Capitalism, Uniformitarianism, and Darwinism is in our social genes and this is what it looks like. We are it; it is us.

I will now enter the jungle of the scientific database and try to put together a story of evolution from a modern perspective.

CHAPTER 8

In the Beginning ...

AMONG THE TRANSCENDENT INTELLECTS of the twentieth century that came from German culture and the tradition of its incandescent genius Johann Wolfgang von Goethe were Albert Einstein and Kurt Gödel. Both fled the German implosion of the Third Reich to that bucolic retreat in America, the Institute for Advanced Study at Princeton. From Albert Einstein came *The Theory of General Relativity*. Einstein was a flash of light, a Halley's Comet streaking across the sky. Gödel was a hard-to-see dark energy. From Kurt Gödel came *On Formally Undecidable Propositions of Principia Mathematica and Related Systems*.

The central theme of Einstein's General Theory is the fusion of space and time. This avoids the paradox of Parmenides that Aristotle and Newton were willing to accept: a cosmic evolution from the void, being out of non-being, the laws of nature out of nothing. In Einstein's General Theory a paradox is replaced by an incommensurable reality. Along these lines we recall Herakleitos: "This world, the same for all, no god or man made, but it was ever and is and will be an ever-living fire, kindled in measures and quenched in measures."

If space itself is relative and joined with time, cosmic evolution will seem to occur in a curved space-time manifold. As Fritjof Capra puts it, "All events in it (timeless space) are interconnected, but the connections are not causal. Particle interactions can be interpreted in terms of cause and effect only when the space-time diagrams are read in a definite direction. When they are taken as four-dimensional patterns without any definite direction of time attached to them, there is no 'before' and no 'after' and thus no causation."[1] This is called relativity.

Evidence that might confirm the curvature in space predicted by the General Theory has actually been observed by astronomers. Curved space was also predicted 2,500 years ago by Xenophanes, who saw the cosmos as a spherical body, living, conscious, and divine, the cause of its own internal movements and change. From this point of view linear time is a bookkeeping artifice introduced for the purpose of determining mortgages, who shall become king, and how history shall be written to justify it all. The linear perspective of the cosmos only works for relatively short distances of the local planetary system.

Gödel demonstrated mathematically that whatever axiomatic system is used, there will be true statements that lie outside its domain. Adding additional axioms is fruitless. As Stuart Kauffman at the Santa Fe Institute put it in a Zen-like koan, "To be is to classify is to act, all of which means throwing away information. So just the act of knowing requires ignorance." Although the mathematical proof to reach this conclusion is an intellectual triumph, all it really means is that the universe will always bigger than our ability to rationalize it. This inherent limitation of rationalism applies all the way across the domains of science, from the cosmos to personal consciousness. Metaphysicians have realized from time before history that the gods have not revealed to mortals all things from the beginning. New levels of technical sophistication and hubris, new insights about creation and evolution, do require that this insight be reached over again.

In Greek mythology Tartarus was both a deity and a place lower than Hades. Tartarus was the unbounded first-existing entity from which the light and the cosmos were born. From the dark came the light. Out of chaos came order. The study of mythology, a sort of pipe dream compared to everyday consciousness, gives us rich material for analogy, metaphor, and allegory and has been used for inspiration by our intellectual giants like Pythagoras, Copernicus, and Newton. Mark Twain summed it up when he said that history doesn't exactly repeat itself, but it does rhyme. Those only interested in dialectical disputes—good versus evil, truth versus fiction—consign themselves to secondary roles. They will be the apprentices to the metaphysician.

Georges Lemaître, a Belgian physicist and Roman Catholic priest, revived an ancient idea for an initial condition for the material universe called the primeval atom. From this first cause the universe expanded forward in time. This is in the tradition of Genesis in the Bible and of Aristotle and Newton. The entire universe originated in a cosmic explosion about 15 billion years ago, and it has been undergoing a spherical expansion ever since. Lemaître assumed the universality of the physical laws of nature and the Cosmological Principle. The Cosmological Principle assumes that the universe is homogeneous as well as isotropic: the same no matter which direction it is measured. This stems from Copernican Principle, which states that there is no preferred or special observer or vantage point. The same cosmological laws would be derived by any observer on any planet, in any solar system, in any galaxy, anywhere in the cosmos.

This reminds us of what Giordano Bruno said in 1600. "This entire globe, this star, not being subject to death, and dissolution and annihilation being impossible anywhere in Nature, from time to time renews itself by changing and altering all its parts. There is no absolute up or down, as Aristotle taught; no absolute position in space; but the position of a body is relative to that of other bodies. Everywhere there is incessant relative change in position throughout the universe, and the observer is always at the center of things." For this relativistic or existential view of the universe he was burned at the stake by the Inquisition. Generally we merely abuse or ignore people who hold such views.

Edwin Hubble estimated the distances to the galaxies in 1929 and found a correlation between the distance of galaxies and the Earth and their recession velocity, measured by their redshift. The redder a galaxy was, the further away it was. He found galaxies receding from his Earthly perspective anywhere he looked in the night sky. The Copernican Corollary and therefore the Cosmological Principle were seen to be proven. It looked as though we were at the center of an explosion that had happened long ago but not far away. For a moment the cosmos seemed to make sense, but it didn't last long.

Fred Hoyle coined the phase "Big Bang" and made scathing references to a cosmological he found he couldn't subscribe to. According to Hoyle, the Big Bang theory simply imported Genesis: 1-3 into astronomy:

In the beginning God created the heaven and the Earth. And the Earth was without form, and void; and darkness was upon the face of the deep: and the Spirit of God moved upon the face of the waters. And God said, Let there be light; and there was light.

Anomalies to the Big Bang/redshift theory began to appear soon enough. Determining how far away objects are in order to correlate distance with redshift in the universe was always chancy at best. Halton Arp found extremely high redshift quasars physically located in low redshift galaxies thought to be close by. He had to leave the country to avoid the prejudiced contumely that was heaped on him. That was just the beginning. The existence of dark matter had to be postulated in order to explain the gravitational effects observed between and within galaxies. In fact, in order to explain the level of gravitational effects observed, the universe had to be mostly composed of dark matter. In order to explain the fact that the universe seems to be expanding at an accelerating rate, dark energy is invoked. While we can neither detect nor do we understand the nature of these invisible hands, one thing we know: they do not naturally fall out of the theory of the Big Bang or of laws of nature defined by us on our little planet. Most recently a very old or fossil galaxy has been found buried within our own. Fossils once again have become a stumbling block for an oversimplified ideology. The Big Bang theory can't account for many things, including the evolution of galaxies and, as we shall see, life.

Following Hubble's discovery there was a broad contraction of opinion toward Lemaître's theory of a primeval atom and a cosmic explosion, to the point where an overwhelming consensus of the academic community now supports the Big Bang theory with all its warts and blemishes. John Horgan in *The End of Science* says, "The Big Bang theory does for astronomy what Darwin's theory of natural selection did for biology: it provides cohesion, sense, meaning, a unifying narrative. Cosmology, in spite of its close conjunction with particle physics, the most painstakingly precise of sciences, is far from being precise itself."[2] William Corliss at Science Frontiers has been collecting information on this brouhaha for decades. He keeps a running tab on astrological anomalies reported in scientific journals, and the list is very long and getting longer every day. Corliss observes, "the Big Bang, it seems, is one of those 'politically correct' paradigms, which one criticizes at his peril." We are at the center of the Big Bang that is moving away from us in all directions—say Hallelujah!

That any cosmological theory should receive such a majority in our present state of knowledge is more a comment on the culture of the academy than anything else. Those gentle giants of Germanic science meditated on the origin of the universe in their decline while they walked the country roads of Princeton. Einstein argued for the Big Bang thesis that most cosmologists presently adhere to. Gödel argued for a steady-state universe that has no beginning or end. In a delightful profile of Einstein and Gödel, David Berlinski says that the steady state vision of Gödel awaits another time to be fully understood, or perhaps another universe. In the larger culture there are two choices: the Young-Earth "And then there was Light" cosmology of Creationism, or the Old-Earth, "Big Bang" cosmology of the Darwinists. Which side are you on? In the first case everything begins with a flash of light that we may be ever seeing, but never hearing; in the other case the cosmos begins with an explosion that we may be ever hearing, but never understanding.

Beyond the continuing revelation of the astonishing beauty of the cosmos, it is amazing how little the disciplines of astronomy and physics have advanced our understanding of creation, the human condition, or the evolution of life. Some of this is due to the overwhelming confusion created by the evidence and the limitations of consciousness; some of it is due to the imposition of theories that are politically correct. We are just as likely to proceed further down the path of enlightenment listening to the vibrations of Bach's Chaconne in the Partita # 2 that plays on the Trinitarian formula, "From God we are born, in Christ we die, by the Holy Spirit we live again," as we are from discussions on the red shift. Wisdom is more likely to shine through a meditation on the Zen riddle, "First there is a mountain, then there isn't, then there is," than will fall out of mathematical inflations based on flawed or shortsighted assumptions leading to no particular quantitative predictions.

Once we understand the metaphor behind the poesy and aphorism of myth and throw off our binding prejudices about metaphysical tradition, we can appreciate the metaphysical axiom what is above is below: the same laws of nature operate in the macrocosm as well as the microcosm. In the face of an increasing level of confusion in astronomy, we can still appreciate that the profound and timeless contrast between the bright little world of Apollo the sun in contrast to the awesome, terrifying spectacle of Hermes' nighttime sky has been extraordinarily stimulating and productive. It has led to wisdom from time immemorial. It provokes the insight that culture guides our thought process for its own secular purposes through the language we speak. We don't see things as they are; we see things as we are, as Anais Nin put it. In general, astronomers find the messy business of biology disturbing and the social sciences beneath contempt, but this is exactly where some of our everyday confusions in cosmology will be resolved. From the experiences of life we may actually reverse engineer cosmology so that it connects with what we see in biological evolution.

The first lesson to learn is that evolution at all levels is balanced by devolution. In the Indian allegory Lakshmi came out of the cosmic ocean of milk when it was stirred by the gods and the demons. Lakshmi represents self-knowledge. She was a mark of good fortune, and among her attributes was that of the destroyer, Shiva. Shiva is both female and male, wears a necklace of skulls, and is the dance of time. In the words of Ananda Coomaraswamy:

> *In the night of Brahman, Nature is inert, and cannot dance till Shiva will it: He rises from His rapture, and dancing sends through inert matter pulsing waves of awakening sound, and*

lo! matter also dances, appearing as a glory round about Him. Dancing, He sustains its manifold phenomena. In the fullness of time, still dancing, He destroys all forms and names by fire and gives new rest.

The metaphor of the creator/destroyer comes easily to Indian tradition. It stems from the monsoon in tropical India that is clearly perceived to be both destructive and fructifying at the same time. Light versus dark is an elementary discrimination performed by the nervous system. Good versus evil is the rudimentary understanding of the unconsidered life. In temperate climate the best weather is separated from the worst weather by the seasons. In the tropics, on the other hand, it is easier to transcend this instinctive intellection and to see creation (evolution/devolution) flowing from the same hand.

Fred Hoyle, Geoffrey Burbidge, and Jayant V. Narlikar proposed a cosmology in 1993 that was better suited to explain things that are anomalies to the Big Bang theory. They supposed that the universe is continuous and eternal, with local "Minibangs" of creation going on all around. The cycles of evolution and devolution that we can also see in all directions in space are local manifestations of ongoing creation. Constructed to answer some of the problems of the Big Bang theory, the Minibang theory is flawed in its own ways, but as we shall see the conventional Big Bang theory is simply incapable of being the Godfather of creation because by the simple rule of expansion, it is evolution without devolution and therefore can be neither.

We are all too familiar with theories of evolution that gloriously come to a stop with us the chosen people. From the Pali Canon of Buddhism we have a wonderfully concise statement of devolution—the Aggañña-Sutta. It is attributed to the Buddha (around 400 BC), and it is an attempt to show that the ugly results created by high civilization that are deviations from Dhamma, the Truth:

> *In the past we were mind-created spiritual beings, nourished by joy. We soared through space, self-luminous and in imperishable beauty. We thus remained for long periods of time. After the passage of infinite times the sweet-tasting earth rose from the waters. It had color, scent, and taste. We began to form it into lumps and to eat it. But while we ate from it our luminosity disappeared. And when it had disappeared, sun and moon, stars and constellations, day and night, weeks and months, seasons and years, made their appearance. We enjoyed the sweet-tasting earth, relished it, were nourished by it; and thus we lived for*

a long time. But with the coarsening of the food the bodies of beings became more and more material and differentiated, and hereupon the division of sexes came into existence, together with sensuality and attachment. But when evil, immoral customs arose among us, the sweet-tasting earth disappeared, and when it had lost its pleasant taste, outcroppings appeared on the ground, endowed with scent, color, and taste. Due to evil practices and further coarsening of the nature of living beings, even these nourishing outcroppings disappeared, and other self-originated plants deteriorated to such an extent that finally nothing eatable grew by itself and food had to be produced by strenuous work. Thus the earth was divided into fields, and boundaries were made, whereby the idea of "I" and "mine," "own," and "other" was created, and with it possessions, envy, greed, and enslavement to material things.

From a Christian perspective, this is the Manichaean heresy that took centuries to suppress. Darwinism inherently follows the perspective of Genesis assigning to the chosen people a special god-like creation associated with language that results in the cessation of evolution. We, according to this theory, transcend evolution and the laws of nature. The war of the light against the dark in human affairs is highly motivating, but I will not take sides. From my generally positive, democratic frame of reference I will also embrace devolution in the search for creation. I see life as the best of all possible worlds, but not with the ironic undercurrent that Voltaire gave to that sentiment.

The word life, from the Proto-Indo-European root *Leip, means that which "sticks" or is "continuous". It comes from blood that is sticky and without which life quickly expires. This will be obvious when we get into the labyrinthine maze of biological evolution: the capacity for growth, reproduction, functional activity, and continual change preceding death are but the echoes of earlier stages. In his attempt to imitate life with automata, Christopher Langton at the Los Alamos National Laboratory asks appropriately, what is the lowest level of interactions that must be simulated to create life? To what interactions between cells, molecules, atoms, or the elementary particles of matter must we go? We have already seen the boundary for life pushed back to microscopic levels since Charles Darwin limited his natural history to the organisms that he could see with the unaided eye.

I will say, in deference of the practical need for categories, that wherever the line is arbitrarily drawn for life, whatever is outside this line is life-giving and life-supporting. This is close if not the same as the most primitive views of life, where the observable aspects of nature were personified and life was given a much broader and more integrated reality than we do today. Today everything outside our little pods of life is seen as the dead letter of physics and chemistry. I will read the story of evolution from the cosmic level and go forward rather than start from the multiplicity and try to work my way backward against the spines and thistles of cultural tradition.

Following the theme of Lakshmi, we suggest that the First Law of Creation is exactly the same as the First Law of Thermodynamics. The First Law of Thermodynamics is the law of conservation: no energy is gained or lost in the universe as a whole—$E=mc^2$. In fact, matter is just a low-level form of energy. Evolution in the broadest possible sense is neither rising, nor is it falling. What is given is taken away in equal measure. I will not abide any variations on the flat-Earth theme—being out of non-being or the Big Bang theory—that tend to end up being a just-so story justifying the ideology of the chosen people. I will live with the infinity of an incommensurable reality rather than a comfortable paradox. One extrapolation from my approach is that if something called intelligence appears anywhere in the universe it is something that is imminent in the universe everywhere. Self-knowledge, I assume, comes directly out of the ocean of milk stirred by both gods and demons as assumed in Hindu tradition.

The Second Law of Thermodynamics states that energy tends to move from a heated body to a cold body until they equilibrate. The conventional point of view of Western science is that life is a special blessing; it is orthogonal to the laws of physics, especially the Second Law of Thermodynamics, and it is generally treated as an independent process. In opposition to this I propose that the Second Law of Creation involves the dissipation of heat energy as well, but the physical reality of the cosmos constantly diverts this equilibration process to do work, the work of elemental, organic, and biological evolution along the way. No additional ingredients need to be added; no special pleadings are required.

When we plumb the depths of matter down to its elementary molecules, we return to the threshold of creation and the birthplace of space and time. Unable to conceive of a primordial seed, I can only start with hydrogen, helium, and a trace of lithium. From this elemental trinity the rest of the cosmos will be reborn. Hydrogen, the first and the simplest element, means "water forming," but its initial role is to create the energy for stars burning bright. Young stars are the forge for the nucleosynthesis of more complex elements up to the element iron. The star creates heavier elements by nuclear fusion. With the conversion of energy to matter comes an increase in gravity. Gravity counters the expansive energy of hydrogen burning. Gravity in this high-energy field is also a force that fractionates the matter created in stars.

Elemental iron still in its nascent form of a gas settles out to become the core of a middle-aged star.

Another element forged by nucleosynthesis in the early life of a star is carbon. Carbon is the fourth most common element in the universe after hydrogen, helium, and oxygen. After it leaves the energetic environment of a star, the sixth element is as black as black can be. As graphite it is one of the softest elements, and it is used as a lubricant. Under other conditions carbon can be formed into a diamond. The heat and pressure of a major impact event on Earth creates a sparkling of diamonds. Because of the symmetry of its four electrons that form an equal-armed cross, it can form the hexagonal molecules of organic chemistry as well as the strings of molecules that take on the shape of life, providing the basis for metabolism. Carbon is the basic Lego of organic chemistry, with over ten million compounds so far described. Carbon is the backbone for organic chemistry and life long before DNA ever appears on the scene. In Hopi tradition the equal-armed cross was the sign of the Earth Goddess, Spider Woman, long before Christianity ever appeared with its crucified deity. In the underworld, the abode of the gods, she dwelt with Tawa, the sun god. There was no living thing until they willed them into being with song. "I am Kokyanwuhti," the Spider Woman crooned. "I receive Light and nourish Life. I am Mother of all that shall ever come." From knowledge of the complexity underwritten by carbon, we can reverse engineer our way back to the hearts of creation in our imagination.

When the supply of hydrogen begins to be exhausted, a old star, now called a red giant, can no longer support its expansive stage against the attraction of gravity. The red giant begins to collapse into its iron core; it begins to die. The higher heavier elements of creation are created under the even greater heat and gravitational pressures now applied. If the star is five times the size of our own, the collapse punches through the scrim of the visible universe to create a black hole, it is said, where everything vanishes including the light of reason itself.

For smaller stars the collapse finally reverses itself when the energy of nuclear repulsion between elemental particles takes over, producing the visible event called a supernova. What is left behind is condensed, neutron stars, some of which are rotating. These produce very regular impulses of energy and are called pulsars. The first extrasolar planets were found orbiting a pulsar by Aleksander Wolszczan in 1992. The discontinuity of a planetary system around a star may trigger the unique rotation of a pulsar. Pulsars may be lighthouse beacons scattered about the universe, memorials to the past lives of planetary lives.

Following the emission of light from a supernova is a high-energy pulse of cosmic radiation. This is followed more slowly by a pressure wave that can trigger the compression of dense clouds of cosmic gas and dust called nebulae to give birth to new outbursts of stars. Finally a wave of matter from the supernova expands through the surrounding interstellar medium in all directions. With the intense outburst of new stars, in the dirty nebular environment of older exhausted stars, the final stages of chemical evolution occur. We recognize all these elements by their characteristic frequencies. Even the densest matter has a note to play in the music of the spheres.

Less dramatic than the formation of new stars are the lesser winds of gas and dust that produce enough heat for the simple chemical reaction of hydrogen with oxygen to create water. The vast Oort cloud beyond the orbit of Neptune of our planetary system is the icy rime left over from of the creation of our sun. The Oort cloud also includes ices of ammonia and methane. Inside the Oort cloud is the Kuiper belt. This is an icy rubble of planetoids outside of the orbit of Neptune. The Kuiper belt is thought to be the repository of the periodic comets. Clube and Napier identify the Taurid Stream that is on an Earth-crossing inner orbit. The Taurid Stream may be the remains of a giant comet of galactic origin. The Taurid Stream is responsible for regular displays of falling stars in the nighttime sky; the concentration peaks every 3,000 years or so and may have been the source of Typhon and, in the next peak, the Star of Bethlehem.

Supernova debris rolling over the remains of older supernovas helped to determine the composition of our solar system 4.5 billion years ago. These events made the chemistry of life as we know it possible. The accretion of denser clouds of matter around a star created planets through gravitational attraction. In between Jupiter and Mars Johannes Kepler indicated there should be a planet. All we have found is an asteroid belt, perhaps the result of uncompleted planetary agglutination. Objects in this belt occasionally drop into lower orbit and impact the inner planets. Earth is one of those inner planets where the flotsam and jetsam of expended stars, dirty snowballs, and chunks of unaggregated matter can make an impact. In the Gnostic interpretation of Jesus' Parable of the Sower, life is not in the seed alone, nor the sower alone, nor in the soil alone. Life was created when the appropriate brew of elements fell at just the right distant from a young star, which we now call the Sun.

The energy of aggregation, gravity, is created in the molten iron core of Earth that was too hot at first to do anything but consume and vaporize the matter raining in on it. As it cooled a crust of lighter, siliceous material formed around the central iron core. As a planet of sufficient size and gravity to hold an atmosphere, as a planet sufficiently removed from the sun to have a long, cooling decline, Earth met the requirements for an elaborated organic evolution. Iron was the blood-red soil that the ancient gods would spit into to create life. Primitive microbes eked out a living breathing the iron out of bedrock in the presence of a sulfur catalyst. Iron would become the element centered in the hemoglobin molecule that carried oxygen to the cells in the multicellular body. It was the hemoglobin gene that was mutated by the cosmic rays of a supernova some 30,000 years ago, creating our different blood types and acting as a marker for that high-energy event. Around this time we began to paint pictures of life in Ice Age caves, invented a calendar, and deduced the existence of the seed: the basis for civilization. The impact events at 13,000 and 9,000 years were secondary events that followed that supernova. It is told that it was Gaia's last son, Typhon who brought on the Age of Iron, when we learned to forge iron and shape it in our fury and terror about what God had wrought. Iron is the color of the blood-red dust of Mars that they say all living planets eventually return to, but I get ahead of myself.

In 1953 Stanley L. Miller, working for Harold C. Urey, put ammonia, methane, and hydrogen in a bottle along with water and ignited electric sparks inside the bottle. A week later it was found that amino acids, the building blocks of proteins, had been created. Miller and Urey won a Nobel Prize for this experiment. The experiment was enshrined in high school textbooks as an example of how the origin of life could occur under natural conditions. In the original experiment five amino acids were created. After the death of Dr. Miller in 2007 it was discovered that another experiment was done where steam was injected into the bottle. From this experiment twenty-two amino acids were found. In all living organisms there are twenty amino acids encoded for in DNA for the production of proteins.

The challenge presented by this experiment is that amino acids have mirror images of themselves. When amino acids are produced in the laboratory this way, the result is a racemic or 50-50 mixture of a right-handed form of the molecule and a left-handed form of the molecule. In all living organisms, however, the amino acids are left-handed. Louis Pasteur pointed to this as the critical distinction between the dead chemistry of the laboratory and the living chemistry of life. Whenever there is wrinkle in the mosaic of rationalization like this, it attracts snake-oil salesmen, miracle workers, and scientists—and this was no exception.

The existence of two forms of a molecule is disadvantageous for living organisms. The efficiency of enzymes to promote chemical reactions depends upon chemicals that are exactly suited stereochemically one to another. The machinery of life would have had to be prohibitively complex to start with, if it had to adjust constantly to left- and right-hand forms. The pharmaceutical industry finds it useful, for instance, to make only one form of a drug that can have different chiral or handed forms. Otherwise half of the product will be useless. In the worst case, one of the forms may actually have a negative effect.

So pharmacologists use catalysts that will steer the product in one direction or another, but how did creation perform this miracle?

Manfred Eigen and Ruthild Winkler dealt with the dilemma of life's biochemical bias in their book *Laws of the Game* in 1981. With great enthusiasm they claimed that the left-handedness of life's amino acids was the determinative discovery that elevated Darwinism from a theory to a natural law equal to gravity. As they saw it, the selection of one chiral form over another made Darwin's theory a natural law that could be traced back to the fundamental principles of physics. Darwinism, they proclaimed, was established as a law that permits no alternatives. Eigen and Winkler began their deterministic biology with the assumption that the evolution of life, even in these early stages, was the result of competition, selection, and change. They stipulated that even though they could see no selective advantage of one form over another, "All we can say is that even if two players start with exactly equal chances, one will always end up winning."[3] They attempted to solve the dilemma using the rules of zero-sum game theory. This was an invention by mathematicians to predict a game where winners win at the expense of the losers: survival of the fittest.

In practice Darwinists have not been able to find a physical analog that can reverse the Second Law of Thermodynamics, as life seems to do with its selection for left-handed amino acids. A demon has to be postulated that will fill up the tank to do what life seems to do: collect energy or information instead of simply dissipating into a state of disorder. Without such a demon, as Eigen and Winkler candidly point out, the axiom of the survival of the fittest is merely a tautology and Darwinism is strictly a religion.

Eigen and Winkler adjusted their limiting assumptions until the equation proved their theory to be correct, but we know from endless practical experience that zero-sum games are about as rare as proofs of genetic damage producing new species. Losers come back on another day; on other days the former losers win and the former winners lose. In fact it has often been noted that winning is the first step toward losing, since the losers change and the winners rest on their laurels. In the long haul everyone wins and everyone loses: the meek do inherit the Earth, only to lose it again. The game goes on as long as the environment that supports it survives. When the environment gets old, the game dies—winners, losers, and all. Game theory has its value, but like Tarot card reading it gets much better results in the hands of a skilled practitioner who is attempting to make predictions, not ideological points.

It is wonderfully curious to see Darwinists using the religious terminology of "demon" to finesse their intellectual stumbling block. It was in the early stages of the development of Christianity as a religious institution that original sin had to be so invented. Sin was the demon required to explain the absence of an inborn sense of Christian morality in God's own creation. This became especially important after Christianity became the state religion and life continued on much as it had before. Before that point Christians were a small minority, and Rome could be the demon. Eigen and Winkler's demon is what causes life to appear out of a heartless, pointless, lifeless physical cosmos. Manfred Eigen had been awarded a Nobel Prize for his work on fast chemical reactions to short pulses of energy, but the response to his solution to the dilemma of handed chemistry was lukewarm. Perhaps it was seen as a rather too obvious intellectual fabrication in defense of a beloved principle.

The real solution to the left-handed chemistry of life has turned out to be much more interesting than the demonic manipulations of game theory. It began with a big crack in the edifice of the Central Dogma that had become obvious with the Nobel Prize awarded to Stanley Pruisner in 1997. Pruisner identified a malformed protein called a prion as the causative agent behind mad cow disease. He proposed the idea of a species jumping spongiform encephalopathy, mad cow disease, caused by a malforming protein. A prion protein became a designated demon. There was considerable discomfort with Pruisner's theory because it involved proteins giving a maladapted shape to normal proteins. According to the Central Dogma of genetics this was not supposed to happen. Only DNA could do that.

Unlike most chemistry students, Pruisner had an active interest in philosophy, history of architecture, Russian history, and economics. This helped to enable him to skirt the steady drumbeat of convention that impounds the minds of typical chemistry students. Also, he did his research in San Francisco during the time when the siren call of free thought was loud and clear. It was in the Bay Area that Luis Alvarez broke the back of Uniformitarianism. It was in the Bay Area that Steve Wosniak and Steve Jobs elevated the discipline of engineering to unheard of levels of creative fashion. These sorts of innovations were frowned upon in the environment of Harvard or Yale; they were actively suppressed in the environment of Oxford and Cambridge.

It was the mad cow panic that allowed Pruisner's theory to breach the conservatism of the Swedish academy, which is generally very mainstream. As with some other epidemic panics of the twentieth century, mad cow disease was based on very flimsy evidence fueled by a flare-up of public fear concerning the reappearance of the plague. Millions of cattle were sacrificed, creating a situation where several important governmental and academic entities had much face to lose if this speculation was wrong. There were other ideas about the cause of the disease in cattle, however, and the hero of this story was Mark Purdey, an organic beef farmer. After a long, intensive study that included some of the same ethnographic literature that had influenced Pruisner, Purdey concluded that the agents that caused the disease were multiple and acting in concert: diet, the use of organophosphate pesticides, and ultraviolet radiation. He concluded that the appearance of transmissibility from cattle to humans was forced from an imperfect database, but it was too late for a change of mind.

With time we find that not only have the large number of mad cow deaths predicted in humans at the start of the panic not materialized—not only has the Centers for Disease Control reduced the risk of mad cow disease to virtually nothing—Pruisner's prion has turned out not to be an exception. It turns out that proteins called chaperones do normally give the active shape to the proteins produced from a DNA blueprint. DNA doesn't even have the information necessary to give proteins their active shape. Proteins adopt their active shape naturally in the appropriate environment, or they are further adjusted by chaperones. Pruisner's discovery has become part of a complete revolution of the Central Dogma of DNA called epigenetics, but for the moment this brings us back to the left-handed chemistry of life before DNA has even appeared on the scene.

In 2003 Donna Blackmond at Imperial College London mixed left-handed and right-handed versions of the amino acid serine in water. Whatever the quantity of right-handed serine was used, the final mixture became left-handed; left-handed serine acted as a catalyst to produce more of itself by itself. This is essentially a case of pattern recognition. Not only can proteins act as a catalyzing agent to reproduce their active forms, the amino acids that they are made up of can also act as catalysts to replicate their own chiral form. This had been predicted fifty years earlier by F. C. Frank, but it was never followed up on.

Amino acids have been discovered riding onto the planet on meteors. At least fifty of the amino acids found on the Murchinson meteorite, which fell north of Melbourne, Australia, in 1969, are not present on Earth. They cannot be contaminations from Earth based life forms, yet they show a dominance of the left-handed form. Ronald Breslow at Columbia has shown that the polarized light from a neutron star can cause the destruction of one of the chiral forms. Perhaps, in other sectors of space where the light is polarized in a different direction, right-handed amino acids dominate and right-handed forms of life have evolved. By a sort of filtration of the original mixed batch of amino acids, life as we know it can now proceed. We only have to invoke the idea of chance if one ignores, for some reason, the environment of the cosmos. Even those few right-handed amino acids that made it through the filter would have been auto-catalyzed by the dominant left-handed form. It was not a gladiatorial contest between left-handed and right-handed amino acids conducted under the banner of Darwinian natural selection; it is what was what normally occurred in this particular space and time.

Investigation of amino acids auto-caltalyzed into proteins has produced even more amazing results. M. Reza Ghadiri led a research team at The Scripps Research Institute looking at the possibilities for auto assembly made up short proteins of thirty-two amino acids called peptides. These peptides naturally fold into a long helix without the help of chaperones. When they mixed left-hand and right-hand versions of these peptides with left-hand and right-hand amino acids, the peptides correctly linked up and formed exact copies of the template molecules. They further discovered that if they added mutant peptide templates to a mixture of amino acids, templates with a single incorrectly handed molecule, the peptides would not make exact copies; they automatically corrected the mistake. "That is astonishing," says Ghadiri. "Based on [our] understanding, polypeptides can self-replicate, form complex networks, error correct, form mutual systems, they have all sorts of emergent properties, and they can now do homochiral amplification."

In short, it is not necessary to force the issue of evolution with the intellectual fabrication of an Eigen/Winkler demon, but the pride and prejudice surrounding the dilemma of asymmetric abiogenesis involving left-handed amino acids, mad cow disease, the sacrifice of millions of cattle, the Central Dogma of Watson and Crick, the Nobel Prize, auto-catalytic proteins, pesticide producing companies, social panics, and government cover-ups is a bizarre and sobering tale of the tortuous path that must be traveled in the search for a single missing link on the path of evolution, but the work goes on.

A group of scientists led by Louis Allamandola at NASA Ames Research have shown that a mix of water, methanol, ammonia, and carbon monoxide ices zapped with the kind of energetic rays emitted by hot young stars produced oily organic molecules in an environment imagined to be like the young planet Earth. When dipped into water, these oily compounds naturally formed multi-membrane chambers similar in size to living cells. Such protocells would have performed an essential service. Protocells create an inside and an outside. Some say that the first biological process was created with the formation of protocells—osmosis. Being what they were, these membranes were semi permeable. Some chemicals could pass through the membrane; others could not. An osmotic as well as an electrostatic tension would therefore exist across the membrane. These were differences that would make a difference. These protocells created energy gradients that could lead to work being done, and they provided the staging area for interactive linkage between the potential building blocks of life. A life-nurturing ocean full of the organic molecules and the elements necessary for life would become an ocean full of life. DNA, the big memory molecule, is still nowhere in sight at these stages of biological evolution. DNA is far too complex and delicate a molecule to exist at this level of background energy, but so much has already happened without it.

The Russian biochemist A. I. Oparin pointed to the natural formation of protocells in Earth's primeval environment as the point when natural selection began. Oparin followed the remarkable tradition of Russian biologists Pavlov, Kropotkin, Timiriazev, Vavilov, and Vernadsky. Unlike some others he went along with the embarrassing science of Stalin's favorite, Trofim Lysenko. Taking the party line helped his career. Fortunately his subject did not threaten the party line. Oparin made a contrast between life and machines that simulate life like Vaucanson's digesting, shitting duck in his classic *Life: Its Nature, Origin, and Development*. In a machine, he says, "only the energy source or fuel undergoes chemical change, while the actual structure remains materially unchanged … and the less it is changed the longer the actual machine will last." On the other hand, as Oparin puts it, "the elements of construction of the living body themselves take a direct part in the metabolic reactions which serve as the source of that energy which is transformed into mechanical movement. Organisms can only exist for any length of time as a result of the continuous accomplishment of chemical transformations."[4]

Life and mechanisms that simulate life are very different kinds of machines. What is clever about machines is that we take a dense form of matter, iron in the classic example, in a state too ordered and elemental to be part of a biological organism and we shape it into parts including an enclosed chamber for explosive expansion. Then we power the movement of a piston with a form of energy, concentrated fossil fuels for instance, that is too energetic to be used by a biological organism to get work done. It is somewhere between these states of order and disorder that creation naturally assembles its living organisms on the third planet from a young star. If make a graph of the electromagnetic spectrum from highly-energetic x-rays, to ultraviolet energy, to infrared energy, to mild radio waves and plot it against the process of nucleosynthesis in stars, to organic evolution, biological evolution, embryological evolution, and to social evolution we find a simple correlation between increasing molecular complexity against a background of decreasing cosmic energy. The milder the energy, the greater the complexity allowed—up to a point.

Life falls out of dusty chaos at a certain distance from a dying star in what J. B. S. Haldane called the pre-biotic oceans of Earth that were a hot, dilute soup of self-replicating organic molecules. Life is the thin red line somewhere between heaven and Earth. Life goes on without effort, without practice, without care. Life does not abrogate the laws of physics. If there is anything that abrogates the laws of physics, it is the closed system of the man-made machines. Such machines are unsustainable, cannot reproduce, and reach a natural dead-end in our own lifetime as rust, carbon dioxide, and water.

The classically defined thermodynamic machine is a closed system that goes to equilibrium. Life avoids going to equilibrium by the flow of metabolic chemistry. Oparin compares the open system of life "with the organization of a musical work, such as a symphony, the actual existence of which depends on determinate sequences and concordances of individual sounds. One has only to disturb the sequence and the symphony as such will be destroyed, only disharmony and chaos will remain."[5] Individual lives disintegrate when the natural capacity for error correction starts to fail and the harmony cannot be maintained. Music is a more appropriate analogy to the larger picture of life's evolution than game theory. Music is a symbiotic act. Music is endlessly improvisational and diverging. Game theory is a rigid algorithm of mathematical steps. Oparin was trained as a biochemist, and his theories about the creation of life and evolution are much more reasonable than the chess games of Western Darwinists. Some have called Oparin the twentieth-century Darwin.

As we recall Watson and Crick were greatly influenced by the rationalism of Germanic science, especially the presentation of it by Erwin Schrödinger in *What is Life?* Thermodynamics uses probability theory to deal with the challenge of describing the physical state of a molecule at any given moment. Probability theory gives a way to deal with the apparent randomness of the physical world at the atomic level. Watson and Crick introduced random molecular mutations as their venue for change in biology. The problem with Watson and Crick's theory is that it is a typical physicist's view of biology. Of course we often use statistics where we don't understand how things work or don't have sufficient data even when we do, but we don't use probability theory to explain how a muscle contracts. We don't need to. We use it to predict how long a person is likely to live, since the details necessary to make a perfect prediction are beyond our ken.

Schrödinger was a Nobel laureate in physics and close friend of Einstein, the godfather of the First Law of Thermodynamics. Schrödinger's monograph was actually an attempt to dispel the supposition that life is a contradiction to the Second Law of Thermodynamics and, therefore, of the laws of physics in general. Citing Max Planck, he tried to show that the order-from-disorder dialectic that was posed as the critical distinction between life and non-life was a specious construction. Although Schrödinger continued to lean on Descartes' widely followed metaphor of life as a mechanism the principle that he proposed of order from order was a fundamental revolution in the concept of life. It is exactly what we have seen with increasing clarity so far.

To appreciate Schrödinger, we have to understand that German science in the late nineteenth and early twentieth centuries was coming under the influence of what Western Europe has generally seen as oriental philosophy: anything east or south of the Glory that was Greece around 300 BC. Richard Wilhelm was making translations of Chinese classics such as the *I Ching* into German. Heinrich Zimmer was making important contributions with his studies of Indian philosophy and history. Carl Jung and Sigmund Freud consumed world cultures not for oddities to put in their collector's cabinets, although they did that as well, but with the aim of getting to root of human culture and consciousness.

From this broad integration of human perspectives, what Europeans called laws of nature looked more and more like cultural inventions suited for a particular culture. The union of modern physics with oriental metaphysics was the beginning of the most profound leap in intellectual evolution in European history. It portended an eclipse of the revolution achieved five hundred years earlier with Copernicus and Galileo. It opened the possibility of a revival of Europe's own Pre-Christian metaphysical tradition reflected in the knowledge of the Dominican who became a Hermetic philosopher, Gioardano Bruno, among others.

Schrödinger was influenced by the study of Indian practice of Vedanta. The Vedanta principle is rooted in the Upanishads, which are about the same age as the Torah. In the Upanishads, the Atman is the personal self that equals the omnipresent, all-comprehending eternal self. Vedanta is devoted to the principle that human life is divine. It is similar to the spirit of Virgil's *anima mundi*: "the all-informing soul, that fills, pervades and actuates the whole." Unconcerned with the battle between good and evil that Judeo-Christian tradition capitalizes on, Vedanta strives for understanding.

This perspective removes the obstacle that defines human life as orthogonal to the laws of nature in Judeo-Christian tradition, including its secularized upgrade. Schrödinger observed, "Again, the mystics of many centuries, independently, yet in perfect harmony with each other (somewhat like the particles in an ideal gas) have described, each of them, the unique experience of his or her life in terms that can be condensed in the phrase: DEUS FACTUS SUM [sic] (I have become God)."[6] Realizing that divinity is personal because it is universal is actually a trivial conclusion once one claws one's way through the delusion of a special creation with all its institutionally administered benefits: we are the chosen ones, we are the fittest, we are redeemed. The Pope now supports a détente with Darwinism, but he will excommunicate anything resembling Vedanta, putting it under the heading of relativism. It lacks the moral commandments that are at the heart of Judeo-Christian culture.

The First Law of Thermodynamics or Creation is the law of conservation: no energy is gained or lost in the universe as a whole—$E=mc^2$. The Second Law of Thermodynamics or Creation states that energy tends to move from a heated body to a cold body until they equilibrate. Ilya Pregogine showed that dissipative structures far from equilibrium in an open environment with which they exchange matter and energy are irreversible; they cannot be restored without an expenditure of energy. By this he determined that there was an arrow of time in the physical universe just as there is in biological evolution and in human imagination. This is really a Fourth Law of Thermodynamics or Creation.

The Third Law of Thermodynamics or Creation states that as heat is dissipated the disorder of active molecules tends to an ordered, more stable materiality—water vapor becomes water and then ice. Finally all processes tend to cease and go to absolute zero, but the Third Law also states that for technical reasons the process cannot reach absolute zero. In the final analysis, matter cannot exist without some energy; energy cannot go so far as to free itself from gravity. This is reflected in the Principle of Tai-Chi and the Taoist image of the black tadpole chasing the white tadpole inside a circle. In the black tadpole there is a dot of white. In the white tadpole there is a dot of black. No matter how dark a situation becomes, there is always a soupçon of the light remaining, and vice versa. The tension between repulsion and attraction is the inertia that keeps absolute equilibrium from being reached, keeps the wheel of life turning in a particular direction, makes the cosmos and creation a steady-state, open, living system. We recall the thoughts of T. S. Elliot, who said, "We shall not cease from exploration and the end of all our exploring will be to arrive where we started and know the place for the first time."

From this I decide in favor of Hoyle, Burbidge, and Narlikar's cosmology that the universe is continuous and eternal with local "Minibangs" of creation going on all around, all around. Supernovas expend energy and motivate the evolution of the elements, organic chemistry, and life. Energy and matter are interacting at all levels. Evolution goes hand and hand with devolution under the icon of Shiva. If a Big Bang had been the be-all of the cosmos, and all that happened was the dissipation of energy with matter flying away from its starting point at a greater and greater distance, there would have been no creation. The Big Bang theory turns out to be an accomplice to the Uniformitarian geology where nothing happens and nothing can happen.

With the collapse of civilization in Europe in the first half of the twentieth century, a brief flame of intellectual and spiritual inspiration on the Continent went out and the extinguished torch was passed back to the Anglo-Saxon academy that had followed the principle of applied science for well over a century. When the Nazis took over Germany, Schrödinger left for a teaching job at Oxford, but he was denied access to the country by the English government because of his exotic lifestyle—he had a wife and a mistress. America denied him access for the same reason. It is likely that the Irish accepted him only because he had been rejected by the English. Schrödinger's tolerance of open systems was considerably beyond what a Christian culture could cope with. Watson and Crick completely failed to notice or at least to be influenced by Schrödinger's central thesis. It was and it remains far too metaphysical. Although there are a few who would follow Schrödinger's lead, such as Stuart Kauffman, there is nowhere near a quorum in the modern academy to create a conversion to such a paradigm.

At this point I would like to recall Erasmus Darwin's revolutionary evolutionary thesis at the end of the eighteenth century, before he was slandered and his writings buried. From his *The Temple of Nature* comes the most profane section of his theory of evolution, a section left out by his grandson Charles:

> *Ere Time began, from flaming chaos hurled*
> *rose the bright spheres, which from the circling world;*
> *Earths from each sun with quick explosions burst,*
> *And second planets issued from the first.*
> *Then, whilst the sea at their coeval birth,*
> *Surge over surge, involved the shoeless earth,*
> *Nursed by warm sunbeams in primeval caves*
> *Organic Life began beneath the waves ...*
> *Hence without parent by spontaneous birth*
> *Rise the first specks of animated earth.*[7]

CHAPTER 9

Microbiological Evolution

THE EARLY STAGES OF life have altered the original environment in dramatic ways, so it is difficult to put together a coherent story about the abiogenesis of life on Earth. We simply have great difficulty imagining such an alien landscape. The atmosphere, for instance, is presumed to have been composed of methane, ammonia, water, hydrogen sulfide, carbon dioxide or carbon monoxide, and phosphate. One thing is for sure: life was a conjuration of what the environment had to offer. So far we have seen heavier elements forged in the furnace of the stars and organic molecules formed in an environment of reduced energy fields in the wider universe. The ambient energy level on the planet, the radiant energy of the sun plus the heat produced by the Earth's core, had to have been low enough so as not to ionize or vaporize the organic elements of life.

Sidney Fox, director of the Institute for Molecular and Cellular Evolution, showed that amino acids would naturally form short protein strings called peptides under conditions that might plausibly have existed on Earth. We recall the work of M. Reza Ghadiri. His research team showed that peptides in solution can self-replicate, form complex networks, error correct through pattern recognition, form mutual systems, and reform right-handed proteins into left-handed proteins. Enzymes are proteins that have an attraction for a specific molecule that is part of an entirely different chemical compound. Weakly bonded to such an enzymes, the chemical compound will be slightly reconfigured. The enzyme assists chemical reactions to occur that otherwise would only rarely occur or would not occur at all. Their other special characteristic is that enzymes continue to exist after the reaction. Like a transistor they are not destroyed by the molecular reaction that they participate in.

The creation of the memory molecules, RNA and DNA, is still largely a mystery. As recently as 1999 leading researchers supposed that the creation of the nucleotides that are the basic constituents of the memory molecules on the primitive earth would have been a near miracle. Yet John D. Sutherland and colleagues at the University of Manchester have found a recipe that allows the auto-assembly of an RNA nucleotide. By reverse engineering it is assumed by many that prion-like proteins, proteins that self-replicate, had something to do with the formation of the first memory molecule, RNA. RNA has the capacity to mimic the order of amino acids, thereby enabling proteins to be copied. Edward N. Trifonov at the University of Haifa has identified the oldest DNA segments by looking for the common elements across a wide range of microbiological organisms. The earliest gene segments were complementary pairs that had the highest degree of thermal stability. This was organic chemistry on the brink of life.

Some would say that the identification and naming of the cell by Robert Hooke while peering through his microscope marked the birth of modern biology. Most biologists mark the appearance of the cell in the oceans of Earth as the beginning of life. As Louis Allamandola and his colleagues have shown, oily molecules will form into protocells when dipped into water. These protocells created a microenvironment within an environment. The cells would create the conditions for feedback to occur with changes in state, a change in permeability for instance. Cells became a staging ground to attract enzymes.

So far this is still the natural domain of organic chemistry. The change to biochemistry comes when arrays of enzymes line up on a cell wall. This becomes the basis for metabolism, a series of reactions where the product of a reaction becomes the substrate for the next reaction under the influence of chemical state of the cell. The evolution from cells to human civilization will be the story of an increasing isolation from nature with an increasingly sophisticated necessity for pattern recognition and feedback control. To all appearances life is a weed that establishes itself on any favorable exposure at this stage, like moss on the north side of a tree trunk. It is now thought that the earliest forms of life appeared as soon as the planet was cool enough hold water, around 4 billion years ago.

Not long ago the most primitive single-celled organisms known were called bacteria. The discovery of the extremophiles in recent decades has opened the possibility of an even more primitive group called archaebacteria.

Some make a complete taxonomic separation: Archaea evolving into Bacteria. The geological eon occupied by Archaea extends from around 4 billion years down to around 2.5 billion years ago. Extremophiles are organisms that can live in conditions that would be ruinous to life such as ourselves. The list of extremophiles is getting longer every day. The extremophiles express a remarkable range of chemical transformations, sucking up virtually every available energy source, but we know very little about them so far. They don't survive in our Petri dishes. An exception to this is the curious bacteria *Deinococcus radiodurans*.

Deinococcus radiodurans was discovered by A. W. Anderson in 1956. Anderson discovered that a high blast of gamma radiation to attempt to sterilize canned food worked for all but this extremophile. A dose of ten Gys of ionizing radiation is sufficient to kill a human. *Deinococcus* can survive a dose of 5,000 Gy. Originally it was supposed that effective DNA repair mechanisms were the explanation for effective damage control. *Deinococcus* has several copies of its genome. Michael Daly et al. believe that a high level of intracellular manganese protects proteins rather than DNA from being oxidized by radiation, and that this is the difference between *Deinococcus* and sensitive bacteria. The implication is that it is the proteins that are the basic core of memory and through them damaged DNA was repaired by reverse engineering. We are very close to the Adam and Eve relationship of DNA to proteins in this organism.

Sampling from the Moon's surface suggests that it was once part of the Earth. It is supposed that the Moon was ejected from the Earth following a huge impact event around 4.5 billion years ago. It appears that a heavy bombardment by asteroids the size of Ireland around 3.9 billion years ago, but they did not wipe out Earth's primordial life forms. Nearly 3.5 billion years ago an asteroid that was significantly above the average hit. This one released ten times more energy than the Alvarez meteor. The upper ocean boiled off into steam, and giant tsunamis coursed around the world's oceans, scouring the early land mass or land masses. The only life that could have survived would have been those extremophiles, according to Gary R. Byerly, a professor of geology at Louisiana State. The fact of the matter is that the extremophiles were probably all that were there at the time. Even if they were all killed off, those organisms would have quickly reassembled and repaired themselves once things settled back down and their preferred ecological niches had returned.

We are still completely dependent on the health of these metabolizing microorganisms. In fact, we are ransacking the far-flung recesses of the ecosphere to find microorganisms that can help us recycle our mountains of sludge, neutralize our oceans of toxic waste, and release a sustainable green energy so we can continue to stave off the consequences of our own excesses. They can evolve into these situations much faster than we can. As the energy runs out, the atmosphere fails, the basic resources diminish, and the high bloom of life begins to fail, these microorganisms will still be here.

Many scientists believe that the extremophiles that thrived around the volcanic vents in the deepest oceans wrote the first chapters of our personal ancestry. These volcanic vents, discovered by Robert Ballard off the Galapagos Islands, spew out mineral-rich fluids including hydrogen sulfide. Early Earth must have smelled like a rotten egg, which solves an old conundrum. The smell of a rotten egg precedes both the egg and the chicken. Early Earth was one bad place. Günter Wächtershäuser, a German chemist turned patent lawyer, proposes that it was actually a naked form of metabolism that evolved around these vents, predating genetically imprinted metabolism. The first vent extremophile to have its genome sequenced was *Methanococcus jannaschii*. It survives in waters as acidic as lemon juice, with temperatures approaching the boiling point of water, at pressures sufficient to crush a submarine.

These early organisms made heat-shock proteins to survive high temperatures. Such proteins spontaneously form into thick-walled, fat-barrel shapes with small openings. All forms of life from bacteria to humans still make heat-shock proteins, even though they are no longer used in the same way. Today hydrogen sulfide-oxidizing bacteria still form thick mats around volcanic vents and are grazed upon by amphipods and copepods. Anaerobic bacteria also live symbiotically in the tissue of giant tubeworms that wave like banners in the marine currents around the smoking volcanic vents. Tubeworms contain hemoglobin that combines with hydrogen sulfide and transfers it to their bacteria the way oxygen is carried from our lungs to our body cells. In fact, tubeworm hemoglobin also carries oxygen, as a waste product, when there is no hydrogen sulfide around to compete for the space. Tubeworms form the base of a unique food chain that includes segmented worms, huge clams, mussels, and crabs. The oxidation of hydrogen sulfide by bacteria is enough to allow this food chain to temporarily reverse the Second Law of Thermodynamics without the need for sunlight or a rain of organic detritus from the upper levels of the ocean. Life is modifying and using the energy from the molten core of the Earth. Even though we have taken a different path of late, we are all descendants of those sulfur bugs.

There was one entire group of archaebacteria that I want to take note of before we move on to the bacteria. Halobacterium live in an environment with high concentrations of salt like the Dead Sea. This bacterium has a purple protein that is light sensitive. This bacteriorhodopsin absorbs sunlight and using the energy pumps protons out of the cell. The resultant gradient across the cell membrane, when reversed, was used to drive the synthesis of adenosine triphosphate, ATP. ATP is like a rechargeable battery that is used to power metabolic activities along with enzymes. Later rhodopsin would become the light-detecting pigment in all primitive, light-sensitive spots in multicellular animals as well as camera-like visual systems such as our own.

Bacteria were the main actors in the Proterozoic eon from around 2.5 billion years to around 500 million years ago. To the casual observer, very little had changed. Bacteria display two types of behavior. There is the planktonic behavior of cells floating independently of one another, and there is a colonial state of bacteria. In the colonial state they are able to switch their metabolic behavior when a quorum is reached. Whole suites of genes are switched on or off. In this form they attach to surfaces at first by weak van der Waals molecular forces and then by more permanent cell adhesion structures such as pili or hairs on the bacterial surface. These colonial organisms are closely packed and enveloped in a polymeric matrix or film called a biofilm. Biofilms help protect the colony from toxic organic compounds. It is in this colonial form that we find them in Western Australia's Pilbara region. They are dated at around 3.5 million years ago. They were closely associated with ancient volcanic vents located in tectonically active sites.

The volcanic energy that heated the early oceans of Earth would remain an important factor in the heat budget of the planet to the present day—part of the energy gradient that allows life to exist. The extraterrestrial energy input that allowed life to expand beyond the volcanic vent were those radiated to the planet from it's relatively cool star, but more than that, a spinning planet. There can be no exactly right distance from a star that would favor the evolution of life where one hemisphere is constantly facing the sun and the other is facing the blackness of space, as in the case of the planet Mercury. It was the flicker effect that kept life from being either curdled on the one hand or frozen on the other. A spinning planet also buffers life from small eccentricities in orbit or minor variations in the sun's output that could also be disastrous. Life is a whirligig spinning within an atmospheric whirligig with a temperature gradient of 5800 Kelvin when facing the sun and 2.7 Kelvin when facing the blackness of space.

Anaerobic metabolism defined a narrow range for the expression of life. The capacity to move beyond the vents probably occurred soon after a very simple change in the Earth's atmosphere. Up to this point the planet was almost certainly a cloud-enshrouded planetary body like Venus or Titian the moon of Saturn. I propose that Earth had cooled enough to allow the appearance of the Sun. This was the cause for an exaltation of blue-green algae. This was effected through the evolution of a new metabolism called photosynthesis. Photosynthesis used ultra-violet radiation from the sun combined with carbon dioxide to produce organic compounds, especially sugars. Sugars became the energy source for the production of ATP. ATP would become the replacement for hydrogen sulfide as the molecular currency to be consumed to run cellular processes like biosynthetic reductions, motility, and cell division—with the aid of enzymes.

Life that was originally the fallout from the cooling inferno of cosmic creation, huddled around its volcanic vents, is lured out into the regular pulsing of sunlight so it can soak up its ultra-violet radiation. Some organisms would lean toward metabolisms that thrived in the sunlight; some organisms would lean toward secondary metabolisms that could function better in the dark. The algae and the green flagellates, to a great degree, would be the ultimate source for vitamins A and D. The algae have saved us many times in times of great trial. When there was absolutely nothing else to eat, they have scummed over the water like manna from heaven.

A waste product of this new form of metabolism was oxygen. The iron that was dissolved in Earth's acidic oceans reacted with the oxygen to produce oxides. This precipitated out to produce our present-day sources of iron ore, and it reddened the soils of the planet. The Earth had to become red first before it could become green. Iron was also an oxygen sink that kept anaerobic life from being poisoned at first. Eventually the elemental iron was largely used up. The buildup of free oxygen became an ecological disaster for anaerobic forms of life.

The Great Oxidation caused major geological changes in the Earth's air and water. The oxygen surfeit sponsored the next major biochemical shift in life. The nitrogen-fixing bacteria that took inactive nitrogen out of the atmosphere to make nucleotides (DNA) and amino acids (proteins) were susceptible to oxygen poisoning, for instance. Without them life was finished. They could only continue by remaining in anaerobic conditions by binding oxygen. This was a defensive response to the pollution problem. An aggressive response came from new life forms that could burn the almost unlimited supply of oxygen for energy. There would be an incredible radiation of these new aerobic life forms. These were great days in the evolution of biochemistry. Nothing approaching the innovation in basic metabolism that was going on would ever occur again at this level of evolution.

The 3.4-billion-year-old stromatolites, also found in Western Australia, are the next forms of life that have left fossil remains. They will dominate the fossil record for billions of years to come. They are presumed to have been colonies of blue-green algae. These biofilm colonies would become havens for other bacteria, protozoa, and fungi, each performing specialized metabolic functions. They would also give safe harbor to bacteriophages, which are basically free-living RNA molecules. They are viruses that enter bacteria and take over their metabolic machinery in order to reproduce themselves.

The first virus identified by Wendell Meredith Stanley in 1935, the tobacco mosaic virus, takes the form of a long, hollow rod in which ribonucleic acid molecules (RNA) form a helical spine around which proteins are attached. The structure of the virus reveals something of the nature of the early cellular environment. The protein coat of the virus protects it from cellular enzymes. The activity of a virus also points to the special circumstances of the enclosed environment of a simple cell. When released from the cell by its death, a virus is incapable of any metabolic activity. It only returns to an active form capable of reproducing itself when it is absorbed, it is ingested, or it invades a new cell. Stanley showed, to everyone's shock and amazement, that the tobacco mosaic virus continued to retain its potential for biological infection even after it had been crystallized.

Viruses can act as messengers, picking up a part of the genetic material of one cell and carrying it over to another cell. This is called transduction. The most complex of the viruses, the bacteriophage, infects bacteria and kills them with enzymes. This is a sort of inside-out digestion. Phages, as they are known more familiarly, look like space probes that we might launch to investigate the planet Mars. They carry RNA or DNA that is a double-stranded version of RNA. Phages are thought to be the most widely distributed and various entities in the biosphere. This is a reflection of the antiquity of their presence in the biosphere and the distribution and variety of their prey, bacteria. Some phages kill their host immediately; others move in and live with their host until the host begins to deteriorate. Then the phage becomes actively destructive. The highly interwoven nature of life with near-life; with the inorganic environment of hydrogen, oxygen, carbon, and the like; and with the ultra-violet energy cycles from the sun is obvious from very early on.

The appearance of memory molecules, RNA and DNA, confirms that a certain stage of local stability in the environment had been reached: organic chemistry had evolved into biochemistry. With the appearance of the memory molecules we see the most machine-like aspect of life: a molecule that is turned on and turned off without being metabolized by the process, but which does wears out with use. The associations of protein and RNA would iterate into longer sequences of genetic material called Deoxyribonucleic acid, DNA. DNA is a double helix of RNA-like strands cross-linked through complimentary pairs of bases with the backbone of deoxyribose sugars discovered by Phoebus Levene. It was the loss of a hydroxyl group from the ribose sugar in RNA that allowed DNA to be more flexible and therefore to be a large molecule. Because of the left-handedness of the amino acids, only a right-turning helix of DNA exists. Because of its potential size DNA would become the boilerplate for vast arrays of metabolic sequences. The large DNA molecule is only stable within the cellular environment; it is incapable of surviving the death of the cell unlike RNA.

The Central Dogma of Western science requires complete autonomy of DNA. DNA is considered to be source of all influence by analogy to a puppet master. However, laboratory studies show that DNA is incapable of faithful reproduction of itself unless the appropriate protein enzymes are also on hand to help. The error rate of DNA reproduction without the environment of appropriate proteins is one in a hundred. The error rate of DNA reproduction within the appropriate environment of proteins is one in a billion. This is clear reflection of the essential role of protein in the evolution of and the continued nurturance of DNA. As Barry Commoner put it in his 2002 article in *Harper's Magazine*:

> *DNA did not create life; life created DNA. When life was first formed on the Earth, proteins must have appeared before DNA because, unlike DNA, proteins have the catalytic ability to generate the chemical energy needed to assemble small ambient molecules into larger ones such as DNA. DNA is a mechanism created by the cell to store information produced by the cell.*

Looking at the issue from a different perspective, Mario Vaneechoutte of the department of Clinical Chemistry, Microbiology, and Immunology at the University of Ghent contradicts the reigning Darwinian paradigm of self-replicator RNA or DNA molecules that are competing for the fastest rates of replication and therefore are the fittest to survive: the Selfish Gene theory. Adopting an information-theory perspective, Vaneechoutte points out that DNA and RNA are actually not self-replicator molecules at all. They are information storage molecules that are replicated by the process of life. Only the cell system as a whole can enable the act of self-replication. DNA does not reconstruct the environment within which its copied proteins will operate anymore than a dictionary creates the writer that will use it to write a book. As Vaneechoutte sees it, the protometabolic chemical community that started to use nucleotide sequences as memory, followed by nucleotide sequences, and then chromosomes, might be compared to the interactions that became possible when humans started using spoken symbolic language, resulting in the use of symbolic, encoded information, printed, and electronic texts.

As Claude Shannon, the godfather of *A Mathematical Theory of Information*, further clarifies it: information is utterly divorced from meaning. Meaning is determined by the user. Therefore, as a practical matter, meaning depends upon the degree to which the transmitter shares a common environment with the receiver. The information storage molecules are a machine aspect of life. Genes may be mixed, recombined, merged, and, of course, copied over and over again, but they work best if they are not accidentally mutated by the process of metabolism or by any other inadvertent occurrence. As Vaneechoutte restates the nature of evolution at this stage: it is not why some genes replicate more successfully than others, it is why certain information is successfully replicated. He says that the natural selection of Darwinism must be set aside for this stage in the evolution of life. Symbiotic-auto creation that is a reflection of the ambient energy level is the name of this game.

Bacteria have their genetic material intermingled with the rest of the cells' metabolic chemistry in the form of a large, circular DNA molecule with a few scattered bit and pieces called plasmids. The genetic material was not isolated from the rest of the cell inside a nuclear membrane. Here was the testing ground for feedback between the various steps in the process of metabolism: the modification or control of a process or system by its results or effects. The presence of enzymes that trigger a particular metabolic process with the help of ATP to supply energy also influence the triggering of subsequent steps in the process. Feedback is what keeps an individual biochemical process from becoming the sorcerer's apprentice.

Feedback is the modus operandi of the integrated or ecological view of life first mooted by Ernst Haeckel a century ago. Feedback was Nickolay Vavilov's answer to his concern with the infinite and unpredictable variability in living forms implicit in Darwin's thesis of random mutation. Feedback was left out of the Central Dogma by Watson and Crick, not because it was considered, tested, and refuted, but because it was simply ignored. Watson and Crick were chemists, not biochemists, and certainly not biologists. Also, feedback had a Lamarckian aroma to it. Feedback entered the field of molecular biology as an afterthought when Jacques Monod discovered the lac operon. The lac operon consists of three genes. One acts as a switch to turn on the production of a lactose enzyme, but only when lactose is present in the environment.

In order for life to survive even simple changes in the environment, feedback was required. Some bacteria automatically respond to untoward conditions by forming a new, thicker cell membrane. Unable to pass food molecules through this heavier membrane, the organism becomes inactive. In such a resting state the bacteria is called a spore. Bacteria can exist in this suspended state for decades, centuries, or even eras if need be. Russell Vreeland of West Chester University successfully revived a bacterium after it had laid dormant for 250 millions years in a salt crystal. When conditions improve and energy, raw materials for digestion, or water are again available, bacteria come out of their hardened state and resume normal life. By this simple method of metabolic control, the spore is capable of spreading out in space and of occupying areas with untapped resources, but at first the main advantage was that the spore phase enabled the organism to survive a decline in vent activity.

Going through phase changes was where memory molecules would have acquired their great value to life as a passive, machine-like memory able to survive hibernation only to reconstitute the active metabolic process with the addition of water, light, or food. Without the stable capacity for memory storage represented by DNA, the more complex forms of life could never have occurred. With a metabolic resting state as a possibility, bacteria would be able to move out of the mothering ocean onto the land, where the challenges for self-control would be even greater. Life would eventually be allured into all of Earth's recesses with the capacity of metabolic control. This is an adaptive capacity that all life forms benefit by having. On the highest stage of biological evolution, something resembling the spore state would still be the method by which malaria would jump from animal to animal, on the wings of the mosquito.

At the stage of evolution reached by the first free-living cells, reproduction would be achieved by binary fission—cloning. Cells simply divided up all their parts. DNA was separated in two single strands with the aid of an enzyme. Each identical daughter cell would end up with the same genetic library as the parent. By this method of reproduction, and with sufficient food and water, one billion bacteria can be produced in ten hours. With a textbook knowledge of genetics, we would assume that each and every bacteria would be the same. This turns out not to be the case. As Carl Zimmer wrote in a *New York Times* article reviewing the evidence for individuality of cloned microorganisms, there is a wide range of behavior and capacity. He says that in times of stress, some members of a colony respond by producing toxin molecules. These bacteria sacrifice themselves by bursting open, killing off the unrelated *E. coli* (*Escherichia coli*) around them. Some of their fellow clones survive, however, and thrive in the state of reduced competition. He cites several other examples, including the case of *E. coli* under attack by antibiotics. Most cells die, but a few, called persisters, which seem to be in a state of metabolic hibernation, survive. From the persisters will come the new hordes of bacteria, including new persisters. There is no genetic difference between the persisters and the majority of *E. coli*, but that is not to say there isn't any chemical difference at all. In fact there is a whole world of chemistry outside of DNA that is important.

Among free-living microorganisms a sort of genetic epiphany may occur with ingestion of dead cells or from viruses that may, themselves, be escaped bacterial plasmids. More commonly, new genetic information will be passed directly between two living cells at any time by direct contact. This is called conjugation. Abigail Salyers, a microbiologist at the University of Illinois, says that instead of creating a new gene the hard way—through mutation and natural selection—the bacterium can just stop by and obtain a resistance gene carried by some other bacterium. Promiscuous transmission of new genetic information throughout a population by this method can be extremely rapid. As we do more and more genome sequencing of the microbial kingdom we find more and more evidence of lateral gene transfer between what had been thought of by taxonomists as distinct gene families.

E. coli is a bacterial organism that has found in the human gut, which offers the ripe, primitive conditions that it loves. *E. coli* reproduce by fission and acquire new genetic information by conjugation. Mating partners find each other without regard to size or shape, and mating between a cell of one so-called mating type and another is common. Genetic information is shared between members of over twenty so-called genera. *E. coli* normally divide every twenty to thirty minutes. They can sustain a conjugal act for up to two to three hours. Recovery from conjugation is almost instantaneous. Conjugal relations in *E. coli* may occur in the soil, the nodules on the roots of leguminous plants, and in the intestinal tracts of fish, poultry, rodents, cattle, and humans. The Marquis de Sade couldn't hold a candle to this level of primeval debauchery.

Microbiologist Bruce Levin did an interesting experiment following his own intestinal *E. coli* over the course of a year. Sampling and culturing the bacteria on his toilet paper he found an ebb and flow that was constantly altered by the ebb and flow of external events. His intestine with its inhabitants came to be known as the Levin Archipelago. Curiously, when another member of his family became sick and was treated with antibiotics, the bacteria in his own intestines were also affected. So much for the washing of hands.

During the bacterial conjugation, the cell that transmits the new information is conventionally called the male: the cell that receives the information is conventionally called the female. This is, of course, a recrudescence of the spermish view of life going back through Darwin to Aristotle, in which sperm produced by the male is the seed. In the moment it makes sense, but in order to adhere to the definition, the cell called the female may have to become a male in the next conjugal relationship when it becomes the donor of information. Since mating can occur between twenty different so-called genera of *E. coli* the definition of the word genus is violated. Genus is supposed to describe groups that are reproductively isolated from one another. If biologists could only turn their microscopes around and take galactic perspective, *E. coli* would be a galaxy with no clearly definable edge. The main reason this doesn't happen is that organizing life into sharply defined, centrally dominant genomes is the only way to make some sort of congruence with the theory of Darwinism. The fact is that the mano-a-mano struggle to impose one's own superior genes to the extinction of another has very little relevance in the prokaryotic world.

Just as it is impossible to define the exact genetic territory of one type of bacteria from another it is impossible to isolate them from the rest of creation. Bacteria suffuse themselves through biological space and time as an absolute necessity for all life. All living things require nitrogen, but only certain bacteria that live on the roots of certain plants can transform elemental nitrogen from the atmosphere into a usable form. Bacteria break down dead bodies. Without this constant recycling, the vital elements that are necessary for life would soon be locked up in the corpus of death unavailable for further use. Life would cease. Of the thirteen vitamins required by the human body for normal growth and health, five are produced within the human body by the bacterial flora that live in the intestine. These bacterial fellow travelers may have been adopted to replace vital genetic information lost due to mutation, or they may have been with us from the beginning, providing genetic information that has never existed in the human genome.

The ease with which genetic communication practiced by the bacteria has been turned to our favor reveals another potential way in which evolution could have occurred in the past and might happen in the future. The exhaustion of genes in the human pancreas that make insulin causes diabetes. Doing what comes naturally, the human gene that produces insulin can be spliced into the genetic plasmids of *E. coli* by recombinant DNA technology. Such altered *E. coli* become factories for producing the insulin enzyme. Interferons are produced by multicellular body cells like our own to protect us against viral infections. Biologists have figured out a way to make *E. coli* produce interferons. With the aid of bacteria we are still finding new ways to mediate with the viral world, a world we generally cannot tolerate.

As we saw in the case of bacteria that catch viruses, not all of our bacterial fellow travelers are friendly. *Staphylococcus aureus* is another strain of bacteria. This is the infectious bacteria that is constantly on the verge of turning hospitals into killing zones. Because of this we know a good deal about its capacity to survive and evolve. *Staphylococcus's* capacity to swap genes with other bacteria enables it to develop resistance to antibiotics very rapidly. Keiichi Hiramatsu, Professor of Bacteriology and Director of the Department of Bacteriology at Juntendo, has shown that staph may acquire genes from a wide range of other organisms up to and including humans. James Musser at the National Institute of Allergy and Infectious Disease has shown that different strains of staph have picked up the same resistance to antibiotics independently. The gene for methicillin resistance has been acquired by five different strains of staph in five different places. The war against bacteria using antibiotics, following our nineteenth-century understanding of biology, has been shortsighted at best.

As microbiologist Lynn Margulis would observe in her excellent book *Microcosmos*, life had a common language. "By the middle of the Proterozoic Aeon 1.5 million years ago," says she, "most of biochemical evolution had been accomplished. The Earth's modern surface and atmosphere were largely established. Microbial life permeated the air, soil, and water, cycling gases and other elements through the Earth's fluids as they do today. With the exception of a few exotic compounds, such as the essential oils and hallucinogens of flowering plants and the exquisitely effective snake venoms, prokaryotic (bacterial) microbes can assemble and disassemble all the molecules of modern life."[1] This was a life form that could cleave off, separate, and ramify, but then rejoin, digest, and renew. This is how life was birthed in the primordial soup, to eventually become metazoan (multicellular) animals, until finally we turn around and discover that a normal human body contains ten times more bacterial cells than body cells. Before I get to metazoan organisms, we have the evolution of the bacteria to the protozoa to address.

The point of distinction between bacteria and protozoa is that the protozoa are cells with nuclei. The memory molecules are isolated from the rest of the cell. These include the green, golden, brown, and red algae; the ciliates; the sporozoans; the sarcodines; and the flagellates. These organisms are also marked by a profusion of new cell organelles such as cilia and flagella, although they are not unknown in the bacteria but exist in a rudimentary form. In fact, these organelles are actually formerly independent bacterial organisms that have developed symbiotic relationships with a host cell, fully evolved cellular organisms that have been ingested but not digested.

It was in 1910 that Konstantin S. Mereschovsky, at the University of Kazan, published the first essentially modern view of the origin of protozoa cells by the mutual confederation of various kinds of bacteria. It took the rest of the century for Western biology just to catch up to the idea. Darwinism stood in the way. The principle was called endosymbiosis. There is a critical factor implicated at the major branch point in evolution among the bacteria. According to Stephen Winans of Cornell University there are bacteria that create a waste product labeled Al-2 and there are bacteria that don't. Al-2 is a chemical signal or autoinducer. When enough of it is present, it induces a in the bacterial colony coordinated behavior, such as luminescence. Initially quorum sensing, as this is called, is done in a language that is unique to the variety of bacteria. Al-2 is a kind of bacterial Esperanto.

When Bonnie Bassler first discovered chemical communication between bacteria, the academic mainstream was shocked and disbelieving. Darwinism is too flimsy a theoretical construction to support this as well as much of what we have come to know about microbiological life. Nevertheless, many different varieties of bacteria can read autoinducing chemical signals. Bacterial colonies encased in an exopolysaccharide matrix of biofilm facilitated and required such communication. Water channels have been found that are the circulatory system to distribute nutrients as well as signaling molecules. The photosynthetic activity of biofilm colonies such as the Australian stromatolites, made up of blue-green algae, fuels biochemical processes and biological conversions of the total community.

It was from the line of bacteria that didn't colonize and didn't produce AI-2 that would come protozoa: single-celled organisms that ingested other single-celled organisms. It was bacteria that would increase their genetic potential by ingestion rather then colonization that would be our forefathers. Dennis Searcy and Lynn Margulis at the University of Massachusetts believe that it was a protozoa resulting from a consortium between a thermoplasma-like archaebacterium that provided hydrogen sulfide and a eubacterium-like *Spirochaeta* that craved hydrogen sulfide to protect itself from oxygen. Searcy was convinced that our own cellular cytoplasm would still retain a vestigial capacity to produce hydrogen sulfide from sulfur from this early ancestor. Most people thought he was crazy. No one had any reason to expect this to be the case, since hydrogen sulfide is toxic to nearly all forms of metazoan life. Not only did Searcy find this capacity in his own cells, he found it in protozoa, fungal, plant, and other animal cells, where it normally lies dormant.

The concept of symbiosis covers three possible outcomes: parasitic (advantage to one at the expense of the other), commensal (advantage to one with no expense to the other), and mutualistc (advantage to both sides). In retrospect, a thermoplasma consortium would look like the archetype for mutually advantageous metabolism to weather the changing environment. I don't know how the high court of Darwinian fitness, with its cockfight mentality, judges something like the thermoplasma consortium or whether can recognize it at all. It certainly has no basis for understanding why our cytoplasm continues to carry the capacity to metabolize hydrogen sulfide.

While endosymbiosis would become the main venue for the creation of protozoa, bacteria would continue to come along for the ride after the fact. The ciliate *Euplotidium* has a band of bacteria in residence on the outside of its cell wall. These bacteria protect *Euplotidium* from ciliates that preys on them by shooting out dangerous ribbons. In the absence of these defensive bacteria, *Euplotidium* are ingested, digested, and evacuated by their predator. *Euplotidium's* defensive bacteria do not go through all their stages of development unless they can smell, as it were, that the predator is in the environment. In this ectosymbiotic relationship they are completely dependent on one another for survival, yet they remain genetically separated.

The colonizing Al-2 bacteria are still around doing well. They are found everywhere. Dental biofilms that create plaque on our teeth are composed of 600 varieties of intercommunicating Al-2 bacteria.

The magnitude of the change created by endosymbiosis is mirrored by a tenfold increase in genetic material in the protozoa over the bacteria. This is a reflection of the fact that the memory molecules would now be clearly separated from the metabolizing areas of the cell. The memory molecules would be enclosed within a nucleus. Metabolism is coordinated from a distance, across a cell membrane. I presume this was successful because it avoided confusion; it allowed greater control on what was being expressed at any given time. On the other hand, no longer would genetic information be so freely shared around. Conjugation would become less a matter of swapping the latest biochemical news and more a matter of conserving a tradition. Whatever possibilities there might have been to reverse engineer DNA damage using protein as a model were reduced if not lost entirely. But as we shall see, some genes in the genome will make proteins that control or block the expression of other genes in the genome.

Although capable of carrying a greater array of functions, the protozoa are less flexible so far as genetic identity is concerned. Because of this, they can actually be put into more defined categories and identified as species. What they give up in genetic flexibility, they recoup with larger on-site memory banks that sponsor a higher degree of behavioral capabilities. Such organisms require more energy more of the time than bacteria. The overall theme of evolution so far is enlarging the field of play to more and more of the inorganic world, the organic world, and finally to pile on its biological self. Creation goes on without effort, without practice, without care, and in retrospect in a general direction.

The main exception to the centralization of the memory function is to be found in the memory associated with energy production. Some protozoans have chloroplasts with their own genome; others have mitochondria with their own genome. It was these energy-producing organelles that first suggested the idea of endosymbiosis since they looked like nothing more than ingested bacteria. Energy production is so central to life that it must have a mind and an evolution of its own, an evolution that is completely stable and reliable. The organisms that contained chlorophyll and emphasized photosynthetic activity would create the basis for the plant kingdom. It would be the organisms that were dependent upon plants and were limited to mitochondrial activity that would found the animal kingdom. Many have called this the most radical change in the history of biological evolution. It is the most significant branch point in taxonomy.

The protozoa are short-lived and fragile beings in their vegetative form.

Continued survival for these organisms depends upon the capacity for cyst formation that is a resting stage reminiscent of the spore stage of the bacteria. Because of their higher energy requirements, protozoa must build up a reserve of carbohydrates before encystment. Cyst formation involves dedifferentiation and the loss of all the specialized cell organelles. Cilia and flagella stop moving, and a wall begins to form around the organism. The cyst encloses a larval form of the cell line. Although metabolism is reduced, it does still go on. The factors that trigger this stage are desiccation, the absence of food, temperatures too high or too low, decreased oxygen, the accumulation of waste products, and in the case of *Trichonympha*, which occupies the intestinal tract of the termite, a large meal is sufficient to cause it to go into a state of hibernation. It is the first couch potato.

Among some protozoa there is a separation between the paired strands of the DNA double helix controlled by special enzymes. In higher organisms this is the process that creates the germ cells or haploid gametes for sexual reproduction. The process is called meiosis. Among protozoa that do this there is a significant reduction in metabolism leading to a resting state. The organism becomes a cyst, which is even more like the spore state of bacteria. I propose that the first appearance of the meiotic division of DNA, the basis for sexual reproduction, was for the purpose of acting as a metabolic switch. This gave it immediate utility. The cyst stage makes the organism capable of survival through hard times as well as giving it the possibility of wide dispersal.

A variation on the meiotic division of DNA in the protozoa is found in the alga-like *Chlamydomonas*. *Chlamydomonas* is like an animal in its freeform lifestyle—it has a pair of flagellum—but like a plant it has chloroplasts. The vegetative or growth stage of this organism is performed by what appear to be sexual or haploid gametes. One might suppose that the haploid gametes, the separated strands of DNA, are more vulnerable to genetic damage than is the case when they are joined together. If this is the case *Chlamydomonas* is vulnerable during its active stage. It is after pairing or fusion that the fertilized diploid zygote encysts. In this case the organism is more thoroughly protected against genetic damage during its cyst stage. The critical factor remains the capacity to shut down metabolism under unfavorable conditions.

To the protozoologist it is the ciliates that stand at the top of the evolutionary ladder with their remarkable cellular organization. The ciliates have two types of nuclei: macronuclei and micronuclei. The macronuclei are large. They are the active, in-use files for the control of the everyday activities of life: nutrition, metabolism, regeneration, and growth. The micronuclei are responsible for creating and recreating the macronuclei. They are the record of inheritance, a backup system as it were. The chromosomes in the micronuclei are the double-stranded memory molecules, DNA. They separate by meiosis and become haploid in the manner of sexual gametes, although there are no chromosomes that can be identified as X and Y, male and female. It is during conjugation with another organism that the micronuclei divide. They join with a haploid gamete of the conjugal partner. There is no crossing of genetic variations as a result of this conjugation. Two ciliates come together; two ciliates separate. After conjugation and the restoration of the micronuclei, the old macronuclei degenerate and disappear into the cytoplasm as amino acids. They are then duplicated by the recently conjugated micronuclei. Ciliates with totally restored genetic material can now clone themselves by the simple process of fission just as bacteria do.

Sexual reproduction by way of meiosis is a great mystery to Darwinists. It is the cornerstone of their method, but they don't understand how it came about. When Richard Dawkins asks what is the good of sex, he is echoing the question posed by John Maynard Smith, who pointed to the high cost to individuals who reproduce sexually compared to the parthenogenetic female who produces exact copies of her self. Dawkins is attempting to answer George Williams at Princeton, who pointed out emphatically "that nothing remotely approaching an advantage that could balance the cost of meiosis [sexual reproduction] has been suggested. The impossibility of sex being an immediate reproductive adaptation in higher organisms would seem to be as firmly established a conclusion as can be found in current evolutionary thought."[2] Even the long-term advantages of sexual reproduction have been questioned by Steven Stanley of Johns Hopkins University. He noted that many asexual species are neither less well adapted, less variable, nor shorter-lived in geological time than sexual species. In short, the Darwinian explanation for origins of sex as their gene-sorting-and-sharing device following mutation can only be supported by a careful selection of the evidence and a stiff shot of belief.

From the view of the protozoologist, meiosis is a metabolic switch that leads to cyst formation. Most protozoan species also form true sexual gametes, sperm and egg, which fuse to form a zygote just as the metazoan organisms do. The sexual gametes are dimorphic: the male forms are small and provided with flagella to facilitate the searching out of females; the female forms are larger and provided with stored nutrients. This is another variation on the theme of hibernation. Among the higher organisms, plants can survive years, even decades, in the reduced, inactive form of a seed. Without the capacity for hibernation it is unlikely that life would ever have gotten beyond the bacterial stage, let alone the protozoa.

Protozoa are able to survive the effects of profound physical mutilation through the regeneration of lost parts. This regeneration occurs at the time of fission or asexual reproduction of the macronuclei when new organisms are cloned. Regeneration, however, shows the effects of aging over time. If only asexual reproduction is allowed, protozoa get weaker, the reproduction rate slows down, and they eventually die. It is with the ciliates that the genetic basis for aging has been most clearly understood. It turns out that the copying or cloning process that occurs during asexual reproduction requires an end piece of nonsense on the gene called a telomere. The telomere is the attachment point for the RNA that makes the copy. The attachment point is not copied, but with every new copy a piece of the telomere is clipped off until there is no more telomere. At this point a section of the functional gene becomes the attachment point, and information is obscured. In short, the gene no longer works. Only a certain number of copies can be made before this degradation introduces disorder. Herein is the cause of the aging process.

In ciliates, at the time of conjugation when the macronuclei disperse into the cytoplasm and the micronuclei restore them along with their telomeres, the hourglass of aging is turned over again in this newly reborn individual. The genome of the micronuclei that restores the macronuclei of the new individual contains fifty chromosomes. This number explodes to 350 chromosomes in the macronuclei, reflecting all the rearrangements necessary for the actual expression of the information.

Micronuclei can also age through destruction wrought by random mutation. Ultraviolet mutation is a possible mutant agent, but under normal conditions it is the corrosive effect of free radicals created by oxygen reactions that are a major factor in biological aging. Damaged genes are out of shape for the musical score of which they are a part. How do the micronuclei repair mutational damage? They rejuvenate themselves by the process of sexual reproduction or meiosis. By this overall process of error correction the macronucleus ciliates is made up of a 70 percent coding fraction that is effectively involved in making proteins for cells parts and metabolic enzymes.

Lynn Margulis pointed out that, "Even careful cloning of single plant cells produces highly variable offspring." Margulis stated that, "Meiosis, especially the part that involves the pairing up of chromosomes side by side and special DNA, RNA, and protein synthesis, is, we believe, like a roll call or a taking of inventory. It ensures that all the sets of genes, including those of the mitochondria and plastids, are in order before the multicellular unfolding that is the development of the embryo begins."[3]

How is it decided which is the clean gene and which is the damaged gene? The decision cannot be arrived at by the genes themselves. Genes that are damaged are removed by RNAs whose job is to recognize and remove by pattern recognition that which is out of tune or out of shape. This is the reverse engineering of billions of years of biochemical tradition that has already transpired to reach this point in evolution.

In an early notebook Darwin proposed that sexual reproduction, or interbreeding as he called it, was a device to keep variations in check, "the interbreeding of individuals carrying opposite variations destroyed these variations."[4] This was a supposition gained from the practical knowledge of breeders. Unsurvivable mutants were weeded from the breeding population by this method. Later on, when it appeared that asexual reproduction played a more important role in the plant kingdom, Darwin downplayed sexual reproduction entirely and prudently avoided mentioning the purifying role of sexual inbreeding in his magnum opus. This left him with no understanding of the meaning of sex, especially since the female role in conception was not yet known.

When Darwinism got a gender upgrade from Gregor Mendel, sex would then become the genetic selection and mixing device for mutated genes that gave variety to evolution. This became the *sine qua non* of evolution, after Morgan, Muller, Watson and Crick. There may be a very subtle refinement of the genetic memory by mutation (the creation of alleles)—it would be hard to eliminate the Central Dogma completely—but the only well-proven role for sexual reproduction among the protozoans is to reverse the degradations of use and abuse on the genetic memory.

Regarding the question of genetic variety providing a continued basis for evolution at this level, *Paramecium aurelia* is an interesting case study. *Paramecium* are another group of unicellular ciliates, called Lady Slippers because of their shape. Tracy M. Sonneborn has shown that there are nearly 200 genetic alleles or alternate forms of a gene associated with *Paramecium*. Yet the only definite way to positively identify the different varieties is by the antigenic response created when their blended remains are injected into a rabbit. Antigens from the rabbit can recognize, attack, and kill only that variety of *Paramecium* for which the antigen was created. The different varieties are marked as different mating types and are also identified by different behavior. Symbiotic bacteria of different types have been found living in the cytoplasm of *Paramecium*. In one case the bacterial symbiont provided the host with folic acid, and the *Paramecium* no longer had an external requirement for this vitamin. In another case it was found that stocks that carried a particular symbiont bacteria killed *Paramecium* that did not have them. This is certainly a choice parody of the Big Dogma.

Although *Paramecium* have been observed for over a century, all around the world, and for millions of generations, they follow the rule of isolated breeding populations. Different mating types never or rarely conjugate or reproduce sexually with one another. They rarely out breed. They have been isolated from one another for countless billions of generations, yet it is still virtually impossible to identify any other difference outside the antigen differentiation. Even though *Paramecium* establish isolated breeding populations, there is no genetic drift that has caused them to become different species. Sonneborn has determined that genes, cytoplasm, and environment all combine to create each race of *Paramecium*. He believes that the different races are simply a reflection of their geographical circumstances.

This reminds us of Niles Eldredge's groundbreaking thesis of punctuated equilibrium for higher life forms. After a careful study of fossil collections of Cambrian-age trilobites he found that for long periods of time there were very minor and what appeared to be random variations with no particular direction. Very occasionally Eldredge would find an abrupt species change in the fossil record, but cosmic impacts, atmospheric catastrophes, ice ages, and the like seem to have been involved. These events may have been a matter of indifference in Paramecianna. Although we can't really tell for sure, it is widely agreed that the antiquity of their works suggests that evolution for them was finished long ago. If there have been punctuating genetic events in the life of a *Paramecium,* it has been lost in the confusion.

Following the principle that a flea hath smaller fleas that on him prey, the ciliates have many species that on them prey. Here we see again the theme of bacterial cells invading or being ingested by other bacterial cells, but in the protozoa we see the invaders become parasites. The attraction for such an invasion stems from the storage of energy resources by protozoa in order to maintain their life style. Ciliates parasitize other ciliates in order to feed upon the stores of carbohydrates the host has accumulated in order to reproduce sexually, but they must do so discreetly so as not to significantly injure the host species.

Reginald Manwell, in his classic and fascinating text *Introduction to Protozoology,* grapples with the dilemma of parasitism from a Darwinian perspective. He wonders at the long step, or the coordination of many small steps, that must have occurred in the evolutionary process to achieve such parasite-host relationship. He speculates that perhaps "a fortuitous combination of mutations in both the potential parasite and its prospective host made the establishment of the host-parasite relationship possible." This is such an irreducible complexity that he finally decided that random mutations in nature "seem not to account for the remarkable degree of adaptation."[5] Manwell suggests that encystment was the key that led to the unintended consequence of parasitism. The host ingested the cysts of the future parasite, but for some reason neither the cyst nor the restored protozoan was digested, attacked, or rejected by the host. I suggest the cyst was benign as far as the host was concerned because it was metabolically inactive. When the cyst did finally hatch, the host related to it as if it were its own.

Many attempts have been made to forcibly produce genetically distinct species of protozoans in the laboratory. It has never happened. Paramecia do survive various deleterious agents or chemical mutations, but the mutations are only maintained in subsequent generations by fission (asexual reproduction). Such mutants either lose their new trait upon conjugation (sexual reproduction), where it is corrected, or they never sexually reproduce and die of old age. An exception to this is in the case where the genome is doubled. This is called polyploidy. Polyploidy is an error that occurs with meiosis, and it occurs at all levels of evolution where meiosis occurs. In spite the huge increase in genetic material, the mutant polyploid genome is reproduced by fission and by conjugation at the normal intervals in paramecia. Over time there is a loss of these extra genes, but genes with a high rate of expression disappear at a slower rate. There are no apparent innovations in biochemistry or in morphology involved with this mutation. *Paramecium tetraurelia*, with three copies of its genome, is still nothing more than a strain of *Paramecium aurelia*.

In metazoan life polyploidy is a rich venue for innovation. The first effect of polyploidy is an increase in size. The larger-than-normal lemon on a lemon bush is a polyploid fruit. In the plant kingdom polyploidy is pervasive, and the estimates are that as much of 80 percent of living plant species are the result of polyploidy. The lower animals such as the flatworms, leeches, and brine shrimp benefit from polyploidy. A few higher polyploid animals, such as the lizards, are sterile and have to reproduce by parthenogenesis. The triploid mole salamanders (three copies of their genes) are all female and reproduce by kleptogenesis. The females steal spermatophores from the diploid males of a related species to trigger the development of their eggs without incorporating any DNA in the process. Size of the embryo becomes an issue in metazoan organisms, where a non-polyploid parent creates the enclosing birth vessel, such as the chicken or us. Polyploid human fetuses are naturally aborted very early on. It seems that the protozoan capacity for self-reliance, efficient feedback, and effective gene correction has precluded the need for innovative usage of polyploidy events. It appears that it is only some metazoan organisms at appropriate stages of their evolution that will be able to risk taking advantage of this reproductive accident.

Beside the increase in size, a secondary long-term advantage for polyploidy is as a backup against genetic damage. Humans show evidence of hexaploidy. We have six copies of hemoglobin. At least one mutation event may have happened in our protozoan ancestry. Our Hox genes, which are the blueprint for the bilateral body, are tetraploid. This is also true of all vertebrates starting with fish, so the mutation event certainly goes back to our earliest metazoan ancestors. Since normal doubling follows the rule of 1, 2, 4, 8, I suggest that the six copies of hemoglobin reflect endosymbiosis of a diploid organism with a tetraploid organism. Some of the pseudogenes uncovered by the Human Genome Project appear to be the wrecked remains of once-functional Hox gene copies. As with the case of self-correcting polypeptides, or oily molecules forming into protocells in water, Hox genes multiplied by polyploidy are a preadaptation for more complex possibilities to come. This is the way creation works: as it unfolds it creates scaffolding for a greater elaboration in the future.

Is there any wisdom we can distill from the collected knowledge of creation at this point? Unfortunately wisdom has vanished from the philosophical map in modern culture and all we hear is the chaos of crows. My favorite symbol for creation is captured a seed pot that I purchased from Bernadette Ascencio, who lives at the Acoma Pueblo in New Mexico. The traditional seed pot is a sphere that has only one tiny opening in the top. This small opening is designed to keep pests away from the stored seeds. Around the opening there is what is called the starburst pattern. This is created by drawing groups of straight lines from the opening. Each group of ten lines or so radiate out in parallel following a central line that aims in the direction of the opposite pole. As these groups of parallel lines are drawn from the center moving around the axis, they overlap previous groups of lines. The pattern stops about half way down the pot which is where the lines stop overlapping.

Looking at the starburst again we see that another pattern has appeared coincident to the drawing of the straight lines. There are circular patterns at a certain distance from the center. If the pot was a perfect sphere, the induced patterns would have been closed circles, but because of the slight imperfections in the surface of the pot spirals and even a branch point are created. The

starburst pattern need not have produced its spiral patterns as result of prescience in the first instance; it was a creation out of simpler elements transformed by energy unique to space and time. This goes beyond simple pattern recognition. It breaks the mold, but rebuilds it in a more complex form after the fact, or in retrospect. It is evolution and devolution under the heading of a creative event.

Before creation there was no living thing until Tawa and Kokyanwuhti willed them into being with song. "I am Kokyanwuhti," the Spider Woman crooned. "I receive Light and nourish Life. I am Mother of all that shall ever come." The sign of Spider Woman in Hopi tradition is the two-armed cross that is the most rudimentary starburst pattern and is the skeleton of profound philosophic speculation from time immemorial. The cross represents the intersection of light (energy) with dark (matter). From the cross comes the calendar, the wheel, and at the opposite end of creation a revolution in linguistics. In the late nineteenth century Ferdinand de Saussure proclaimed that communication was a good deal more than stringing together words out of a dictionary. Claude Lévy-Strauss elaborated by defining a horizontal dimension that was speech read left to right like a musical score and a vertical dimension that defined the structural relationships: the context within which communication occurs.

The base of the Acoma seed pot below the starburst pattern has a different design. This is a design made up of seed-shaped lozenges that enclose symbolic cornfields. This reflects the knowledge of the seed and of the calendar: the basis of civilization. The tic-tac-toe pictograph of the cornfields is the same as the Chinese pictograph for "well." In the I Ching, the Book of Changes, the Well stands for that which does not change. Political structures change, architecture changes, fashion changes, but the well in the center of the field or the plant that draws water up out of the ground does not change. The common inspiration behind the Acoma seed-pot and the Chinese pictograph are not random coincidences. Carl Jung called such relationships an archetype. There are no immediate linear connections that would satisfy the geneticist or the taxonomist. It is the synchronicity of the advancing horizon of life itself that we see. Like a pulsar we give off a signal that is unique to our own particular planetary creation, yet it is common to all possible creations. The Acoma seed pot is as succinct a statement of creation as I have found. Now that we have a clear direction for evolution, we are now in a position to give it shape.

CHAPTER 10

The Shapes of Things to Come

AT THE END OF the Proterozoic eon around 500 million years ago life was composed of bacteria, algae, sponges, and a few other enigmatic fauna. Lynn Margulis and Dorion Sagan sum up this stage of biological evolution in their latest book, *Acquiring Genomes*. "The language of evolutionary change is neither mathematics nor computer-generated morphology. Certainly it is not statistics. Rather, natural history, ecology, genetics, and metabolism must be supplemented with accurate knowledge of microbes. Microbial physiology, ecology, and protistology are essential to understand the evolutionary process. The behavior of microbes within their own populations and in their interactions with others determined life's winding, expanding evolutionary course."[1]

Suddenly there was an explosion of metazoa over the course of the next 5 to 10 million years, but possibly as short as a million years, according to Bowring et al. During the great efflorescence from the single-celled protozoa to metazoan life in the Paleozoic era, there would be a pronounced increase in the total amount of DNA. There would be one hundred times more DNA in the new metazoan organisms on average. All of the major phyla of animals appear at this time. The Cambrian period has been one of the great mysteries and embarrassments of paleobiology since the nineteenth century. Charles Darwin saw this jump in evolution as one of the most serious objections to his theory, so naturally it has received a lot of attention. Creationists point to biology's Big Bang as typical of Darwin's crank thesis, but they can't take much advantage of it since it fits their belief even less. They prey on it only because Darwinists make such a fuss. If we clear away the prejudices and look only at mass of data now accumulated, a solution auto-assembles itself.

Evolution for the first few billion years occurred on a horizontal level. All the metabolic chemistries of life were evolving piecemeal in their special niches. The sharing around of information by the bacteria turned into the libraries of information collected by the protozoa. Even though this was all happening at the microscopic or electronic microscopic level, it doesn't mean it was trivial. Just because most botanists and zoologists look down their noses at and are largely ignorant about the importance of microbial life, it doesn't mean that these were empty eons. On his Science Frontiers website William Corliss cites Kurt P. Wise, the young-earth creationist with a degree in paleontology from Harvard who studied under Stephen J. Gould. Wise observes that actually the Cambrian Explosion did not see the greatest increase in biological innovation. The earlier Archaean Explosion produced seventeen new phyla of bacteria employing an extraordinary range of different metabolisms. Although some thirty-eight new phyla did emerge from the Cambrian Explosion, they utilized only one type of metabolism.

With this correction in mind, we have to remember that everything has to wait its turn. Without bacterial colonization of the planet, the Cambrian explosion could not have occurred. Without the first recycling emergency when the original forms of life poisoned themselves into minor players with their own waste product, oxygen, the Cambrian exposition would not have happened. Oxygen totally changed the face of life. The planet had to turn red before it turned green. It took hundreds of millions of years to evolve into an aerobic planet. Without the collection and alignment of metabolic chemistries in the nucleus by the protozoans, the metazoans were impossible; furthermore, the energy gradients had to be low enough before the Cambrian explosion could occur.

I believe the Cambrian explosion happened because evolution had reached a point where it could happen. It was triggered by gross climate changes and environmental havoc. The foundering of a huge chunk of sea floor was coincident with the Cambrian period. Perhaps that was the trigger. Devolution gave evolution another waterfall in the inevitable downhill course of the Second Law of Thermodynamics, a diversion that allowed work to be done by way of the Second Law of Evolution. Once the Hox genes were established, life began to evolve vertically instead of just horizontally. Life was given new options in a new environment, the embryonic environment of metazoan organisms. This produced something that we could see—fossils. Over the next 500 million years, these fossils increasingly began to look like us.

The great efflorescence from the single-celled protozoa to metazoan life in the Paleozoic era produced a pronounced increase in the total amount of DNA, but without any new metabolisms. There would be no correlation between animal or plant evolution and the number of chromosomes—a *Paramecium* has many more chromosomes than a human. Even on the issue of the total amount of DNA, there is no correlation between animal and plant evolution and what we think of as complexity. Plants tend to have more genes than animals. As John Watson put it in his classic textbook, "While higher plants and animals have much more DNA than the lower forms, no one anticipated the findings that certain fish and amphibians would have 25 times more DNA than any mammalian species. And as more and more plants were examined, closely related species were sometimes found to vary in their DNA contents by a factor or five to ten."[2] In short, there is little correlation between the size of the genome and the resulting organism. This was yet another discovery that had not been anticipated, nor can the Central Dogma of Darwinian evolution explain it.

The fungi represent a very early stage of metazoan development. Many of the fungi practice a form of sexual reproduction, but, as in the case of the ciliates, the sex cells are exactly the same; there is no way to identify one as an egg, another as a sperm. Each contributes an equal number of genes to the new individual and equal amounts of food reserves to nourish the new organism. This kind of sexual reproduction is called isogamy. Isogamy resembles parthenogenesis or cloning in that there is nothing that can be construed as out-breeding. There has been no gender specialization in this organism genetically or morphologically. This is a very stable and effective approach to survival with spore formation, giving the organism the capacity to survive cold or drought or conditions of poor nutrition for very long periods.

The fungi are responsible for mold, mildew, rot, smut, rust, and blight, as well as jock itch and athlete's foot. The fungi are also the metabolic engines that give us bread, beer, and cheese, as well as sliced mushrooms that go very nicely with chopped liver stir fried with some Marsala wine. Lichen are a fungi that live in a commensal relationship with algae. They retain their own genetic integrity and cannot live one without the other. The algae provide energy; the fungi provide raw materials. Whereas bacteria and protozoans ingest before they digest, fungi digest before they ingest. They secrete digestive enzymes on their food before they ingest the byproduct. This puts a hold on their evolutionary possibilities. Ingestion without digestion has been a major theme of evolutionary diversity so far. The fungi were created endosymbiotically in the first place, but this is no longer an evolutionary venue because of their method of digestion. And this is a good thing for us since the fungi are major recyclers of worn-out bodies on the living planet. Their rock-solid stability is an essential basis for the rest of evolution.

A dramatic, intermediate stage of evolution between free-living cells and metazoan organisms is found in the slime mold *Dictyostelium*. Slime mold is considered to be a social amoeba. *Dictyostelium* normally browses in the manner of undifferentiated, free-living amoebae, dining on bacteria as well as cannibalizing its own. It divides every three to four hours by simple fission. Upon starvation there is a release of a chemical signal by one of the amoebas. The other amoebas aggregate in response to this chemical signal to form a sausage-shaped slug that crawls about, orienting in an appropriate fashion to heat and light. After awhile the slug erects itself into a vertical stalk that stiffens itself with cellulose. Cellulose is a complex carbohydrate that forms the main constituent of plant cell walls. The remaining cells—they are now cells in a colonial relationship, not amoebae—stream to the top of the stalk where they form a small sphere. Each of these cells produces a spore that is released. The stalk dies, and the life cycle is complete. The spores lie dormant, waiting for conditions to improve. Therefore, as far as *Dictyostelium* is concerned, the chicken comes before the spore, or at least the reproductive organ of the chicken.

The question for the Darwinist is, what good does this do the slime mold? What utilitarian function does this social activity serve that could not have been done with fewer resources by free-living amoebas? John Tyler Bonner, emeritus professor at Princeton University, sums it up by saying there is no certain answer because we do not even know the evolutionary significance of the aggregation of the amoebae into cooperating cell masses in the first place. This is the embarrassed answer by Darwinists to the appearance of multicellular life—the metazoa—in general.

We have already seen ectosymbiotic behavior by bacteria in order to create colonial biofilms. We have seen endosymbiotic behavior creating a new level of organismic complexity in protozoans. We have seen ectosymbiotic relationships of bacteria with protozoa and metazoa. Now we have evidence of behavior in amoebas that continues the theme of communal activity. Amoebic midwifery has been observed by Elisha Moses at the Weizmann Institute of Science in Israel. Amoebas normally multiply by non-sexual fission. Commonly, however, the two new cells fail to completely separate. The distressed cells release a chemical attractant that signals nearby amoebae to come to the rescue. Amoeba gather around and nudge the stalled fission to completion. While this is going on, so-called cheaters can take advantage of the riches that accrue to socialization. The possibilities for parasitism are numerous. Countermoves, however, generally lead to optional cheating: cheating only when conditions allow it. However, the use of the word "cheat" is really inappropriate here. It shows that the scientists using the word have a dog in that race. The demands of socialization are never without their drawbacks.

More can be learned about the advancement of social behavior by watching the Tour de France than from reading either the Bible or textbooks on Darwinian evolution. The relationship between the pelaton (body) of the bicycle tour and its individual members is a fluctuating equation that is solved based upon internal chemistry; the multiplicity of awards for different strengths and conditions; the different colors of the teams and their differential budgets; and the environment, always with the possibility of special chemicals available to some but not others. Each individual solution is unique, although the overall solution will nearly always come out of a fairly predictable leading group, at least for those who understand the race. The wild card of the outlier, the persister, or the optional cheater is a surprise, but adjustments are made soon enough. Creation cuts an ever-deepening valley that assures that random perturbations will not stray far off stream. Incomprehensible levels of unconsciousness mediation help us make our way through the slipstream of creation.

Contact through chemical signaling has been an evolving method of communication between bacteria from the beginning. Direct contact offers even more opportunity for linked, metabolic communication between individual cells. More communication goes hand and hand with more complex physical structure. An outstanding attribute of metazoan organisms is, as Bonner put it, "not just that individuals are associated, but there is an alignment of their life cycles. In colonies the timing of the cycles of the individuals is under fairly strict control: all the individuals arise simultaneously so that the cycle of the colony coincides with the cycles of the individual."[3] Those life cycles are the accumulation of timed sequences of metabolic activities that take us from conception to death.

Bonner supposed that natural selection, in the original sense that Darwin used it, operates on life cycles not just on an individual at a given moment, say at the time of mating as Darwinists would have it. "Evolution then becomes the alteration of life cycles through time; genetics the inheritance mechanisms between cycles, and development all the changes in structure that take place during one life cycle."[4] From a Darwinian perspective there is no selective value for any of this sort of evolution at the time. That would suggest precognition or divine intervention.

Bonner made a curious discovery about the slime-mold slug involving cell specialization. If a slug was stained and the front end cut off and grafted to the rear of an unstained slug, the stained section migrated rapidly up the unstained slug to the front. Having once been front cells, they became genetically altered to always be front cells. Were the priorities for such cellular specialization already rudely established by a time-sequenced linearity of metabolic process developed in the protozoans? Is the fore and aft knowledge of the slime mold an expression of such an inner organization? In any case, this property will be well developed in the metazoans. The ability to induce a large population of *Dictyostelium* cells to proceed synchronously and with invariant timing through a developmental program has attracted an enthusiastic group of researchers. It is embryology externalized.

With the metazoans, understanding the development of the adult from the seed or egg is an entirely new and challenging discipline. This is the discipline of embryology. From the fertilized egg, cell division produces a ball of cells. This opens up to create a cell lined with cells called the blastula. New cells will form within the blastula that will become the embryonic disc. Embryologist Rosa Beddington showed that the fore-aft orchestration of body parts of the metazoan organism is triggered from the blastula, which was once thought to be only a protective cloak, surrounding the embryo. From the outside these cells secrete a transforming growth factor that induces the primitive streak. The establishment of the body axes begins to take place prior to and during the very early period after conception when the embryo begins to ingest itself in a process called gastrulation. The shapes of things to come is determined by the very localized embryological environment.

Later in embryological development Jill A. Helms, an orthopedic surgeon, showed that exchanging cells destined to become beaks in quails and beaks in ducks resulted in quails with duck's beaks and vice versa. At this level of embryological development, cells are now programmed, but they are still capable of persuading neighboring tissue cells to join with them to produce the bizarre result. At a later stage of development mature kidney cells from dissociated kidney tubules reform themselves into recognizable kidney tubules. Dissociated cartilage and kidney cells mixed together resort themselves according to their original identity. At this point, once a kidney cell always a kidney cell.

It is the Hox genes in the DNA that orchestrate the details for embryological development in the metazoan organism. They are a biological archetype. The genome at large is a welter of genes scattered helter-skelter in no apparent order. The Hox genes are a tidy little island of good order. They are a broadly conceived musical score that directs the order of development. They do not encode for proteins that are used in cell construction or for enzymes for metabolism; they switch on cascades of other genes. They define a sequence space for embryological development that leads to the phenotype, the physical body that we see. Every cell has a full set of Hox genes, modified by their local chemical environment, so they know what role to play and, equally important, what roles not to play. Cells orchestrated by Hox genes at an early stage of embryological development acquire a profession and a mind of their own.

Hox genes are just as much a reflection of fundamental metabolism of kingdom Animalia as its mitochondrial inclusions. The plant kingdom has its own Hox genes along with its chloroplast inclusions. These genes and inclusions are part of the fundamental difference between plants and animals: those that stay and can easily orient to gravity and/or light and those that must mature while changing their orientation to gravity or the sun. Those that must evolve and survive in a moving environment are more irritable by nature; they evolve a nervous system.

Hox genes are found in evolutionarily more primitive organisms, such as *Velella velella* that is radially organized, as well as fungi and *Dictyostelium*. In all bilaterally organized vertebrates such as us, Hox genes are so rudimentary that a fruit-fly Hox gene that is deliberately knocked out by mutation can be replaced by a human equivalent without the organism missing a beat. Hox genes are interchangeable between species, genera, families, orders, classes, and even phyla, but not kingdoms. Hox genes were at the heart of the explosion of animals in the Cambrian period evolving separately along fifty to one hundred genetic lines. For those of us still looking for the light, Walter Gehring and his collaborators have identified the common origin of eyes in evolution, the *Pax-6* master control genes. It is a set of Hox genes.

The primary tool for genetic research in this area is knocking out a gene or replacing a gene in the laboratory and seeing what happens. Although generally useful as a part of a larger picture, this research can also turn into a mathematical flea circus; we are after all looking at things that can only happen in a laboratory. The most exciting discovery is when we uncover such processes in the wild, which will also reveal how adaptations occur naturally and what adaptations can occur. A most astonishing example is a case of ectosymbiosis that served as a replacement for the loss of one of the basic elements of the *Pax-6* gene set. This is the relationship between an animal, the Hawaiian bobtail squid, and a bioluminescent bacterium.

Euprymna scolopes has a light organ in its belly. In embryonic and juvenile squids, it bears a striking resemblance to the eye development in mammalian embryos. Several of the same genes are involved, but it is missing the retinal tissue that gives eyes their photosensitive capacity. The part of the genetic record that is missing is one of the Pax-6 Hox gene set that was originally acquired by endosymbiotic ingestion, the part that included the genome of the halophilic bacteria that contained the photosynthetic pigment rhodopsin. Since the squid still has highly functional eyes in the normal position in space, this anomalous eye must be a copy of the normal that is out of the normal embryological sequence. Because it is normal in other respects, it seems to be an eye in search of a retina, as we shall see.

Soon after the squid hatches from its egg, the eye-like organ is occupied by bacteria. Out of the great multitude of microbes in the environment, it is the bioluminescent bacterium *Vibrio fischeri* alone that enters the special crypts designed for their occupation, encouraged by its ciliated lining. Here they establish an ectosymbiotic relationship. *Vibro fischeri* show typical colonial behavior. When sufficient numbers of individuals occupy the crypt, they produce a chemical death signal that kills the ciliated epithelium of the crypt. The bacteria, on their side, lose their flagella. Isolated or in low population concentrations, bioluminescence is almost zero in *Vibro fischeri*. It is only in the state of great population density that autoinduction triggers a simultaneous and hyperactive output of light. It was this symbiotic relationship between squid and bacteria that revealed the autoinduction process in the first place. In exchange for their light, the squid provides sugars and amino acids to the bacteria. This is the long-lost relationship that the squid was missing, albeit with a different outcome: luminescence rather than photosensitivity.

The color of the light produced by the bacteria is adjusted by the squid by a lens, and intensity by a shutter-like mechanism so as to match the background of moonlight or starlight falling on the water at night when the squid are feeding. The eye-like light-producing organ produces an electrical response in the squid that indicates that the complicated light and color response is under neurological control. Without this bacterially induced camouflage, the squid are quickly spotted by predators from below and consumed. This ectosymbiotic relationship lasts for the night, the bacteria are flushed out, and the process begins again the next day. There is more to be learned about this amazing relationship, but so far we can see that an ancient endosymbiotic genetic tradition has been lost, to be replaced by a cycling ectosymbiotic relationship with a distant bacterial cousin.

The *Pax-6* set of genes produces the true and the pseudo eyes of the Hawaiian bobtail squid, the pinhole eye of the Nautilus, the simple eyes found in larva, and the compound eye of the fruit fly, as well as our own exquisite visual system. A review of this subject by William A. Harris at the University of California, San Diego, shows how the development of the eye of a cephalopod differs from the development of the eye of a vertebrate like us. In the cephalopod the eye is induced entirely from embryonic epidermal tissue. The vertebrate the eye is also induced from embryonic epidermal tissue, but it also employs embryonic neural plate tissue that has specialized to eventually become the nervous system. Because the vertebrate eye shows a prolonged embryological development, it incorporates a later level of cellular specialization within which to create its effect. Because of its position in embryological space, however, it is still an eye.

The curious thing about the difference in timing between the cephalopod and the vertebrate eye is that in the vertebrate eye the photoreceptive cells appear to be installed backward. The optic fibers come through the eyeball to the side where the light is entering. It has been suggested that this is an evolutionary mistake, since the nerve fibers block and distort the light and the photocells are pointing away from the source of light. On the other hand, the cephalopod lens is focused by a muscle across the opening of the eye ball to which it is attached. This muscle, although transparent, still tends to distort the image and reduce the light. The vertebrate lens is free floating and clear in the opening of the eyeball. It is the eyeball that flexes to change the focus. There are several ways to skin a cat, as they say, and whatever way you chose there will be a downside. The main point is that it all falls out from the same genetic tradition, the *Pax-6* genes.

The gatherings of bacteria that form a quorum and produce light at an early stage of evolution, and the symbiotic relationship of those bacteria with other colonial bacteria that would protect and nurture them begin to reduce the irreducible complexity of the eye—the argument made by William Paley against Erasmus Darwin's theory of evolution—to manageable proportions. It reduces it to an auto-assembling creation following laws of physics that create the laws of chemistry that create the laws of biology within a predetermining state of the environment. Even though complexity is reducible, it doesn't mean that it wasn't an intelligent creation; it certainly wasn't random. Only the fate of individual cells or the sparrow in its windward flight is to some degree unpredictable.

Bivalves are early recipients of the *Pax-6* genes, but scallops are the only members of the group with fairly well developed eyes that include a lens. Other bivalves have poorly developed eyes or none at all. A functional visual system has made scallops active swimmers. Indeed, they are the only migratory bivalve in creation, but whether they have acquired any advantage over their eyeless cousins is a moot point. In some ways they have an advantage; in other ways they have created a problem by attracting attention to themselves. The potential for vision would be an invaluable preadaptation ready for development for mollusks who had lost or would lose their protective but constraining snail-like shells—the cephalopods. The squids, octopi, and especially the cuttlefish have maximized the use of vision, a large brain, shape shifting, and color change to create a dynamic, self-conscious organism capable of astonishing feats of camouflage or visual display.

There are eight Hox genes that determine the fruit fly. The first affects the mouth, the second the face, the third the top of the head, the fourth the neck, the fifth the thorax, the sixth the front half of the abdomen, the seventh the rear of the abdomen, the eighth various other parts of the abdomen. This fly has only one complete set of Hox genes, unlike vertebrates, which have four. This means that knocking out a single Hox gene will have an obvious effect on the corresponding body segment orchestrated by that gene, a deadly effect. The famous example is the *Antennapedia* mutant fly that has legs growing where antennae normally appear. This is a macromutation that is not survived.

Fruit flies and all insects share an arthropod ancestor with crustaceans, such as shrimp. In this ancestor the Hox-gene cluster that is responsible for the abdominal segments was duplicated. This was not genetic damage that turned out to be new and useful information; it was the duplication of useful information. The duplication appears to have created a backup that allowed different subsets to go off on their own while the traditional embryological function was still being supported. These new genes, although producing the same proteins as before, could be expressed in different regions or at different stages of embryological development, producing different results. The insect branch would lose limbs in the lower abdominal region, but what they would lose in legs they would gain in wings. The macromutation of the abdominal Hox gene cluster set the stage for the class separation between crustaceans and insects.

The loss of an abdominal Hox gene in shrimp has turned out to be just barely survivable; it created a degenerate group of prawns called the *Lucifer* genus that has many fewer appendages than normal with no claws and no gills. This is an evolutionary dead end waiting to happen. The fruit fly is missing the Hox gene at the end of the line that in crustaceans and vertebrates is responsible for the expression of a tail. The fruit fly has no tail. This was a macromutation that was not only tolerated, but it conveyed an advantage. It eventually allowed the insect the ability to utilize a new medium, the air. Crustaceans dominate life in the ocean, but insects are the most diverse group of animal life on the planet, with the number of species now in existence estimated to be between six and ten million. Part of this explosion of numbers of species in Insecta comes from the loss of the tail.

Vertebrates with four sets of Hox genes have to have between two and four of the homologous genes knocked out simultaneously to get a visible mutation, depending upon how many have already been knocked out of service. The polyploidy that created four sets of Hox genes created a genetic backup, but it is not as simple as that. When all four of the Hox-9 vertebral genes are knocked out, it also changes the expression of the vertebrae both upstream and downstream of the mutation. Claudia Kappen uses the analogy of different numbers of pigments that may be used to reproduce the color spectrum: a box of crayons with ten pigments or a collection of fifty colored pencils, and so forth. She suggests that different repertoires of Hox genes account for different embryological orchestration and offer the possibility of increasing complexity and/or increased resolution of other genetic principles found in the genome.

Darwinists are at a loss concerning the Hox genes. "The incredible conservatism of embryological genetics took everybody by surprise,"[5] said Matt Ridley in his book *Genome*. With its *idées fixes*, Darwinism cannot take advantage of this amazing principle of biology. It is a fundamental theme that is reincarnated in a multiplicity of forms. From a Darwinian perspective it is a miracle, but there is an even greater miracle to report.

Sidnie Manton, highly respected biologist at Cambridge, suggested separate polyphyletic origins for the three groups of arthropods: the centipede family, the insect family, and the shrimp family. She suggested separate origins from worm-like ancestors. She came at the anomalies of classification in the arthropods from the perspective of comparative anatomy. It was simply impossible to impose a branching tree-of-life model on these groups of animals. About the same time in the 1970s, Donald I. Williamson of the University of Liverpool suggested the ultimate in saltatational evolution. He argued from the perspective of a planktologist that in the invertebrates that have a larval form as one of their developmental stages have acquired the genome of a larva-like animal. This is not bacteria sharing genes or protozoa ingesting one another, but rather metazoans combining genomes by hybridization. Williamson suggested that this may have occurred in eggs shed in water fertilized by another species or by penis-aided entry into a female, the wrong female. He emphasized that both adults and larva evolved by the familiar method of gradual descent with modification, but with hybridization superimposed.

A case in point would be the hydrozoa, *Velella velella*, commonly known as by-the-wind sailor. In its obvious form when it washes up on shore, the by-the-wind sailor is a highly modified vegetative polyp. In this stage it is similar to its Cnidarian cousin, the sea anemones. *Velella* doesn't attach itself to the ground like an anemone; it is upside down by comparison, and it throws up a sail. In another stage of its life the by-the-wind sailor produces tiny medusae out in the center of the Pacific Ocean, unlike their cousins the sea anemones that have no medusa stage. These free-swimming medusae produce sexual gametes similar to their Cnidarian cousins, the jellyfishes. From the fertilized egg of the *velella* medusa comes a planula, which is a larval form that will develop into a by-the-wind sailor. The jellyfishes have no polyp or vegetative stage. The Williamson theory supposes that all the Hydrozoa that produce two body forms during their life cycle like the by-the-wind sailor are hybrids of Scyphozoa (jellyfish) and Anthozoa (sea anemones and corals) neither of which have alternating generations. Taxonomists, attempting to impose a conventional tree of life on the Cnidarians, are at complete odds as to how to classify them.

Another likely suspect in the hybrid genomes scenario is the velvet worm. Onychophora is one of the most primitive of animals. It was one of the first to crawl onto the land in the days of the Cambrian explosion. They are part of the primitive group that gave rise to the arthropods, which gave rise to the crustaceans and the insects. Like the crustaceans but unlike the insects, they have all their appendages. Rigidity in the appendages of the velvet worm is achieved by hydrostatic pressure. There legs are like balloons. Movement is obtained by stretching and contracting body segments. The legs operate on the principal of resistance to movement, like crutches, while the body lifts itself up and over them to move forward. A pair of simple eyes is found at their anterior end. Simple eyes are like vertebrate eyes without a lens or aperture control. These eyes are good for detecting the source of light but are not useful for forming images. Generally they are only found in larval forms of insects.

The velvet worm uses its simple eyes to avoid the brightness of daylight. It forages at night. Two species live in caves to which they are well suited by preadaptation: the essential requirements for this habitat were largely present before they moved in. The velvet worms are renowned in modern zoology for their curious reproductive behavior. Some lay shelled eggs, while in others the fertilized egg attaches to the mother's innards by way of a placenta. The worms give birth to live young. This is a very early appearance for a method of reproduction that has long been considered to be a very advanced innovation. It is the method of insemination that one group of velvet worms practice that is of particular interest to us here. There is an African species where the male simply leaves a packet of sperm on the body of the female. After the packet dissolves, the sperm are absorbed through the skin. Inside the body of the female they swim to the ovary, where they fertilize the egg.

This informal approach to fertilization had the potential to produce a hybrid among different animals that were still fairly closely related. Since there are no caterpillar-like larval forms in the marine environment, I suppose that it was a randy female of one of the early insect descendents recently broken from the line of Onychophora, not long after it went on land, that mated with a velvet worm not long after it came on land. From this union there came two: the miracle of the caterpillar with all its legs (the velvet worm genome), biologically deconstructing itself in a pupa to metamorphose into a butterfly with wings (the insect genome). It is hard to imagine how such an animal could have evolved without such a hybridization. We will see this in greater detail when we formally get to the insects in evolution. It is one of the wonders of creation.

Another hybrid possibility is found among the Ascidiacea commonly known as the sea squirts. They remind us of the two behavioral states of bacteria: the planktonic state of free-living individuals and the colonial state. The free-living larval form of the sea squirt resembles a tadpole. This organism is not capable of feeding. Its role is dispersal. This organism has a notochord, a dorsal nerve cord, and a ganglion of nerves that serves for a brain. It is therefore counted as a primitive vertebrate. Once the tadpole finds an appropriate rock to occupy using light sensitivity, gravity, and tactile senses, it attaches and metamorphoses into the radially organized, invertebrate sea squirt. The sea squirt is a sac-like filter feeder that creates colonies of itself. Closely related to the Ascidiacea are the Larvacea. The Larvacea look like the tadpole form of the sea squirt, except they do feed and they never change.

The conventional explanation for these similarities is that Larvacea are examples of a well-known aspect of evolution called neoteny. "Neoteny" means retaining juvenile traits into adulthood. We will see this later on, but here it is unconvincing. When the tadpole metamorphoses into the sea squirt, it eats its own brain. This is not delayed development; it is a radical conversion. A dramatic reorganization called ooplasmic segregation can even been seen in the egg following fertilization. The fertilized zygote ends up in what is called the animal hemisphere while the endoplasm ends up in what is called the vegetal hemisphere. Williams' thesis seems the much more likely explanation for this bizarre behavior than neoteny.

Although evidence for Ascidiaceas can be found in the Cambrian period, Williams supposes that these sorts of hybridizations occurred after the greatest mass extinction of life, the Permian extinction, which occurred 250 million years ago. It is associated in time with the greatest period of volcanic activity in the last half billion years in the area we now call Siberia. It may be associated with the breakup of Pangaea, the original continent. It took 50 million years for life on land to fully recover its diversity. It took 100 million years for marine life to recover. In contrast to the Cambrian explosion that produced thirty-seven new metazoan animal phyla, nothing of the sort would happen at 225 million years or at any other time down to the present. That stage of evolution was over, but hybridization was occurring.

The hybrids proposed by Williamson are counted as subphylum variations at most, not new phyla. The Darwinian rules become quite blurry at the microbiotic level. Some cases of endosymbiosis are accepted. The endosymbiosis of blue-green algae to create the plant kingdom and the endosymbiosis of mitochondria to create the animal kingdom come to mind, even though that is a violation of principle. Williamson's hybridization of genomes is another story. In 1940 geneticist Richard Goldschmidt proposed that there could be complete changes in the primary pattern or reaction system of an organism above and beyond the point mutations of Watson and Crick. This was received with raucous contempt and labeled "the hopeful monsters of evolution." The laughter still echoes down the long hardened halls of academia, and there are very few that would care to be on the receiving end of such contempt.

Darwinists are also loath to come to terms with the reality of yet another recent development in molecular biology, epigenetics. If the story of polyploidy, Hox genes, and hybridization is the story of punctuated macroevolution on a grand scale, epigenetics is the story of detailed refinement, the microevolution of the individual genome in the stream of evolution. The story of epigenetics really begins when the Human Genome Project discovered that there are only around 30,000 genes located on twenty-three human chromosome pairs doing the work of building and running the body, and we must also keep in mind that less than 2 percent of the total genome is involved. That number is now down to around 20,000. This was far less than the 100,000 genes that were thought to be necessary. It is half the number of genes that have been counted in the bacteria found in the human gut, called the colon microbiome.

Part of the solution to the dilemma of DNA shortfall came from an unexpected direction—RNA. RNA may have begun life being matched to the shape of protein molecules in order to make new ones and then being collated into DNA to perform the role of memory. RNA then took on the intermediate role of messenger, taking the stored message of DNA back into the cell to produce proteins. All of this made it well suited for the role of a virus moving inertly through the environment, infecting organisms to take over cellular chemistry for its own purposes, but this is not the only role for RNA. RNA also has the role of error correction of damaged DNA. However, now we turn to a group of specialized RNAs called spliceosomes.

Spliceosomes cut apart and rejoin messenger RNA such that they end up making different proteins than the DNA is actually coded for. The current record for different proteins produced by spliceosomes from a single gene site is 38,016. With spliceosomes in the house, the actual protein produced by DNA is only fully specified outside the nucleus in the environment of the cell. Spliceosomes are a reflection of the environment of the cytoplasm that, in turn, is in communication with the outside of the cell. Hormones from the brain and from other organs in the body, or pheromones from another animal or from completely non-biological sources can induce the formation of proteins for which there is no genetic record in the DNA. Alternative splicing continues the theme of non-centralized control of the cell's metabolism.

If spliceosomes augment the expression of DNA, blocking the expression of DNA is just as important. The simplest control mechanism to block DNA discovered so far uses one of the most primitive of organic molecules in the nature, methane. The attachment of a methyl group to a section of DNA blocks expression of that gene. Bacterial DNA has been found capped with methyl groups. These help to assist in gene repair by marking accidentally repeated sections of the genome. Methylation helps to time gene expression, and the accumulation of such marking gives a bacterium its unique identity. Methylation differentiates native DNA from introduced DNA, which, in turn, enables a primitive immune system to operate against infection from bacteriophages.

But how do these methyl groups know where to go? RNA of course. Craig Mello and Andrew Fire won the Nobel Prize in 2006 for discovering transcription RNA. These non-coding RNAs block the transcription of RNA messages into protein. Similar non-coding RNAs identify the placement of methyl groups to block the production of proteins in the genome itself. The Encode project at the University of Queensland in Australia has begun to identify exactly what the genes in the human genome actually do. They find that a staggering 93 percent of the genome produces RNA transcripts whose role is to control normal DNA activity. A sizable proportion of the so-called junk in the genome is only junk if you follow the dictates of the Big Dogma.

Orchestrating the activity of DNA is at the heart of metazoan specialization. In the nucleus, DNA is studded with proteins that determine which genes can produce transcripts and which cannot. Timing the order of biochemical events is one of the outcomes. Embryological development appears, at a distance, to come from a point source, DNA, but upon closer inspection it turns out there is a symbiotic orchestration of genetic traditions—preadaptation—mediated through unique environmental conditions—epigenetics. The embryonic epidermal layer that will produce skin, nails, the epithelium of the nose, mouth, and anal canal; the lens of the eye; the retina; and the nervous system all have the same basic DNA. Those specialized tissues are shaped by the epigenome. From the nervous system will come the brain, but the brain itself would be nothing more than a congregation of reflex reactions were it not for the complex architecture of the brain to do specialized activities, shaped by peculiar patterns of experience by methylation into memory.

Evelyn Fox Keller, science historian and professor emeritus at M. I. T., has said that the parlance of molecular biology is in need of a complete overhaul. Watson and Crick assumed one gene equals one messenger RNA, equals one protein, period—no exceptions. There may have been a time in evolution long ago and far away where things like that actually happened, but no more. The baggage that we have to throw off, according to Keller, is the expectation that if we could find the fundamental units that make things happen, if we could find the basic atoms of biology, then we would understand the process.

Keller's criticism is broadly justified, but her solution is an empty promise. The invention of neologisms is the standard evasion for institutions that are incapable of real change. The dilemma we face goes right back to a founding father whose truth-seeking expedition was a hopscotch voyage around the globe observing mud flats here, mountain ranges there. He brought his pickled, stuffed, and poisoned specimens back home, where he attempted to put them back together again like Humpty Dumpty, using a theory that wouldn't greatly offend the gentlemen and ladies of the town. The value of such a theory has long passed. We need a perspective derived from a new founding father whose truth-seeking voyage involves going to a foreign place and staying long enough the record the whole structure of the environment as well as culture of the natives who live there, including the language they use to make sense of it. Claude Lévy-Strauss offers such a model. What we will lose by this is our inherited sense of entitlement: the burden of superiority that we drag around that our incumbent theory of evolution is careful not to violate. We will also lose its dialectical opposition, the hopeful myth of a wholly beneficent ecology and the noble savage that comes with it.

Hox genes and epigenetics are just beginning to be understood, but they already remind me of two Continental versions of evolution in the nineteenth century. Étienne Serres coined the epigenetic proposition early in the century. He made a link between comparative embryology and a pattern of unification in the organic world. Étienne Geoffroy Saint-Hilaire suggested life could have had environmental causes making transformations on the embryo rather than the adult. This was supported by, among others, Robert Edmond Grant at Edinburgh. Charles Darwin was one of his students, but Darwin was still in the thrall of Creationism at the time.

Under the influence of Goethe's *Urbild*, Ernst Haeckel restated the thesis of "ontogeny recapitulating phylogeny" in 1866. This is the version we generally remember: embryological development of the individual organism (its ontogeny) follows the same evolutionary history of its species (its phylogeny). Although Haeckel was a little too enthusiastic in his cross-species embryological recapitulation—human embryos are arrived at by first by going through the phases of being a fish, a salamander, a tortoise, a chick, a hog, a calf, and then a rabbit—the general ideal of a broad unity across the living world was appropriate. His ideas were marked as foreign in certain places because they were associated with Lamarckism, including the concept of heterochrony, which is a change in timing of embryonic development over the course of evolution.

The English were way behind in microscopic biology and were not easily swayed by such speculations when D'Arcy Wentworth Thompson published his classic *On Growth and Form* in 1917. This book was a beautiful exposition of his thesis that evolution overemphasized selection at the expense of the physical laws of nature to determine the shapes of things to come. His book influenced generations of architects and artists, but it was utterly *sui generis* to the mainstream of biological thought in England. In spite of this, his accomplishment was so breathtaking he was elected a Fellow of the Royal Society, he was knighted, and he was awarded the Darwin Medal.

Twenty-five years before the Human Genome Project, it was already known that there were large sections of DNA that had no apparent coding function. James Watson suggested that apparently inactive sections of DNA might represent a timing function. "For example," he said, "inserting some 200,000 base pairs of spacer DNA between two genes separates the time when they begin to function by at least 100 minutes." This possibility, as he puts it, "is a completely open question."[6] The Cold War was still on, however, and Darwinism was locked in a death struggle with Lamarckism. Nothing would come from this intriguing observation. Epigenetics is still the dark matter of evolution.

Most modern biologists are willfully ignorant of the fact that both Charles Darwin and Jean-Baptiste Lamarck were influenced by Erasmus Darwin and that they both believed in the inheritance of acquired characteristics. Both believed in the "inner need," *besoins*, of an animal to adapt to its environment. As Ernst Meyer pointed out in an essay in 1972, the extension of this to "volition to evolve" was a mistake in translation. This occurred through Lamarckism's association with Marxism for political reasons. At this point he didn't think the difference between "inner need" and "volition to evolve" was worth worrying about. Understanding our own motives and prejudices was more crucial. Old-guard Darwinists still cling to the hope that the inheritance of acquired characteristics is a Communist plot, and that transcription RNA has no purpose and is made only to be thrown away.

For the moment issues like Hox genes, macromutation, and epigenetics can be finessed by never mentioning the word "Lamarck" and by avoiding the evolutionary implications. This is fairly easy since the new generation biological scientist is only vaguely aware of the evolutionary dialectic that developed in the late nineteenth century and the early twentieth century and the deadly competition that developed between them. This generation of scientists is only confronted by the tired old cliché, Creationism versus Darwinism, when it comes to evolution. The harmless neologism used to describe this area while avoiding attack from the cultural immune system is the rock 'n' roll term "evo-devo." Evo-devo manages to wipe away generations of internal criticisms about Darwinism without touching the cause of the dilemma. The other diversion that biologists use is to talk about all the new disease-curing drugs that will become available if we go evo-devo. This was how the Nova television program on epigenetics that aired in 2006 handled it. The whole show was basically about diseases that could be cured or avoided. The program did mention that epigenetics would require a revolution in the theory evolution, but not Darwinian evolution of course.

What went along with the metazoan evolution triggered by the Cambrian period was a complex organism less tolerant to basic genetic change, more vulnerable to attack. Enter the antibody. Antibodies attack viruses, bacteria, and strange proteins found in the body proper, shielding the body from invasions of germs. Although this capacity had existed nearly from the beginning, it now became a fully developed functionality of the body. Since there were so many germs still in attendance doing essential things, there had to be a roll call of friendly germs. The registration of the immune system that produces antibodies comes at an early stage in development. The proteins, viruses, bacteria, and even eukaryotes living with the mother define what will be allowed to pass through the defense of the antibodies and be recognized as friendly. The antibody system came to play such a powerful role that it could become too enthusiastic and attack a perfectly normal body. This is we call an autoimmune disease. We can die from such a disease. Some invading viruses and bacteria learned to hide in the body until the immune system is compromised by poor diet or age, whereupon they come out of hibernation and do their business.

The most effective and least disruptive moment for genome correction in metazoan organisms comes at reproduction with meiosis. We saw this most highly developed in *Paramecium*. In metazoan organisms the role of the micronuclei is played by the germ cells, and the role of the macronuclei is played by the body cells. As in the case of the macronuclei, whose role is to represent the normal vegetative processes of the cell, there is no telomerase to be found in the specialized tissues and organs of the metazoan body cells. In *Paramecium* genetic restoration is done in-house by the micronuclei. In metazoan organisms the whole body is discarded and recycled. It is the role of the germ cells to recombine and reproduce the body proper in order to replace an expired body.

The new animal and plant forms generally have two markedly different sex cells. Sexual gametes are cells that are at least as different from each other as many species are from one another. The gamete that is larger and carries the bulk of the food reserves is called the egg and is traditionally labeled female; the one that is produced in very large numbers, is motile, and very small is the sperm and is traditionally labeled male. The large, enriched egg is an attractive source of food for nearly any other form of life and must be hidden or otherwise protected. The egg is placed in a propitious environment by the adult who is usually called the female. Selecting an appropriate choice for the placement of the egg is one of the foremost skills of the adult female.

In humans the egg never leaves the protective confines of the adult female body, yet the problems connected with fertilization are many. As many as 50 percent of all pregnancies are spontaneously aborted within two to three weeks of conception, before the woman even realizes that she is pregnant. It is estimated that half of these abortions are due to chromosomal defects that were not corrected at meiosis. It is estimated that without these natural abortions, 12 percent rather than 2 to 3 percent of human infants would have birth defects. In the next stage of development, from three to eight weeks, is a period when cell populations are establishing themselves into primitive organs. During this period the embryo is still sensitive to genetic abnormalities, but now environmental influences are paramount. It is during this period that gross structural birth defects can be induced by drugs, alcohol, poor diet, radiation damage, and other physical abuse.

Even among the most complex organisms of the animal kingdom, the capacity for metabolic hibernation, keyed to the meiotic separation of DNA, is still necessary in order that the remarkable space trip taken by sperm from the male to the egg within the mother may be achieved. Even the most elegantly complex animal organism must go back through a reduction phase that harkens back to the independent days of the protozoan. With the DNA necessary for the complex body packed away in the head of the sperm—in a state of hibernation—the male germ cell makes its highly improbable trip up the uterine delta by means of a low-tech sperm tail supplied by mitochondrial ATP. These sperm mitochondria can be seen lined up along the tail in electron microscope pictures. This image alone, fully understood, is enough to rescind the Central Dogma of DNA. The electron microscope image of primitive mitochondria organelles lined up around the tail of the sperm stand as evidence for an ancient macro-evolutionary event of endosymbiosis and a naked condemnation of the hopeful myth of Darwinism as clearly as the animalcules of Hartsoeker stood as a condemnation of Aristotelian Spermism—for those who can see it.

The severe challenge faced by sperm requires that they be produced in large numbers. Up to 10 percent of human spermatozoa have observable defects. The need for a clean genetic record leaves weaker sperm by the roadside. Only the fittest sperm will find and penetrate the egg to complete the next cycle of active, fully expressed adult life. When we see the mob of sperm surrounding the corpulent egg through a microscope, each attempting to penetrate, it is only natural to see males competing for access to the female, each attempting to impress its own set of genes. It is only natural to suppose that the sperm with the best goods will win the hand of the fairest egg. It is only natural for Darwinists to see this as a venue for evolution—sperm competition. But this is not the case. The goods that the sperm carry are in a state of hibernation; the operating system of the metazoan organism is not being read or tested at all. Sperm may be carrying the plans for a body type that is on the verge of extinction, or it may be the next best thing. On the other hand, a perfectly healthy genome may be undone by a deficient sperm.

The thing that is important here is that the sperm are active enough to make the hazardous journey from male to female. Fertility is on the line. And when sperm approach the egg, a chemical trigger opens a calcium ion channel that exists only in the tails of sperm. This turns on a last frenzy of activity that is absolutely necessary in order that the sperm can make that final lunge as it lyses its way through the tough, outer covering of the egg. This is a competition that requires them to be fit. At the other end of the cycle, the sexually mature body preparing its reproductive gametes for conception is equally devoted to the tradition. And as far as the somatic cells in between these two events are concerned, as Colin Patterson states in his book *Evolution*, "Our blood cells and liver cells, for example, have lost the possibility of contributing to the next generation, except vicariously, through the egg or sperm cells that they may help to nourish."[7] This is a necessary but weak connection.

Sperm mitochondria follow a strict endosymbiotic evolutionary pathway entirely of their own. The mitochondria are self-reproducing organelles that are separate from the DNA isolated within the nucleus as well as the much-reduced sperm cell. They have their own replicatable genetic information that works in sympathy with their own system of membranes and enzymatic processors. Whatever might be said about the possibilities for change in the DNA in the head of the sperm, strict conservation is the theme for the DNA of these mitochondria. Point mutations do occur to mitochondrial DNA, so much so in fact that the unique pattern of accumulated damage gives a genetic thumbprint to the organism carrying them. The time of separation between two genetic lines that no longer interbreed can be estimated by the difference in these mutations. But these mutations are neutral in terms of any organic function. Once a sperm successfully lyses its way through the membrane of the egg, it jettisons its mitochondria. The mitochondria for the next generation of sperm will come from the female. This exposes the underlying principle at work here. This is a diploid life form, life constructed on two strands of DNA, which expresses itself alternately as male or female in the same generation. The primary function of sexual reproduction is still gene correction.

Specialization in the gametes will eventually be well marked in sexual behaviors of metazoan adults, male and female. Evolution of the different body shapes of the two metazoan sexes is well expressed in the resulting phenotype. The conception of the next metazoan organism is launched by the sexual dance of life that ends in a frenzy of ejaculation, a metaphor for the actual penetration of the sperm into the egg. Even humans, encumbered by the trappings of high civilization, will heel closely to this ancient desire. Once again we see the basic premise: preserve, repair, and recycle. If there are any other advantages that may accrue to the sexually reproducing metazoans, they come after the fact, in other ways. The most interesting fact is that the metazoan organism is always reduced again to its protozoan roots where the genome is repaired and refreshed. This is survival of the fittest in the service of no evolution. And without it where would we be?

This brings us to a very ticklish issue. It is a "need to know" sort of issue as they say in politics. This need-to-know situation reminds us of the hopeful geology of Charles Lyell. It is another issue on which Charles Darwin was politically correct even though he probably knew better. It is the issue of incest. Creationists and Darwinists are in the same boat on this issue. The realpolitik knowledge on inbreeding is not known to most people, including biologists, especially most evolutionary biologists. The only people who have needed to know the reality of this subject in the past have been animal breeders and veterinarians. In the modern academy, geneticists who are writing textbooks on genetics need to know that inbreeding or asexual reproduction in general is just another approach to life. A healthy genetic condition is not arrived at by any single rule in nature.

Colin Patterson calls a spade a spade in his book *Evolution*. He makes a very concise statement concerning the long-term challenges that fall out from an overemphasis on out-breeding:

> *There are many different lethal and sub-lethal recessive mutations in the human population. It seems obvious that it would be a good idea to try to identify carriers (heterozygotes) of these genes, and to discourage marriages between them, so avoiding the birth of afflicted homozygotes. But the result of such a policy would actually be to increase the frequency of these harmful genes, because they can only be reduced in number by the birth, and selective elimination, of homozygotes; or by preventing carriers from breeding at all.*[8]

What is not understood by geneticists because of the confusion introduced by Darwinism is that meiosis or sexual reproduction is fundamentally a means for gene correction. This is the cause of what is known as hybrid vigor. Hybrid vigor is where out-breeding occurs allowing relatively inbred chromosomes to match up with better versions of themselves and the offspring have qualities superior to either parent. This is a common experience. But there are also at least 6,000 mutations that are not corrected by out-breeding, mutations that are spread by out-breeding. I cannot imagine any way that human civilization as it is presently constituted can even face this challenge, let alone be trusted to come up with a humane solution. In the next chapter we will see how other organisms have dealt with profound losses of genetic information and have evolved because of it.

CHAPTER 11

Plant/Insect Evolution

AROUND 500 MILLION YEARS ago the oceans were teeming with life. The first ammonites and trilobites had appeared. These snail-like organisms were built on a radial body plan. The first chordates that were built on a linear plan were also in existence in tadpole form. Although some of these have not changed since that time, others would evolve into creatures curled up on couches looking at television at images of their fellows taking the first steps on the moon. At that time the planet was not blue, green, and white. The ocean was a deep, dark blue, but the Earth's single continent at the time was a rusty red, heavily eroded, and overlaid with ropy, blackened lava. The planet showed its elements: brilliant white quartz sands and salt flats, yellowish sulfur beds, and purple filigrees of manganese in the shallow waters. Along the temperate margin of land and ocean were shallow bays bluish-green with algae. Fungi were beginning to occupy the beaches in grayish, greenish patches. As plants began to come on land they would carry with them nitrogen-fixing fungi in an ectosymbiotic relationship. These fungi also expanded the water absorption capacity of their plants, increasing the ability to take up minerals that will be necessary for plant life to take up life on land.

The cycads were among the first plants to occupy the land during the late Paleozoic era. Looking like ferns or palms, they are not closely related to either. These gymnosperms would sprout into great landed forests. These were so dense that little light reached the ground. Very few low-growing herbaceous plants were to be seen. The seed of this family of plants was an unenclosed ovule. Gymnosperm means "naked seed." Its seed was protected from ultraviolet radiation damage by its overarching fronds and then by the shade of the great forest itself. The earliest land plants reproduced by spores like the algae from whom they were descended, but the seed-bearing cycads evolved the largest reproductive organs in the vegetable kingdom, including the biggest spermatozoids. They are big enough to be seen with the naked eye. These flagellated spermatozoids needed water to make their trip to the egg. The swimming stage required for reproduction kept the cycads close to water, which was not a problem at that time except in the middle of the great northern continent Lurasia or the southern continent Gondwana, which were probably deserts.

Cycads can reproduce asexually by offsets or pups, or they can reproduce sexually. There are male trees and female trees. It is reported that they can change sex, although it is rarely observed. This reflects the fact the there is very little specialization in the two genders. Reflecting its new environment, the male tree produces pollen that is dispersed by wind or, later in evolution, by insects. The spermatozoid comes from the pollen and swims to the egg. Once the ovule on the female plant is fertilized, the seed takes up to a year to mature. The seed has a marvelous capacity to find new vegetative digs. The seeds float so they are dispersed by water; later in evolution, the nutritious seeds would be dispersed by bats, birds, marsupials, and rodents. Hundreds of millions of years later monkeys would eat the seeds and elephants would consume the whole seed-bearing cone that would be planted with a generous supply of fertilizer. Reproductive flexibility and no competition for resources allowed the cycads to take over the planet for 100 million years or so.

Cycads appear primitive to us because they have survived unchanged for 750 million years. Eventually, however, even their capacity to survive acute downturns by the dormant stage of the seed would not be enough to save them. They lost their major role of greening Earth not during extinction events but as a result of competition from the faster-growing, shorter-lived, flowering plants. Here is where the Second Law of Evolution would begin to run its course for the first great kingdom of the plants that grew on land. They were still dominant during the Carboniferous period around 400 million years ago when insects took to the land in great numbers to become the main representative of the animal kingdom. It was the time of the big bugs.

Naturalists in general are always amazed by the riot of life that is contained in today's tropical rain forests: two-thirds of the world's species are found within what is now a very narrow range of environmental conditions. It was under the feet of the dinosaurs in the Cretaceous period around 130 million years ago that we find evidence of the first flowering plants. They appear much diversified, without a trace of an intermediate stage. Specialists are not sure whether they evolved from the simpler gymnosperms (cycads, gingkos, or conifers) or whether they were a separate development on their own.

Charles Darwin called the sudden appearance of the flowering plants an abominable mystery. As we have seen, the environment and evolution do have to reach a certain point before other things can happen, but the more traditional explanation can also apply: the missing pages explanation. Certain life forms, it is safe to assume, were preparing for cooler times and more exposed environments on the planet before the records of paleontology were able to record it or we are able to recognize it. In the late Paleozoic there was a vigorous mountain-building period. In tropical countries at the present time we still find a remnant of that special environment, the aerie niche of the mountaintop.

It was Frederick Henry Alexander Von Humboldt who made popular the conception of zones of vegetation such as we experience as we climb into the thinner atmosphere of the high mountain. The five zones that he described on Mt. Teneriffe in the Atlantic Ocean were the region of African forms at the base, the region of vines and cereal plants, the region of laurels, the region of pines, and, last, the region of Retama that is unique to Teneriffe. The very top of Teneriffe is rocky and quite bald of vegetation. As we climb up a mountain, we move north, biologically speaking. It is very likely that the angiosperms originally evolved on these archipelagos of life surrounded by oceans of deep green, tropical forest. On these sun-exposed islands gymnosperms had better fertility rates if their seeds didn't dry out, if their seeds were within a protected ovary. These montane niches were not places that favored the preservation of fossils for the convenience of paleontology—in marked contrast to the bottomlands, where legions of cycads, gingkos, and conifers would fall in on themselves to become seams of coal.

It was the conifers that appear to have moved up the mountain furthest to become its early settlers. The conifers had already survived one genetic reversal. They had lost the capacity to produce animal-like, swimming sperm cells that the cycads and the gingkos possessed. With their less effective sperm, conifers resorted to massive amounts of wind-blown pollen to fertilize the egg without an intervening spermatazoid. Pine pollen takes three years to mature, a year for a pollen grain landing on a cone to grow in and fuse with the cell inside, and two more years for the fertilized cone to become mature. I see this as an example of devolution, a time and place where evolution loses capacity and takes a breath before moving forward again.

Such a plant type was not quite as well suited to compete with its cousins the cycads and the gingkos, but having survived the loss of active, well-developed sperm, the conifers did not need to stay in or near the damp, steamy bottomlands. It was both their slowness of growth in general and their ability to fertilize away from the water that allowed the conifers to survive in cooler environments where many tropical plants could not. Tropical plants are too quick to respond to the false spring of February or the warm day or two at high altitude. These plants get nipped in the bud with the return of the chill of night or winter. It was the lethargic pace of the conifer that allowed it to adapt to high-montane niches. The earliest angiosperms have also been found in chaparral-like communities dominated by shrubby conifers.

The willow-leaf podocarp, a conifer and a native to tropical mountains of the Andes, shows an inherent potential within the conifer family to become an angiosperm. The willow-leaf podocarp has an edible, berry-like cone. Here is a gymnosperm with its seed protected like an angiosperm. Here is a seed that is protected by an enclosing fruit. Valentine Krassilov at the Institute of Biology and Pedology Vladivostok thinks that angiospermous characteristics arose in several lineages of gymnosperms that were then accumulated by non-sexual (horizontal rather than vertical) transfer of genetic material.

I would also add that radiation from space causing chromosomal inversions, translocations, and polyploidy was still a likely possibility among the high-riding gymnosperms. Here is a tidy paradox: genetic damage inadvertently causing greater protection from genetic damage and from desiccation and predation at the same time. Eventually the angiosperms would come down from the heights to dominate plant life in the lowlands, introduce fruit, color, and perfume to the Garden of Eden, and leave fossils. The energy put into the production of fancy flowers was too high a cost for the self-fertilizing, wind-blown economy of the gymnosperms to sustain.

The most stunning aspect of the angiosperm is its gaudily exposed reproductive organs, but they are nothing without eyes to perceive them. What were those eyes looking for in the first case? Not the chicken but the egg, not the flower but the fruit, such as the fruit of the willow-leaf podocarp. With insects looking for the rare, special treat of larger fruits, plants with such fruits and slightly modified, more colorful leaves would stand out from the uniform green, relatively low-nutrition gymnosperm forest. Insects are the nervous system that would give the angiosperms their Narcissian mirror. Insects would provide the energy to cultivate the advantages of larger fruit, flower, and seed. These would be genetically expressed through the genetic accident at meiosis of polyploidy, which causes the doubling of size of the genome. The mystery of the origin of flowering plants is solved by the peculiar circumstances of the mountaintop ecosystem. This was followed by an explosive co-adapted evolution between plants and insects.

It was the insect family Hymenoptera, which includes wasps, ants, and bees, that would most actively evolve with the flowering plants. The mouthparts, digestive systems, and life cycles of these insects would shape themselves to the reproductive organs and life cycles of their favorite plants. The 80-million-year co-adaptation of the fig wasp to its desired flower hidden within the fruit is a case of mutual dependency so complete that it could almost be called a case of hybridization across the plant and animal kingdoms. Angiosperms would return the favor of highly directed and efficient insect fertilization by producing perfume for insects not so visually advantaged and nectar, a high-energy drink that can even sustain animals as large as a hummingbird.

Without the angiosperms and their fruits, seeds, grains, and nuts, without their insect agriculturalists, there would be no mammals, and there would be no humans to fall in love with the most stunning member in the angiosperm class, the family Orchidaceae. By attraction to the flower alone, the orchids continue their evolution by their co-adaptation with us. The orchids compose the largest family of plants by far. There is one species of ginkgo; there are over 300 species of cycads known so far; there are 20,000 to 50,000 species of orchids, depending upon whether the taxonomist making the estimate is a lumper or splitter. Orchids occupy nearly all possible ecological niches on land. Orchid comes from the Greek *orchis* meaning "testicle." The name was applied to the European Orchis that has paired pseudobulbs resembling testicles. European folk culture supposed that orchids were the result of animal sperm that had dropped to the ground, with each type of orchid the sperm of a particular type of animal. Charles Darwin's first book following *The Origin of the Species* was *The Various Contrivances by Which Orchids are Fertilised by Insects*, 1877. The orchids seemed to him to be the almost perfect example of his evolutionary thesis.

The orchids show a breathtaking diversity of beautiful adaptations of flower to the insect that fertilizes them. The numerous structures, seemingly created for variety and beauty alone, revealed astonishing uses to Charles Darwin's devoted and careful observations. He predicted that the Star of Bethlehem orchid—Angraecum sesquipedale found in Madagascar that has a nectary buried eleven and a half inches away from the opening formed by the petals—would have to be fertilized by a moth with a proboscis of equal length. Such a moth was subsequently discovered. It was named Xanthopan morgani praedicta to celebrate the wise man's prediction. Darwin did not see that some orchids have flower parts that reproduce the pattern of the reproductive organs of the insects that fertilize them, nor could he smell the insect-like perfumes orchids exude to attract them. These multiple attractions have allowed the orchid to stint on nectar when it is insect fertilized in comparison to bird-fertilized orchids where the quality of nectar must be high. One can freely imagine what it is about the orchid that attracts humans.

Darwin discovered that orchids could and would self-fertilize. Among these is *Coryanthes*. *Coryanthes* has a bizarre puzzle of petals that forces the fertilizing bee to slide down a ramp and get dunked in a stupefying nectar. The bee crawls out of this nectar drunkenly, reeling from side to side, bumping into everything along the way. By this exotic dance self-pollination takes place. Darwin supposed that the self-fertilized orchids that he was familiar with were verging on extinction. Under these circumstances, he made an exception to rule of incest. As he says, "it would manifestly be more advantageous to a plant to produce self-fertile seeds rather than none at all or extremely few seeds." Darwin cried hosanna upon the discovery that other orchids had been able "to resist the evil effects of long-continued self-fertilization."[1]

As early as the 1830s English botanists such as William Hooker believed that the cryptogams, meaning hidden marriage, were totally asexual. The cryptogams are the algae, mosses, and ferns. Modern botanists no longer imagine that the self-fertilization orgy of *Coryanthes* is or needs be a temporary affair. The evil effects of incestuous inbreeding are no longer mentioned by the modern botanist. The neutral word "cleistogamous" is used for flowers that self-fertilize. When botanists talk about evolution, they emphasize the loss of variety that comes from cleistogamous breeding. The cleistogamous ferns have an ancient lineage and hold the record for most chromosomes in creation, but genetic stability is not seen as a virtue. If they were less stable, perhaps they could have become an orchid.

One of the great puzzles that Darwin could not solve concerning orchids was the matter of the seed. Orchids produce an enormous amount of seed. In a northern variety that Darwin counted, there were well over 100,000 from a single plant. From some tropical varieties, well over a million have been estimated. Yet the fertility of these seeds is astonishingly low. The answer for this has come from modern research. We recall that the female sex gamete—the egg—is the one of large size that carries the nutrient for the embryo. Among the orchids, not only does the sperm tend to be infertile, but there is no well-developed female sex cell. There is no cell that is enriched with food to nourish the fertilized egg. This is why there is such a low fertility rate for orchid seed. This is a plant that should not exist, yet the family Orchidaceae rules.

For the orchid to succeed it must find the perfect co-adapted relationship. The tiny orchid seed requires certain kinds of midwiving fungi to begin their growth. Orchids cannot germinate until a specific mycorrhizal fungus infects it. These fungi come from a range of fungi including those that live on dead organic matter and those that are pathogenic. Fungi help the orchid to develop a tripartite relationship with a host plant: fungus, orchid, and host plant. After entering the plasma membrane of the orchid egg, the fungus forms a tight coil or peloton that will connect with the host plant like a placenta. Well settled and connected, the fertilized egg will realize its orientation in the field of gravity, the same force behind the creation of elements in the heart of a star and the formation of planets. Gravity marks the genomes of the dividing egg forming the primordial principle that will lead to roots, stems, and leaves. The fungus receives raw materials from the host; the orchid receives raw materials from the fungus. The orchid is dependent upon the fungus until it can send out leaves and begin photosynthesis. However, the orchid controls the growth of the fungus and continues to support the fungus even after it has become self-reliant.

The most primitive orchids are found in the mountainous areas of the tropical Pacific Ocean at an altitude around 3,000 feet. This is an area of continuously damp, cool, mossy forest. The line of demarcation is very sharp, and it is associated with the level at which cloud formation occurs. There is a profusion of mosses, lichens, ferns, and orchids at the cloud level. These orchids are not the large, showy species prized by collectors. They have small, delicate, primitive flowers. They have survived and adapted here because it was a relatively isolated niche. Orchids may have gone up the mountain to evolve with the angiosperms; they certainly went down the mountain into the fatter environment of high humidity, fungus, and insects.

The loss of well-endowed sperm and eggs in the orchid has resulted in a widespread dependence upon the good graces of life at large. Beside the flower, the single orchid product that we do consume is the vanilla bean that is so bitter it has to be married to sugar to be eatable. According to Totonac mythology, this tropical orchid was born when Princess Xanat fled to the forest with her lover after being forbidden by her father from marrying a mortal. The lovers were captured and beheaded. The vine of the tropical orchid grew where their blood touched the ground. I see this metaphor for devolution/evolution as a basis for a naturalistic theory of beauty and style: the ideal of creation, the natural consequence of creation, but perhaps this is something only a naturalist can really resonate with.

At the other end of the scale of flowering plants from the orchid is the violet. Its simple, five-petaled flower has a beauty of its own, however. The violet propagates by runners, by self-fertilizing (cleistogamous) flowers, and by sexual flowers pollinated by insects. The seeds are often spread by ants. The cleistogamous flowers shoot up from the younger, more vigorous runners. This sort of fertilization provides the opportunity for genetic correction. This may explain why the 500 or so species of this plant around the world all look the same. The sexual flowers, coming from older rootstock, give the plant the possibility variety. They do appear in a narrow range of colors. One member of the violet family, *Viola pedata*, has given up cleistogamous reproduction entirely. This bird's-foot violet produces a large, velvety, pansy-like sexual blossom and is further noteworthy for the fact that it blooms for a second time in the autumn. For a violet this is quiet an innovation. In general *Viola* is widespread, genetically stable, self-reliant, and invasive: the exact opposite of the orchid.

Curiously there is an orchid genus, *Miltonia*, which looks like a pansy, especially *Viola pedata*. Ordinarily one would simply consider this a startling coincidence. However, when we consider how botanists wax eloquent about how the flower attracts, guides, and offers handholds for their vital attendant, the fertilizing insect, perhaps we should reconsider for a moment. Is it possible that insects of similar tastes and physiognomy have intentionally selected for varieties of flowers that are a reflection of their own physiognomy—the pansy orchid and the violet pansy—which are otherwise genetically separate families? Here we return to the idea of feedback that we have already encountered over and over at the biochemical level. Although a feedback process suggested here across the boundary of the two kingdoms is widely appreciated, when the question of such a process comes up around the issue of biological evolution, it is disallowed or ignored. If we were to reduce the Darwinian theory of evolution to an algorithm and run it on a computer, it would branch and diverge forever.

We get a better perspective of evolution from the French tradition of Darwinism. The relationship between plants and insects is better understood by what Nobel laureate Jacques Monod has called the teleonomic performance of an organism: "the aggregate expression of the properties of the network of constructive and regulatory interaction, that is judged by selection; and that is why evolution itself seems to be fulfilling a design, seems to be carrying out a 'project,' that of perpetuating and amplifying some ancestral 'dream'."[2] This idea was coined to stand in contrast with the word "teleology": a creative agent planning for the ends. This Gallic refinement resonates with the Germanic tradition of the unfolding of life, but it is a poor fit in the Anglo-Saxon academy.

Monod's teleonomic performance is in the tradition of Jules Michelet in the nineteenth century and Henri Louis Bergson in the beginning of the twentieth century. Bergson spoke of *élan vital*, where the emphasis was on time not space, on choice not mechanism, on quality not quantity. The debate posed by Bergson is between an anatomical perspective, fixed in space, and a physiological perspective, moving through time. *Élan vital* is impelled by creation from within. As I see it, this is a French version of the Second Law of Thermodynamics working through the Second Law of Evolution: the interplay of energy passing through form, $e=mc^2$—that is, the act of creation. Teleology, teleonomic performance, and French Darwinism in general were addressed from an Anglo-Saxon point of view when J. B. S. Haldane said that teleology is like a mistress to a biologist: he cannot live without her, but he is unwilling to be seen with her in public.

With no grant monies to gain or lose, I will propose that the convergence of *Miltonia* and *Viola* is engendered from an ancient connection to an ancestral dream, the dream of glorious conception, which is best realized and in some cases only realized by symbiotic cooperation. In this case an insect is selecting for its ideal mate in the nectaries of a flower. Is it really all that surprising to suggest that organisms may choose things that are attractive as well as things that are necessary? Is it a surprise to discover that sometimes organisms may be attracted to things at the expense of necessities, and will do whatever is necessary in compensation? Darwinism strangles anything of this sort at birth.

Returning to the mossy, high-altitude, temperate-climate forest of the Pacific, we find showy species of *Rhododendron*, raspberries, violets, and, most peculiar of all, spectacular varieties of *Nepenthes*. *Nepenthes* are the pitcher plants. The feedback loop between *Nepenthes* and the insects has an extra turn in it. As we peer down into the pitcher of *Nepenthes campanulata* in a remote area of Borneo, we find the result of its fatal attraction. *Nepenthes* catches moisture from the dripping forest in its pitcher. In the nutrient-deficient soil occupied by the pitcher plants, it is the insects that get trapped in this pool of water that compensate for the deficiency. The pitcher is a digestive organ for this carnivorous plant.

The diet of this plant consists of ants, cockroaches, termites, crickets, beetles, snails, and scorpions. It even includes the occasional mouse. While orchids depend upon the intimate relationship with a fungus to compensate for the nutritional deficiencies of its egg, *Nepenthes* thrive on the overflow of the cornucopia of insect life that is swept into its pool of forgetfulness to compensate for its lack of an ectosymbiotic relationship with nitrogen-fixing bacteria among its roots. Among the North American pitchers, enormous numbers of ants are entrapped, killed and digested including *Cremastogaster pilosa*. This ant frequently nests in the dead pitchers of a plant whose still active green pitchers are busy killing them off in great numbers not far away, but this isn't the end of the pitcher story.

Back in Borneo we turn to the *Camponotus schmitzi* ants. They have attained a remarkable accommodation with nature and a high degree of notoriety among entomologists. *Camponotus schmitzi* make their nests in the hollow leaf tendrils of *Nepenthes campanulata*. They feed themselves by scavenging insects out of the digestive fluid of their pitcher plant home. A single *Camponotus schmitzi* will swim out to a trapped cockroach, for instance, and drag it back to the edge of the pool, where other *Camponotus schmitzis* will carry the drugged insect up the slippery sides of the pitcher. The arduous ascent of two inches can take more than an hour. The scavenging party then dismantles the cockroach in order to eat it. The leftovers are tossed back into the pitcher.

Charles Clarke, the definitive expert on the *Nepenthes* of Borneo, has concluded that *Camponotus schmitzi* help to keep a member of the *Nepenthes* family, *Nepenthes bicalcarata*, from overeating. Too many insect carcasses in the pitcher can upset the chemical balance of the digestive pool, and this can kill the plant. What goes around, comes around. Whereas the primary stages of biochemical evolution were completed within and between the founding cell lines of microbial life, the powers of cooperation and movement in metozoan life forms opened up amazing possibilities for cross-species interdependencies.

The outburst of the flowering plants is correlated with larger amounts of DNA than are found among the invertebrates and even the higher vertebrates. Polyploidy seems to have been a common mutation that was tolerated. The newly mutated plants were not much different at first; they were slightly larger, especially in the flower and the fruit. By analogy to the case of modern, domesticated plants, I speculate that the larger reproductive organs of those plants were not well suited to their business without help, but they had the potential to be considerably more attractive to insects so help was available.

Most insects know what they like and stick to that, but since they do have nervous systems, they do have the capacity to make choices. I suggest that some of the new relationships with mutated plant forms were simply obvious to all, but other relationships were originally formed by the occasional genius insect. Carpenter bees, for instance, have been observed cutting holes in the base of a flower whose nectaries were too far removed for the bee's mouth parts to reach by normal methods. Honeybees have then been observed following up on the work of the carpenters. Having seen innovations of this magnitude, it becomes obvious that most of the apparently habitual behaviors we see in nature have been insights by individuals at some point of time. Shape-shifting is an essential part of the act of creation as well as in the cognition of ways to get around it.

Orchids show the results of shaping through a symbiotic relationship with insects. *Nepenthes* show the results of shaping in order to prey upon insects. There are insects that clearly show the results of shaping by the predation upon themselves by other insects. Among these are the stick insects, Phasmida. The stick insects are the largest and longest lived of all insects. Unlike the mantises that they resemble, the stick insects are vegetarian. Of all the insect groups kept as pets, they are the most common. This reason for this is that they are large and they are slow. They are curiously shaped to the purpose of mimicking twigs. The young ones even mimic the midribs of leaves or blades of grass. Some stick insects are able to change their color from green to brown to match the twigs they are resting upon. Even the eggs of the stick insects are deceptive. They look like seeds. The deception is convincing enough that ants that collect seeds sometimes collect stick-insect eggs by mistake. Here is an insect whose survival strategy is to disappear.

Some stick insects reproduce by parthenogenesis, that is, the animal version of cleistogamous or cloning reproduction. They have largely or completely dispensed with males, although the male genome has not completely disappeared. Gynandromorphs are common. Gynandropmorphs have both male and female characteristics. The exception to this is the Lord Howe Island stick insect that has been called a land lobster or walking sausage. Once upon a time, on their isolated island, these very large insects were brought to near extinction by the introduction of rats. The Lord Howes are not so invisible. One family survived on Ball's Pyramid, a tiny islet, living under a single *Melaleuca* shrub. The surviving Lord Howes form gender bonds. The male follows the female around and sleeps with her at night with three legs wrapped around her, for the snark is a boojum don't you see.

All aspects of the life cycle of the stick insect are aboveboard and aboveground, and it works. In general courtship routines, with all its competitive busyness, as well as the nurturing of eggs and larva, are reduced to a minimum. The fertile female dumps her cloned eggs on the green world below her perch in an act that makes a sound like rain on a tin roof. They are of no further concern to the parent. Maximum egg production is the thing with the stick insect. The nymph comes out of the egg fully formed like the adult, although small. The nymph goes through a series of molts to reach the size of an adult. There is no evidence of endosymbiotic hybridization at some earlier stage. As far as success is concerned, these insects occasionally get carried away and defoliate whole areas of jungle. Starvation, not predation, is what brings the species back into balance, when there are too many sticks and not enough leaves.

Some species of stick insects have gone back and forth from winged to wingless varieties up to four times over time. This suggests that flight is of marginal value. Some have suggested that flight has been incited by the desire for sight. In animals with fixed eyes, vision quickly fades away in the absence of movement of the eye or the object of sight. Flight brings visual sensation to life, which must be exciting at first. But flight reduces the power of invisibility. In the absence of flight, the stick insects get into a slow, rocking movement in order to see you better, but with less attention being drawn to them since the movement can be mistaken for foliage being blown by the wind.

One of the earliest instances cited as a case study for Darwinian evolution involved another case of camouflage. This was the British peppered moth,

Biston betularia. The typical form of this moth has pale wings with black markings. Resting on a lichen-covered tree trunk, it is almost invisible. In 1849 a single melanic or dark variant was caught near Manchester in England. The color pattern of this moth was the inverse of the typical form and it stood out, making it vulnerable to predation by birds. With the increased population and industrialization of the Manchester area came the death of lichen that encrusted the trees, leaving soot-darkened tree trunks instead. By the 1870s melanic moths were still rare. In the 1880s the dark form began to outnumber the pale form; by the end of the nineteenth century 99 percent of the moths collected in Manchester were dark. They had changed their phenotype or appearance to one that was better suited to survival. So far, so good.

Later it was remembered that the caterpillar of this species was capable of adjusting its coloration. In fact, it can adjust its coloration to from purplish-brown, when perched on a smooth, shiny birch twig; to greenish-gray with faintly granular coloring, when on an oak twig; to mottled, dark gray and white, when on a twig covered with lichen. The caterpillar was able adjust on the fly, as it were. Poulton, writing in 1903, supposed that these were genetic types arrived at by natural selection at some other time. In his classic overview *Insect Hormones*, V. B. Wigglesworth supposed that the different insect forms, larval and adult, were the result of independent evolution. The caterpillar of this species does look like and changes color like a stick insect. What he meant was that the different expressed color forms were latent in the same genome, taking different routes of expression. He was very close to, but did not make the leap to the principle of hybridization.

In any case the color changes in the caterpillar were evoked by switching between different variations of a gene, alleles, in accordance with the light impressions converted to neurological activity mediating a hormonal response. The caterpillar is able to perform a relatively rapid change. Such flexibility is lost in the adult form that can change only from one generation to the next through the gross selection of bird predation. The two adult morphs of the peppered moth are varieties of expression from the same genome. The two varieties were different phases, not a different species. If either of these phases were lost, its survival capacity would be dramatically reduced. Darwinists who are not close to the issue refuse to give up their home-grown but specious application of evolution for fear of handing ammunition to ever-diligent Creationists.

Termites showed up after the Permian extinction and the formation of the great gymnosperm forests. They are the premier metazoan detrivores. They clean up downed wood and other plant matter on the forest floor. They are related to the cockroaches and, more distantly, to the stick insects. Even though the termite genome does have information for enzymes to break down cellulose, their main source of food, they have sub-contracted this metabolism out to symbiotic partners living in their gut as well. Some termites have rudimentary flagellated microbes not much different from sperm cells helping them out; others have more advanced assistance from prokaryote cells or symbiotic relations from a combination of both. The symbiotic protozoa have their own ectosymbiotic relationship, bacteria embedded on their surfaces that produce the necessary digestive enzymes.

The most enigmatic termite symbiont is *Mixotricha paradoxa*. *Mixotricha* has four anterior flagella; these are used only for steering. Movement comes from symbiotic hair-like Triponema spirochetes attached to the cell surface. *Mixotricha* also has rod-shaped bacteria in an ordered pattern on the surface of the cell. In addition it has endosymbiotic bacteria acting as mitochondria and at least one other identifiable symbiont. Margulis and Sagan consider *Mixotricha* to be a composite organism with five genomes. It takes the eukaryote termite genome to gnaw its way through a rotten log; it takes at least four more symbiotic genomes to digest it.

As a group the termites are constantly feeding off of each other front to back, thereby passing on their acquired endosymbiotic bacteria to new generations joining the digestion train. Evolving in the warm, moist, largely untrammeled tropical existence of early life on land, modern termites are extremely challenged by a variable climate. They have solved the challenge behaviorally by building termitaria. Whether they originally built their termitaria to control the environment to a very narrow range of temperature and humidity or to protect themselves from hordes of ants doesn't matter. Both were challenges to be solved. Eugène Marais, who wrote *The Soul of the Ape*, was ahead of his time with his observations of termitaries in South Africa. He concluded that all the lives of its castes and its ranks, in countless number, added up to a single termite organism.

It is widely agreed that all the Hymenoptera are descended from the Xyelidae or sawfly. The sawflies are the earliest Hymenoptera to show up in the fossil record, appearing 200 million years ago. They look much the same as their modern descendents. There is a striking difference between the larva of the sawfly and the rest of the Hymenoptera, however. The larva of the ant is fragile and underdeveloped. It must be attended to. The larva of the sawfly is a lusty caterpillar that most people mistake for the larval form of a moth or a butterfly. These caterpillars can be pests that denude pine forests. The metamorphosis of this caterpillar larva produces imagos that are solitary, weak fliers that only live for a few days.

Something seems to have gone terribly awry in the wasps, ants, and bees. The caterpillar stage of the sawflies that is so self-reliant that the adults needed to be nothing more than reproductives, sperm with wings or eggs with wings, has gone missing. The first stage of this genetic devolution can be seen in the horntail or wood wasp. The adult female has a remarkably long ovipositor. With this fragile looking instrument she drills into an older tree or the wood of a house. She inserts her extremely large egg. From the egg hatches a whitish larva with capable jaws. For several years this larva excavates its way through the wood, ending just below the surface. At this point pupation occurs. From the pupa will hatch the adult. The caterpillar stage is much reduced, but in compensation, the adult female is considerably more skilled than the sawfly female. The females of the parasitic wood wasps take this one step further and lay their eggs where they can attack the larvae of wood-boring beetles.

The horntails and the parasitic wood wasps look like sawflies with a continuous body structure. There seems to be an abnormality connected to the Hox genes in the rest of the Hymenoptera. A 'waist' is formed between the first two segments of the abdomen, and the first abdominal segment is fused to the thorax. Another behavioral solution paralleling this continued genetic devolution is seen in the solitary bees. Among the families of the solitary bee, the female digs a hole in the ground or in wood and builds a nest at the bottom of the hole. She then stocks the hole with food for her immature larva. She lays her egg, and that is the end of her parental involvement. The males in this family of Hymenoptera are smaller and less developed. Like all Hymenoptera males, they have but one role to perform. The halcitine bees act as solitary individuals for the early part of their year of life. The rest of the year a few related females remain in the nest, with one usually taking on the reproductive role. The queen controls the sex ratio, producing only a few males.

The parasitical wasps represent yet another turn in the soap opera of Hymenoptera. Certainly the most astonishing parasitical life cycle involves the wasp *Nasonia vitripennis*. The female wasp lays a fertilized egg (female) on a whitefly that acts as the host to this parasite. When the egg hatches, the larva eats the whitefly from the inside out. It then pupates and metamorphoses into a female wasp. The female *Nasonia* wasp lays an unfertilized egg (male) on a newly hatched sister. The egg hatches and eats the *Nasonia* wasp from the inside out. It then pupates and metamorphoses into a male wasp. Female wasps are produced from being laid on whiteflies. Males are produced from being laid on female wasps of their own kind. Fortunately only a few males are needed to fertilize the females and start the process over again. It gets really dicey now.

John Werren, professor of biology at the University of Rochester, et al. discovered that there are some males produced from fertilized (female) eggs. It was found that these males carried a supernumerary chromosome. This supernumerary chromosome destroys some of the paternal chromosomes, leaving a haploid (male) wasp. Supernumerary chromosomes are ubiquitous in the eukaryotes. They are thought to have once been functional chromosomes. They are more common in out-bred lineages, less common in inbred lineages. Genetic elements such as Werren's supernumerary chromosome in *Nasonia* are called "selfish chromosomes" since they convey no known advantage. They are a home-grown parasite.

The discovery of selfish chromosomes caused a tizzy of excitement in light of Richard Dawkins' "selfish gene" hypothesis. However, a recent study by Tom Price and Nina Wedell, School of Biosciences, Exeter University put things to right. Selfish genetic elements (SGEs) are shown to be frequent sources of reproductive incompatibilities. There is widespread evidence that, "SGEs are associated with reduced fertility in both animals and plants, and present some recent data showing that males carrying SGEs have reduced paternity in sperm competition."[5] The SGEs are not the beginning of a new superior species, but they may be the end of the line for an already challenged species. SGEs produce male *Nasonia* wasps where none are needed.

The next unexpected turn in our story is that these SGEs-damaged wasps can be blocked from spreading by bacteria. It is estimated that 66 percent of all arthropod species are infected by *Wolbachia*, a bacteria that lives in the cytoplasm of the cell. We are used to thinking that bacterial symbionts contribute useful attributes to eukaryote organisms, such as photosynthesis, chemosynthesis, nitrogen fixation, synthesis of vitamins and essential amino acids, methanogenesis, cellulose degradation, and luminescence. The understanding was that even outright parasitism should evolve toward a more benign state over time, since helping the host would help to sustain the invading parasite. Up until recently *Wolbachia* has been seen as wholly destructive. *Wolbachia* had been found to feminize genetic males in isopod crustaceans, to actually kill off males in some insects, and to have been responsible it was supposed for creating all female ant tribes.

The challenge to understanding *Wolbachia* was that we limited our considerations to what it could add, rather then what it could subtract. *Wolbachia* has been found to confer resistance to RNA viral infections in *Drosophila*. It seems to do this by horizontal gene transfer across host species like the sharing of genetic information that bacteria do among themselves in the wild. They also confer resistance in another manner. Research from Werren's lab has shown that the spread of the SGEs in *Nasonia* was blocked by *Wolbachia* creating cytoplasmic incompatibility. Like the different breeding populations of *Paramecium*, *Wolbachia*-infected wasps cannot mate with non-infected wasps. The result is that wasps carrying SGEs will eventually fade out because of their reproductive insufficiency. A kingdom was lost for the lack of a nail; a genome was lost for the lack of an infection. Due to a developing metabolic dependency, some parasitic wasps can no longer even produce eggs without a relationship with *Wolbachia*.

Parasitical wasps represent a secondary stage of genetic devolution in Hymenoptera, and they appear among most of the Hymenopteran families. In this group not only does the male continue to perform his most reduced role, if he even exists, but the female has also lost most of her maternal instincts. A sub rosa relationship with *Wolbachia* protects the parasitic wasp from a rubble of malfunctioning supernumerary genes in a generally collapsing genome. As with all of the Hymenopteran families, there has been a great need for behavioral solutions—or was it the other way around? Were behavioral innovations what allowed for the passing of genetic tradition?

In summary, I propose that the Hymenoptera were founded by a hybrid ancestor. This large order of insects represents a crazy mosaic of genetic interdependencies acquired horizontally and vertically. This accounts for its unique stages of development and probably also accounts for the its poorly defined sex determination. The radiation from the founding group, the sawflies, came about from a disordering of the hybrid genome on many occasions, resulting in the loss of a competent caterpillar stage, which is replaced by larva that are dependent on competent adults. I suppose that a considerable amount of the remarkable social behavior in Hymenoptera is a reflection of lost or confused genetic information.

The ants appear in the fossil record around 110 to 140 million years ago in the Cretaceous period after the rise of the flowering plants. Biologist Manfred Verhaagh of the Karlsruhe's Natural History Museum reports the discovery of a miniature, wasp-like ant like no other ant. It was discovered in the Brazilian rainforest. The oldest fossil ants have been found encased in amber, the resin of the conifer tree, and they look just like today's ants. I suggest that the ants have launched their evolution from wasp-like ancestors on the exposed Mesozoic mountaintops where the conifers were taking hold. Being exposed to cold and cosmic radiation, they learned to build underground nests to protect their vulnerable larva. Today there are officially more than 12,000 different species of ants, with more being discovered every day.

The most primitive of the ants still with us are in the tribe Ponerini. These ancient species are the dominant group in Australia, where other odd relict species have been preserved, such as the monotremes and the marsupials. One member of the Ponerini tribe has a single pair of chromosomes, the smallest number found in any animal. The bulldog ants of the Ponerini are considered the prototype of all ants. They are located on isolated and desiccated islands off the mainland. They build simple nests. The larva of the bulldog ants are more developed and are fed undigested insect parts that are cut up for them by adults, in contrast to the regurgitated food the higher ants feed their young. Bulldog ants pupate in silken cocoons from which they are able to free themselves as young adults, unlike the higher ants. The founder of a bulldog ant nest, the queen, appears to have no privileged position in the community. She is not attended to by a bodyguard, and as the female workers are often fertile, there is little to distinguish between the queen and the workers. The bulldog ant colony has none of the unique specialists of the higher ants. These ants are carnivorous foraging for millipedes and termites. Adult bulldog ants are described as being more like a pack of dogs than the well-ordered armies of the higher ants. Individual bulldog ants, however, act on their own initiative much more so than the ants of higher civilization.

I suppose that at some point ants migrated into the jungle lowlands in their rudimentary social form. As described by Thomas Belt, a naturalist who visited Nicaragua in 1874, "When we see these intelligent insects dwelling together in orderly communities of many thousands of individuals, their social instincts developed to a high degree of perfection, making their marches with the regularity of disciplined troops, showing ingenuity in the crossing of difficult places, assisting each other in danger, defending their nests at the risk of their own lives, communicating information rapidly to a great distance, making a regular division of work, the whole community taking charge of the rearing of the young, and all imbued with the strongest sense of industry, each individual labouring not for itself alone but also for its fellows—we may imagine that Sir Thomas More's description of Utopia might have been applied with greater justice to such a community than to any human society."[3]

Unlike ants that live outside of the jungle, these ants have no permanent abode. The jungle is a movable feast. The jungle itself is their anthill. From the classic text *Ants* by William Morton Wheeler we hear, "Their vast armies of blind but exquisitely cooperating and highly polymorphic workers, filled with an insatiable carnivorous appetite and a longing for perennial migrations, accompanied by a motley host of weird myrmecophilous camp followers and concealing the nuptials of their strange, fertile castes, and the rearing of their young, in inaccessible penetralia of the soil—all suggest to the observer who first comes upon these insects in some tropical thicket, the existence of a subtle, relentless and uncanny agency, directing and permeating all their activities."[4] Sean Brady of Cornell University identifies thirty species of ants with legionary behavior and says they share identical genetic markers, meaning they have a common genetic lineage. From the outside the legionary or army ants are king of the jungle. In the wet tropics nearly all of nature's animal forms flee in terror from these ants if they can. Those that cannot escape will be reduced to their skeletal remains in short order. The only animal form that can hold the ants to a standoff are another social insect, the termites.

Like the termites, the ants have had to adjust to the shrinking wet tropics and learn to adapt to hot-dry zones and then the temperate zone. Some ant societies continued to be carnivorous, some became harvesters and collected seeds, while others became agricultural and raised fungi in their nest. Some ant societies husband aphid eggs over the fallow season of the year and then carry newly hatched aphids out of the nest when the warm, wet season brings the return of their preferred plant foods. These aphids are milked by the ants for the honeydew that they excrete. Like termites, ants construct their nests for effective temperature and moisture control. Ants that store seeds prevent their germination by bringing them up to the surface to dry. Ants go to war with other ants, freely sacrificing their lives for the common weal. Some ant tribes fight huge, grim wars where thousands of individuals tear themselves to pieces. Yet that most excellent observer of ants, Pierre Huber, also saw ants chasing and pretending to bite each other like so many puppies. He is also credited with noticing slavery among ants.

Ants, like humans, are incapable of hibernating to get through seasons of poor resources and bad weather. We are both capable of getting involved in all parts of the reproductive cycle the year round. In a tropical environment this is not a problem. Outside the tropics the availability of food changes the rhythm and shape of the anthill. Wigglesworth states that among some of the tribes of Formicini the determination of the queens is in part due to the season in which the eggs are produced. Winter eggs are large and rich in RNA, whereas summer eggs are smaller and not as rich in RNA because they follow the lean season. Winter eggs will be developed into queens; summer eggs will develop into workers. Among the ants a poor food season results in more workers and fewer queens. What a splendid population control. An untoward disturbance in the seasonal rhythm can produce mutant forms. If the larvae from enriched winter eggs on the way to being queens are fed poorly, or if the larvae from poor summer eggs on the way to being workers are feed too well, intermediate types or intercastes are the result. Here we see obvious epigenomic factors at work. Although the genome remains the same, it is marked differently by the seasonal availability of food. On the other hand, there are some ants where there are genetic differences in the worker castes. This is hard for a Darwinist to explain, since selection is not supposed to occur with these ants.

There is a range of reproductive fecundity between queen ants of different tribes. Some female queens are quite large and well nourished. These queens are capable of self-sufficiency when founding a new colony. They dig the nest and raise the first few generations of workers. This is accomplished out of their own flesh and blood, since the queens do not eat during the entire time. This is the case for the fungus-raising ants of the tribe Attini. The Attini ants have actually learned to create the perfect conditions for raising fungi. They use the fungi to digest leaves that they collect and bring to the fungitorium. Attini live off their fungi entirely, and live well.

In the tribe Formicini the queens are very small, that is, without yolk in a manner of speaking. Like the orchid that produces volumes of seed, such Formicini queens are produced in large numbers in order to compensate for vicissitudes that follow from this genetic deficiency. Rather than establish their own digs, the tiny Formicini queens of *Formica exsecta* and *Formica rufa-microgyna* enter the colony of a related Formicini tribe. They oust the local queen and use the local laborers to raise and nurse their own larvae and pupae. Among the *Formica sanguinea* another tactic is used. Workers go out and raid other nests for workers. This apparent slaving behavior is believed to have evolved out of the habit of Formicini of adopting recently mated queens into established colonies. In long-established slave societies the native workers lose the capacity to do any work at all, like the Spartans of ancient Greece, and become completely dependent on slavery. Only northern species of ants seem to have the need to be slavers.

Sex determination varies widely among the evolutionary lineages of the animal kingdom. One important method is called haplodiploidy. It appears in 20 percent of all animal species and all of the 200,000 species of Hymenoptera, including the ants. In haplodiploid reproduction the male is a haploid organism coming from his father's sperm. The male is essentially cloned. The female is a fertilized diploid organism having two strands of DNA, one from her mother and one from her father. To the extent that her sex is genetically determined, it comes by way of a single, combined location on both strands of DNA. This is called complimentary sex determination. After a male fertilizes the queen, she stores the sperm in her spermatheca. Over the years, depending on conditions, the queen releases the sperm, controlling whether it will fertilize her eggs or not. By this control she determines the sex ratio of the anthill. For the most part she will fertilize her eggs and produce competent females.

Local exceptions to this rule, however, are common, even ubiquitous. Among some ants males have disappeared completely and the females reproduce parthenogenically. The crazy situation of sex determination in the Hymenoptera has been described as an evolutionary aberration and a victim of runaway selection. In their summary on the subject, Michael Mahowald and Eric von Wettberg at Swarthmore College state that there is no model sufficient to explain sex determination in any more than a subset of the group. Sex determination seems to be a reflection of the peculiar life histories and conditions of individual species and a dispersed collection of genetic markers for sex throughout the genome rather than a well-established, sex-determining region such as we find in ourselves.

Entomologists worry about the inbreeding involved here of course. It leads to a lack of biological variety they say. And it does. We would hardly recognize our own ancestors from a hundred million years ago, yet today's ants look much like the oldest ants we have found buried in amber. The bulldog ants of Australia were originally described as a single species since they all look the same, but it turns out to be several distinct species, as far as chromosomes are concerned, having chromosome numbers of 1, 9, 10, 16, 24, 30, 31, and 32. Genetic variety does not seem to be part of the plan for ant evolution. Since it has been estimated that ants represent the largest total biomass of any animal by weight on the planet, I guess the concerns about inbreeding are exaggerated. There are different ways to go on the path of creation and the scenery is different for every one.

That being said the difference between having one set of chromosomes (a male) or two sets of chromosomes (a female) is quite striking. The male takes no part in building, provisioning, or guarding the nest, or in feeding the workers or the brood. The half-wit male has only one role in complex ant society, to impregnate young queens. Here we see fundamental behavior of sperm physically recapitulated in the multi-cellular organism. The male is produced only in the appropriate mating season, and in great numbers. The life span of the male is a week. The life span of the queen, by contrast, is years, as many as fifteen years. It is in the male that we see the ant's sawfly ancestry with its wings, flight muscles, and disproportionately large eyes. Haploid organisms in general go through their life cycles more rapidly, carry a smaller mutational load, and don't adapt as well.

Males appear to be elaborate timing or stimulation mechanisms triggered by the queen who decides to stop fertilizing her eggs. The workers carry the reluctant males up to the anthill openings and push them out into the glaring light. Males fill the air with their clumsy, confused marital flight. It is the ant's version of a Fourth of July celebration. It stirs a fond, habitual memory of competent founding fathers now long gone. Among some ants, queens avoid inbreeding by mating with foreign males. In these ants inbreeding produces diploid males that are sterile. In other cases diploid males are normal, and in yet other cases, as we have mentioned above, the continuation of the species has entirely devolved into the female genome and is sustained by parthenogenesis.

One of the most important pieces of information that is still functional in the ant genome is the gene that produces juvenile hormone in the corpus allatum. The corpus allatum is a congregation of cells in close association with the central nervous system. In animals it is the nervous system that coordinates rapid chemical communication between different parts of the body. The nervous system is also involved with embryological development through its association with hormone-producing tissues. The nervous system coordinates the production of juvenile hormone. There is evidence that a deficiency in juvenile hormone in the adult female leads to eggs with poor yolk, which, we recall, contributes to the production of more workers—the better to feed you with, my dear.

Juvenile hormone has a huge role to play among the arthropods. It regulates the color changes in the peppered moth caterpillar. It regulates diapause, a dormant period in insects that enables them to survive temperature extremes, drought, and reduced food availability. Juvenile hormone favors the differentiation of larval characteristics by blocking the development of adult characteristics. In the words of Wigglesworth, "in the evolution of insects the earliest function of the corpus allatum hormone may have been to delay the differentiation of the gonads until growth elsewhere in the body was sufficiently advanced; and that the control of all the other characters associated with metamorphosis is a secondary effect that was acquired later."[6]

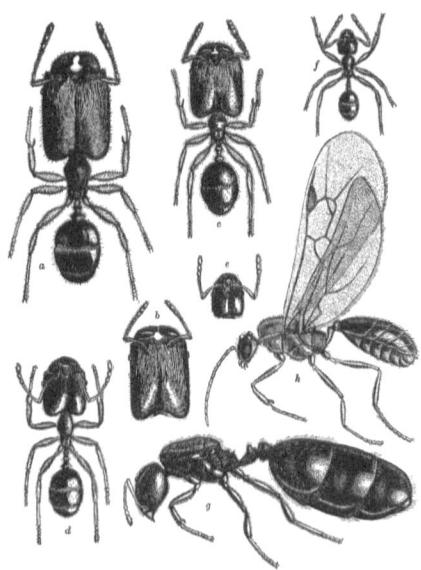

It is the delay in the development of the gonads that allows for the remarkable metamorphosis in the arthropods with hybridized genomes. It is during this plastic stage of development where the ants shape their own development epigenetically. The final factor that distinguishes the queen from her workers is a special food that is given to certain larva by workers. Whether this occurs by chemical election, in a dream-like trance, or by some other method is not known. Once the gonads appear, the die is cast; the possibility for change is over. Just as a sexually mature, light-colored pepper moth will always be a light-colored pepper moth, a queen will always be a queen.

It was the various castes among the social insects that offered such a serious challenge to Charles Darwin and to Darwinists ever since. With mating limited to the queen and the males, how could the different castes of females be shaped by selection or arise in the first place if they are sterile? Darwin said that this situation posed a difficulty that was potentially fatal to his whole theory. Viewing the ant society as a familial organism, he suggested that, "Selection has been applied to the family, not the individual."[7] Darwin was not a Democrat, and selection for the group, especially the Christian Victorian group, was allowed.

Richard Dawkins attempts to solve the conundrum without resorting to familial selection. Dawkins is a democrat, and this is the major difference between himself and Darwin who was a Victorian. For him everything stems from selfish genes that act through the body that they happen to find themselves in. To the question, why would most sisters give their support to a sister who would control destiny through their own genes: it is the relatedness of the individual's genes that determines the cost effectiveness of every individual's participation in the colony. To get a better understanding of what we are dealing with, we have to introduce a new academic discipline to our story.

Ethology didn't exist in Darwin's day. Ethology is the study of human behavior from a biological perspective, looking for analogs to behavior in the animal kingdom, and the pièce de résistance, composing a theory of the evolution of consciousness and behavior. Obviously it is a controversial area. Religion is uncomfortable because it rescinds the special perquisite of the chosen people that God gave to Adam. The scientific academy is uncomfortable with it because it threatens to deconstruct its Victorian operating system.

What did exist in Darwin's day was a set of descartian assumptions that animals were mindless, instinctually driven, windup toys, with the possible exception of the squire's favorite hunting dog. This was acceptable to the Anglican Church, and it is largely acceptable to the modern academy. Karl Von Frisch described honeybee behavior as an instinct, something that was genetically inherited. He got a Nobel Prize for his work. Harald Esch at the University of Notre Dame showed that the information communicated in the hive of the honeybee by the famous bee dance described by Frisch actually involved what they saw on the way. This is too much conscious intention for stimulus-response Darwinists to handle. Esch will not be awarded a Noble Prize.

We have already seen aggregation, agglutination, colonization, endosymbiosis, and ectosymbosis occur naturally without divine intervention or by invoking demons. We see a social tendency of ants already preadapted in the caterpillar stage of the sawflies. These caterpillars tend to group together and to rise up in legionary style when disturbed. Though these creatures are communal by nature, choice still continues to play a daily role in the social life of the ant. Before they set out on the hazardous chore of foraging, young worker ants spend their first weeks in the hive doing household chores while they get their legs under them. Fabien Ravary and Emmanuel Lecoutey of the University of Paris controlled a worker generation for age, size, and so forth to see if they could alter the outcomes of their eventual role in the hive. They found that ants they favored by giving them food continued in the role of foragers. Ants that they didn't favor eventually retired to perform household chores. It is really quite obvious that with the nervous system in play, adaptations will come about much too rapidly for the genome to be anything but dedicated follower of fashion, and the genomic world will have to adapt to this new epigenetic influence if it can.

To explain the complex behavior of ant civilization, and ultimately its shape, we have to look at their unique form of communication. Ants communicate with pheromones, as do all insects. These are chemical signals that do what hormones do in the body. The female peppered moth issues forth a pheromone for the delectation and motivation of the male at night when she is sexually receptive. Ants lay down trails with pheromones to instruct other ants as to the source of food, but it is the ant's so-called "social stomach" that puts them in a social class with the termites. Ants chew up their food and ingest it into crop, which is separated from the true stomach. Food from the crop is regurgitated and shared around with larva as well as other adult ants. The crop is the ant's social stomach.

All animals reflect a concept of self that is measured by ethologists as social distance. Animals get nervous whenever other members of the same species get closer than a certain minimum distance. It is a measure of sociability. It is different for different species and it can change with the seasons. Among the ants in a colony there is no social distance. Ants fondle one another with their antennae, feed each other, and generally share a group smell. An ant colony is as chemically connected as a bacterial biofilm. Mature ants of another hill do not share this smell, but enslaved ants raised from eggs, larvae, or pupae will be initiated and become participating members of their new colony. The word "slave" is actually an inappropriate word to use since it recalls the horrific situation found in human culture that is quite different. There are other dependent relationships in human culture that are less abusive and more appropriate as an analogy.

The absence of a social distance and the presence of a social stomach, enable the ant colony to swarm with a smooth efficiency that to us is telepathic. The social stomach trumps the selfish gene and all those twisted arguments about the degree of genetic relationship of anthill members. The social stomach creates a selfish anthill where different tribes are involved, but the Argentine honey ant has spread around the world in the last century, riding the waves of human commerce, and they all still recognize each other as sisters of the stomach.

The successful queen becomes a dependent of the group once she becomes an egg-laying machine. The workers that she has nurtured take on the burden of feeding her. When she joins the social stomach, all the members of the colony know that reproduction is going on because of the unique hormones that she produces. The whole colony feels rewarded. The successful queen becomes nothing more than the reproductive organ for the colony. It is effectively a sacrifice to become a queen, not that it matters. As we have seen, reproduction for the introduction of variety is not the game plan in ant culture. Adaptation comes out of the behavioral domain. As we shall now see, not only is there a vertical dimension to the ants that become sisters of the stomach, there is also a horizontal dimension. Other species of insects can join this sisterhood.

Ant society embraces a perplexing assemblage of assassins, scavengers, satellites, guests, commensals, and outright parasites. It has an uncanny resemblance to a genome, not the idealized genome of Darwinism before the Human Genome Project, the real genome. Of these ant fellow travelers, most are Insecta, some are Arachnida, and a few are Crustacea. Starting as tolerated or persecuted guests skulking around in the abandoned sections of the anthill, parasites gradually adapt and take on the smell and the taste of their hosts. The ants become despoiled by the relationship and seem to suffer from a kind of social obsession that resembles our own affection for pets. There is no doubt that the fellow travelers of the anthill and of high civilization are taking advantage of a highly developed maternal facility.

Perhaps the most startling relationship of all involves the symbiotic relationship between ants and a genus of Staphylinid beetles. From a paper published in 1897 by E. Wasmann, who studied this situation for twenty years, we have the following observation on the relationship between *Formica sanguinea* and *Lomechusa strumosa* beetles:

> *These Staphylinids are treated by the ants like their own kith and kin, live in antennary communication with them, are cleaned and licked and occasionally carried about, and are fed from the mouths of their hosts, although they are also able to feed independently and frequently devour the ant brood. The ants are especially attracted to these beetles on account of the prominent tufts of yellow hairs on the sides of their abdomen which are licked by the host with evident satisfaction. Not only do these beetles themselves live as guests among the ants, but the same is also true of their larvae. The larvae of Lomechusa are reared by the ants like their own brood; they are licked, fed with regurgitated food and before pupation covered or embedded in cells like their own larvae. When the nest is disturbed they are carried by the ants in preference to their own larvae and pupae*

> to a place of safety. The predilection of the ants for these adopted larvae is all the more remarkable because they are the worst enemies of the ant-brood and devour enormous numbers of the eggs and larvae and their hosts. This brood parasitism, in fact, causes the development of abortive individuals intermediate between the female and worker castes, and these intermediates, which I have called pseudogynes, gradually bring about a degeneration of the parasitized colonies.[8]

The question arises of why the habit of rearing these so-called parasites has not long since led to the extinction of *Formica sanguinea*. Wheeler observed that pseudogynes appeared only after the *Formica* colonies had already reached an exhausted or moribund condition. He observed large, flourishing colonies that had abundant *Lomechusa* larvae and no pseudogynes. It turns out that the degeneration in the ant community hinges upon how the ants care for the beetle larvae. As we have seen, the ants treat the beetle larvae exactly as they treat their own. They bury the beetle larvae as they do their own, which is exactly what beetle larvae also require. While underground the larva pupate in preparation for their metamorphosis into adults. But the ants dig up the pupal beetles too soon, and most of the beetle larvae perish through this act of ill-timed nurturance. It is only the few who are overlooked or are forgotten that insure the survival of *Lomechusa* and, coincidentally, the ant colony as well. It is only when the ant colony begins to age and reach the end of its own resources, and begins to be less industrious in the raising of larvae, that the beetle begins to overwhelm the community with its numbers and its narcotic influence. It is a case of when it rains, it pours.

The Hopi say that they survived the hard times by living with the ant people. The Scriptures advise: "Go to the ant, thou sluggard; consider her ways and be wise. Which having no guide, overseer, or ruler, provideth her meat in the summer, and gathereth her food in the harvest." When Christianity became a state religion, ant culture came to be seen as the kingdom of ants. Aesop told the fable of the ant and the grasshopper to teach the virtues of work and saving. Wasmann called the general situation of cross-species domesticity among the ants an amicable selection, and he was thinking of a model for human civilization. Wheeler observed that ants live in opulence compared to solitary animals, "and are therefore able to support a host of parasites on what may be called their large margin of vitality, without serious danger to the existence of the species."[9] Lewis Thomas observed, "Ants are so much like human beings as to be an embarrassment. They farm fungi, raise aphids as livestock, launch armies into war, use chemical sprays to alarm and confuse enemies, capture slaves, engage in child labor, exchange information ceaselessly. They do everything but watch television."

The anthill stands as a refutation to the theory of evolution by itself alone. The anthill, like a multi-cellular body, is a confederation of sometimes cooperating, sometimes quarreling, often pointless, sometimes even suicidal functionalities identified with many different genetic traditions. None of its specialists or any of the other insects that are fellow travelers is fit enough to survive, competing in the wild as an individual. The real role of the genome in the anthill, outside maintaining necessary housekeeping functions of the various bodies involved, is to produce a central nervous system to enable the various actors to play their roles and arrive on cue. As an arthropodian traveling theater, they survive on a grand scale. It seems to have begun with the loss of important genetic information that was compensated for with a brain the size of a pinhead. And with no apparent genetic changes in the ant for millions of years they have seen the ant-eating dinosaur Xixianykus come and go, and they will certainly outlast the mammalian ant-eater.

CHAPTER 12

The Last of the Dinosaurs

FISH DOMINATED ANIMAL LIFE in the ocean during the Devonian period around 400 million years ago when there was one super-continent called Pangaea. Following the Permian extinction, ferocious marine reptiles had taken over the ocean by the Cretaceous period around 140 million years ago. We might well attribute the behavior and atavistic appearance of the anglerfish to this momentous revolution. In the abyssal depths of the ocean, where no sunlight ever goes, there goes the anglerfish. Life is alive and well in the darkest depths. Oil worms, see-through sea cucumbers, yeti crabs, whale bone eaters and the like work their silent ways, but prey for the anglerfish, is few and far between. Prey may be hard to find, but they need not fear being preyed upon. Some anglerfish have an antenna-like projection that dangles in front of the great toothy mouth, tipped with bioluminescent bacteria to attract attention.

The female anglerfish is a horrid giant much larger than her male consort. In danger of being eaten by the female or of just losing her in the dark, males bite into the body of the female and hold on tight. The male and female become inseparable. The male's eyes are very large, but once he becomes attached to the female, they degenerate and he goes blind. His mouth and digestive system atrophy. His circulatory system merges with the female. In time he is reduced to a set of male reproductive organs attached to the female. And so does this big fish devour the little fishes. The anglerfish gives us a thrilling reenactment of the primordial courtship of sperm to egg. In many ways evolution is the same old story with bigger parts.

Determination of sexual identity by highly differentiated sex chromosomes is the norm in biology, but many fish are hermaphroditic, meaning that they have the reproductive organs of both the male and the female. Such fish have no specialized sex chromosomes. Some fish start off as one sex and then shift to the other in response to stimuli in the environment, including toxic waste. The anglerfish is not straying too far from family tradition with its ectosymbiotic hermaphroditism. What allows the angler fish, and indeed any sexual organism, to play this game is that in every male there is a little bit of egg, and in every female there is a little bit of sperm. It is a chemical signal that identifies the other as part of a common stock. This is what protects the attached male organs from the immune system of the female.

The anglerfish practices polyandry, many husbands for one wife. As many as a half dozen or more males attach themselves to a female. The approach to reproduction practiced by the anglerfish does seem to be an awkward affectation, like the eyes of the scallop or the efflorescence of a bacterial biofilm, but a study by Price and Wedell throws new light on polyandry. Polyandry is not unusual in the metazoan kingdom, although it took us a long time to recognize it, so enamored have we been with God's definition of the family. Price and Wedell suppose that multiple mating allows for post-copulatory selection against "selfish" genetic elements that can lower fertility. Since the female anglerfish can control the emission of sperm to suit herself, she may well be picking the best sperm producer of the bunch.

Polygyny, many wives for one husband, is thought to be the fundamental mating pattern for animals. The argument for this approach is that since the female produces very few eggs at great cost, she picks the best male of the bunch. Polygyny, then, is for females that cannot pick their sperm, but can only pick their sperm carriers. When monogamy arrives on the scene, it is a sign that the male is willing and capable of doing some of the female's work. This can be very attractive, but it doesn't completely eliminate older practices.

In some animals polygyny involves females who gravitate to eggless males who imply maternal fecundity by their accumulation of gaudy feathers, colored beads, and other shiny trinkets. Although we could invoke Amotz Zahavi's handicap principle here, the actual result may be paradoxical as we can see with the cuttlefish. The drab female cuttlefish allows gaudy males to fight it out while another male, as blandly disguised as the female, sneaks in the back door to mate with the actual female. I suggest that a surfeit of males allows the expenditure of risky behaviors that serve as diversions from the real business of reproduction. Since all cuttlefish control their color and pattern neurologically and can choose to be bland or colorful I suppose that gene selection is not the issue here, only the survival of the species. Some males prefer to duke it out; others prefer the nuptial bed, but they all have the same genome.

Some sort of competition seems to be involved with all approaches to sexual reproduction, but nothing is completely resolved, the game goes on, and nature often finds a role for the losers as well as the winners. The gender of the parent, by an evolving definition of sex, is not necessarily determined by the kind of germ cell that it produces; rather it is determined by the parent's relation to the fertilized embryo. Among fish the sperm-producing adult, as often as not, is the one that nurtures and protects the fry. It is the male adult that is on the scene when the fertilization of the egg takes place. It is now proposed that this adult be called the female. In cases of internal fertilization it is the adult that carries the egg that will care for the fry and be called the female. We see internal fertilization among some of the fishes and reptiles. It will become the calling card of mammals.

Neil H. Shubin et al. at the University of Chicago have discovered a fossil fishapod they named *Tiktaalick roseae* dated at around 350 million years ago on Elsmere Island in Canada. It has the neck of a landed animal, a neck that would have allowed it to look around. In the ocean all it takes is a flick of the tail to turn the head in a new direction. The same research group looked at the genome of a paddlefish that has fleshy fins and looks like a fish preadapted for land dwelling. It turns out that the Hox genes that play a role in fin development in the paddlefish are the same ones that induce limbs in land dwellers, and they come with a similar neurological pattern of activity.

Hox genes are the superordinate preadaptation for the evolution of metazoans for what Jacque Monod called a teleonomic performance: the perpetuation and amplification of an ancestral dream. The genome underwrites the potential for behavior through the secondary or epigenetic process of embryology. Plug in a time and a place with the Hox genes as your program along with ecological state of the planet, and you get a general idea what the organism will look like and how it will behave in its particular genomic manifestation. Evolutionary researchers John Tooby (anthropologist), Leda Cosmides (psychologist), and Steven Pinker (psycho-linguist) suppose that complex biological organisms can reverse-engineer to discern what they are "designed" to do. This is one of the fundamental insights of the *I Ching*: know the past in order to predict the future.

Among the higher animals, parents model or mirror useful behavior. When a parent does a particular behavior a neuronal pattern of activity occurs in the brain of the offspring that reflects it, thereby initiating learning of the action. Animals will also anticipate the behavior of other species of animals through the similarity of their neuronal patterning. An animal that is prey will anticipate the behavior of a predator; a predator will have to learn to disguise its intentions. The other important principle of pedagogy is the unwritten principle: thou shall not do what parents did not do. This cuts the torrent of possible information down to a manageable and useful level.

None of this covers creative behavior, without which evolution will cease and prepare to become devolution: that moment when, with a slight adjustment, fins can be used as limbs; that case where hardened mouth parts are first used like a tool to crush seeds that were otherwise unobtainable; that inspiration to use fingers to grasp a straw, to insert it into a termite hill to draw out termites. At some point the rules, the traditions, or the habits have to be broken. Never is there a fossil left behind to record such a moment, but we can see it all around us. Such events may seem to be random fluctuations, coincidental patterns, or mistaken identities, but only in the context of habit. Life is awash with such events. If they solve a dramatic situation for the actor in the moment, they acquire a rational basis. If they fail to be of use, they are accidents. Such events frequently occur in conjunction with a movement into a new ecology where the actor is constantly looking for similes. Insights need not be particularly efficient or foresighted, and unexpected consequences will have to be solved in time.

The Mesozoic era of the terrible lizards came from the opening created by the Permian extinction. In the general delousing of environmentally locked in genomes, fixed behavioral patterns and the general extermination of parasites and predators that occurs with a major extinction event, the survivors are freer to try awkward, untested innovations. During the Mesozoic era there was a rifting of Pangaea into two continents, Laurasia and Gondwana. The Mesozoic is divided into three geologic periods: the Triassic, the Jurassic, and the Cretaceous. An extinction event separated the Triassic-Jurassic boundary around 200 million years ago involving the overthrow of the archaic dinosaurs thereafter to be replaced by more modern dinosaurs. The earliest mammal-like creature has been found in the Jurassic period about 195 million years ago. Zhe-Xi Luo, a paleontologist at the Carnegie Museum of Natural History, supposes that the dime-sized *Hadrocodium wui* to be the earliest known ancestor of the mammals. This nocturnal, shrew-like animal had the middle ear typical of mammals and a precociously large brain. It probably had to eat continuously to feed this brain. The Mesozoic era lasted around 160 million years until the Cretaceous extinction event when large dinosaurs disappeared for good.

From the first moment of their discovery, dinosaurs became the creatures of our worst nightmares. They fascinated our young males and became a test of courage. Through the first half of the twentieth century scientists believed that dinosaurs were slow, dim-witted, cold-blooded animals. The survival of the fittest was naturally suited to be the motto of the most terrible dinosaur of them all, *T. rex*. *T. rex* epitomized the ultimate contrast between righteous Christian, Victorian culture and the reality of nature. In Tennyson's words: "Who trusted God was love indeed/And love Creation's final law/ Tho' Nature, red in tooth and claw/With ravine, shriek'd against his creed." Mothers protected their youngsters from too early an exposure to those natural monstrosities with their ravine shrieks.

A revolution in the perception of dinosaurs began in the 1970s when Jack Horner at the University of Montana put in his thumb and pulled out an egg. Dinosaurs laid eggs and built nests. Fossil evidence left by Maiasaura suggests that parental care continued long after hatching. Large clutches of eggs offer evidence that three types of medium-sized dinosaurs had males guarding and brooding eggs as in the case of large flightless birds such as the emus, the rheas, and the tinamous. DNA research has shown that the genomes of dinosaurs were quite small, about the size of the modern bird. There is a long list of traits that associates birds with dinosaurs, including feathers, elevated metabolisms, and pulmonary innovations. Polyploidy was out of the question for animals being hatched from eggs. The great size of some dinosaurs came from secondary genetic changes keyed to embryological development and the environment in general. Another outrage perpetrated by Horner was his discovery that *T. rex* was a scavenger, not a predator.

During the Cretaceous period four-legged fish were still pulling themselves up out of the ocean that was teaming with marine reptiles. They pulled themselves through the mud onto the land, to breathe the air, to become amphibians, only to find dinosaurs dominating the landscape. While the termites and then the ants were forming their great civilizations, it was from another chordate line that we would find more evidence of the existence of mammalian genes. Welcome the platypus.

When the English naturalist George Shaw received a specimen of this animal from Australia in 1799, he was thoroughly taken aback. The Age of Discovery had trashed the accumulation of centuries of tradition summed up in medieval bestiaries. Gone was the manticore that had a threefold row of teeth meeting alternately, the face of a man with gleaming, blood-red eyes, a lion's tail with the sting of a scorpion, and a shrill voice so sibilant that it resembled the notes of flutes. Gone was the griffin that was a winged quadruped that was vehemently hostile to horses, with the body of a lion and the wings and mask of an eagle. While examining his chimera-like specimen with its duck's beak, webbed feet, otter-like fur, a beaver's tail, and serpent-like spurs on its hind legs, Shaw suspected a hoax.

An educated public marinated in the simple life of nineteenth-century natural history simply could not process what natural history had become by the end of the twentieth century. It became too baroque, and things have only gotten worse with the platypus. From direct observation of their secret life, it has been found that females seal themselves inside their burrows beside streams and lay eggs. These they incubate for about ten days. The hatchlings are the size of lima beans and extremely immature. These neonates are nursed in the mother's pouch from milk-exuding patches for three to four months. By a loose association the platypus has been assigned a position in the tree of life at the branch point that leads to mammals on the one hand and marsupials (kangaroos and opossums) on the other.

When the platypus genome was decoded, it was found that they also shared genes with reptiles and birds. Unlike humans and most other mammals, the platypus has two separate sets of sex-determination chromosomes. It has the XY system that we have, and the ZW sex-determination system found in birds and many insects. In the ZW system the female has two different of sex chromosomes (ZW), and the males have one (ZZ). The platypus has the XY and the ZW systems separated by three other paired sex chromosomes. The problem with assigning the platypus the role of forefather to the marsupials and the mammals is that the platypus does not seem have enough genes to be those organisms; it only has enough genes to be a platypus or perhaps its cousin the short-beaked echidna. It would have had to lose or junk many of its reptilian and avian genes to become a koala bear or a guinea pig. All and all, this is a bit too much to ask of Darwinism when there is another solution to the anomaly of the platypus genome.

The platypus looks very much like a hybrid from very early stages of the marsupial/mammal line and reptile/bird lines. This cannot even be considered by Darwinism obviously, so while taxonomists gnash their teeth over Jungian animus and Australia's platypus, Ogden Nash has one of his koanic epiphanies: "I like the way it raises its family/Partly birdly, partly mammaly." Nature too must like the way the platypus raises its family since *Steropodon galmani*, a 110-million-year-old fossil monotreme, looks pretty much like the modern platypus. Being a jack-of-all-trades, the platypus has been well adapted to live in its environment until now.

The advantage of internal fertilization, which the platypus does in birdlike fashion, is that it decreases the number of eggs necessary for success. They are well protected. With internal fertilization the journey of the sperm is limited to the passage from the male to the female through direct contact, but it still required its ancient, low-tech, mitochondrial engines to motor through the dark passages to the haunt of the egg. The imperative for no change remains the same for the sperm. When fertilization is entirely removed from the external environment—in the monotremes, the dinosaurs, the birds, the marsupials, and the placental animals—the adult female body becomes the sexual body upon maturation and then a vegetative or maternal body after insemination with the aid of hormones. This is the reverse of the stages of development from the caterpillar to the butterfly. The butterfly is strictly a sexual organism.

It falls to the mother carrying the egg to compensate for untoward mutations of the genome as well as she can, and to fend off untoward changes in the environment by intentional behavior. If the male plays any role beyond his one seminal function, communication with the female becomes a necessity. Since the sounds of seduction are only slightly different from the sounds associated with maternity and child rearing—some of the aims are similar—communication can be touchy and potentially ambiguous.

At one time the fossil records of dinosaurs seemed to be the ideal model for natural selection through competition and evolution by the accumulation of small, random changes that happened to provide a benefit, over a long period of time. In 1949 J. B. S. Haldane proposed to measure the genetic rate-of-change of evolution using the dinosaur records. He estimated there was a 1 percent change per million years. The unit of change would be called "the darwin."

Strictly speaking, Charles Darwin saw only one slow, uniform rate of change for all of creation. We have already seen all manner of venues for evolutionary change as well as for devolution. Many of them have been too large or abrupt to be perceived by those using the Haldane method. Nevertheless, we can use the method as very abstract measure to compare dinosaurs with other animal groups during the same geologic period. If dinosaurs were evolving at one darwin, the ants were evolving at the most at one one-hundredth of a darwin, and the platypus was evolving at about the same rate. The mammals did not appear to be evolving much at this stage, but that could change as more pages of the geological record are found.

Jack Horner fostered a change in heart about the lives of dinosaurs, including things that pointed to their eventual extinction. By the Mid Cretaceous the angiosperms made great advances through polyploidy and insect husbandry to became the major component of the botanical ecosystem. Graeme T. Lloyd et al. at Bristol University have shown that the fate of the big dinosaurs was sealed by historical contingency rather than competitive superiority. It was the very large size and long intestines of the big dinosaurs that allowed them to digest the older, low-nutrient gymnosperm browse. The big, herbivorous dinosaurs showed a simple unwillingness to change their diet as documented in their fossilized dung.

Evolution, as we have seen, demands changes in diet through co-adaptation with bacteria and protozoa all down the line. With the metazoans we have to include teeth and mastication techniques along with intestinal flora and fauna. And we also have to include intentional behavior. Much of the genetic drift recorded in the dinosaur fossil record by Haldane was probably evolution in the service of maintaining the status quo: change that represented no change. Behavior that can make the difference between survival and extinction when it is flexible and innovative can also embed a species in a dead-end tradition, especially if the environment is changing slowly enough. The inability of dinosaurs to change would prove that they were dinosaurs after all.

It is hardly ever a single factor that brings down a species. During the peak of dinosauria, volcanic activity was high, the vegetation was lush, and carbon dioxide levels were up to twelve times higher than today. There were no polar ice caps, the land area was more uniformly warm, and the levels of oxygen in the atmosphere were over 30 percent compared to 20 percent today. These were conditions that allowed for the development of very large animals with long intestines and enormous oxygen demands.

Near the end of the Mesozoic era volcanic activity began to decrease dramatically. Plant paleontologist Nan Crystal Arens found a 90 percent drop in the numbers of species of flowering plants in the Hell Creek formation made famous by Jack Horner. This was before the K-T extinction boundary 65 million years ago. This and other evidence suggest the beginning of seasonality on planet Earth, an alternation of dry with wet seasons. As far as life was concerned at the time, the planet was dying. This would have had a devastating effect on the entire ecosystem, although the smaller animals were less handicapped under the circumstances. While the first great co-adaptation of land animals with plants was on the wane, the lizards, crocodiles, snakes, and birds continued to innovate rapidly, eat whatever was available, and change shape. The herbivorous insects continued to play an important role in the new ecology, and the mammals were tagging along taking advantage. The big dinosaurs were already dying out when the coup de grâce came in the flash with the K-T extinction. On top of all the proximal causes for extinction described in Chapter 6, most of the large dinosaurs were not suited to live in the cooler, drier climate of the coming age of the mammals.

The animals that would survive the K-T extinction and occupy the continents that we are now familiar with would be those that had learned to live with the palms, the figs, the grasses, the orchids, and their ilk as well as the pines and cycads of old. It was after the Cretaceous extinction that the bird-like dinosaurs evolved very rapidly into all the forms that we now see. Certainly it was their potential for escaping predation and for covering a wide territory that made a big difference. Flightless birds would only reappear under special conditions: a local concentration of resources and an absence of predators. It is among the living dinosaurs on the Galapagos Islands 600 miles off the mainland of Ecuador that we find flightless cormorants as well as the ultimate test for the theory of Darwinian evolution.

The Galapagos are volcanic islands. They are located at a triple junction of plates that create the Galapagos hot spot. The surrounding ocean floor is pockmarked with volcanic vents flagged by tubeworms and the life forms that are able to build upon ancient anaerobic metabolisms. The islands are around 5 to 10 million years old. Some places are so new that they give an idea of what the red Earth was like when it was first being contemplated by fungi. The Galapagos tortoises quietly grazing on grasses are ghostly throwbacks of the age of the dinosaurs, except for their diet. God only knows how they got there, but it is another dinosaur descendent, the finches of the Galapagos, that are the objects of one of the most intense field studies in the history of natural history.

Darwin noted but was not overly impressed by the finches on the Galapagos during his trip around the world on H.M.S. *Beagle*. He casually observed that one might really fancy that from an original paucity of birds in this archipelago, one species had been taken and modified for different ends. He was referring to the two varieties of mockingbirds. It was after he returned to England that ornithologist John Gould reported that what Darwin thought were blackbirds, gross-bills, and finches were in fact a series of ground finches that were so peculiar as to constitute an entirely new group of twelve closely related species. There were finches with warbler-like beaks; there were long, thick-beaked finches suited to probe in blossoms and the fruit of cacti; some finches had small, obtuse beaks for eating tiny grass seeds; and some had vice-like beaks for cracking very hard grains. Darwin supposed that natural selection had caused the evolution of these birds in order to occupy various ecological niches. P. R. Lowe would coin the term "Darwin's finches" in 1935. David Lack observed about this family tree that in no other birds are the differences between the species so ill-defined, and he supported that the difference in beaks was a result of Darwin's principle of divergence. Darwin's finches have become the standard textbook example of evolution by adaptive radiation under natural selection.

In contrast to Darwin's finches, the marine birds on the Galapagos are largely the same as marine birds around the world. This is explained by the fact that these wide-ranging birds remain an active part of a continuous gene pool. Other species have developed local subspecies: a pelican, a tern, two herons, a duck, a flamingo, and two owls. These have only occasional congress with outsiders and/or they are relative newcomers to the islands. An albatross, a penguin, a gull, and a hawk are designated as local species. A special genus is reserved for the swallow-tailed gull—the only known nocturnal gull—a flightless cormorant, and the Galapagos dove. Local varieties are found between the islands in two groups of birds: the mockingbirds and Darwin's finches.

It is only Darwin's finches that appear to be truly unique to the Galapagos in the sense that there is no cousin on the mainland that really looks like them. Some have thought their closest relative might be the St. Lucia black finch in the Caribbean Islands. Others have suggested the blue-black grassquit. Recently they have been grouped with the tanagers. The discussion may now be over since DNA data has identified an Ecuadorian grassquit as the closest relative. This means that they are of the class Aves (birds), order Passeriformes (perching birds), family Emberizidae (includes sparrows in North America and tanagers in South America), genus *Volatinia*, species *volatinia jacarina*. They are not finches. The confusion over ancestry at the genus level reveals that they have been on the Galapagos longer than other birds about which there is no such confusion. They have been catalogued into as many as five genera and a total of fifteen species.

Although Darwinists are happy to show us the radiation of the different finches into their specialized niches on the Galapagos Islands, they have no reasonable response to the question of why there is a significantly greater adaptive radiation of Darwin's finches than for the other local species, such as the albatross, the penguin, one of the gulls, the hawk, and the Galapagos dove. I would have predicted that all of these birds would have arrived on the islands before or at least at relatively the same time as a weak-flying finch, but what if the founding finches were the survivors of a catastrophe that wiped out most of animal life on the islands? What if the invisible hand of natural history set the initial conditions for the finch adaptations to occur?

Those who specialize in major tsunamis around the Pacific Ocean say that they are not uncommon. For those not wedded to Uniformitarian geology, we recall the impact of the Typhon Comet around 3,500 years ago, the event that is marked in human history by the shift from the Bronze Age to the Iron Age. A huge tsunami inundated the whole of the western coastline of North and South America following that event. Our cousins who had ancestors living on these shores since that time have stories about such an inundation. Among the many recently discovered geological pieces that fit this puzzle is one by Matthew Hornbach of the University of Texas. He proposes that a massive tsunami must have been responsible for dropping some huge boulders and cleaning off Tonga's main island within the last few thousand years. As we have already seen news of this sort of catastrophe does not reach most of the scientific bureaucracy. It is one of those need-to-know pieces of information. It tends to confuse a well-laid ideology.

I suggest that the ancestor to the warbler finch, which prefers the higher altitudes above 1,300 feet, survived a major tsunami event. I suppose that those finches proceeded to occupy a wide range of feeding possibilities left after the inundation as Adam with Eve. Their offspring proceeded to divide up the ecological landscape without competition from other birds or insects that might have been more suited to the task, and there were no predators to worry about. The distance between islands was great enough for these weak flyers so there was not much interbreeding between the islands, allowing local varieties with their own beak and song to develop. I suggest that the other Galapagos birds, most of them stronger fliers, have reoccupied the islands since the extinction/origination event.

For a rough comparison to this situation I refer the reader to Lake Victoria in Africa. Lake Victoria is a pluvial lake that was created following the impact events around 9,000 years ago or so. It is now occupied by 300 species of cichlid fishes, which is a rate of change in the multi-darwins. Darwin said almost nothing about the actual origins of species in spite of the title of his book. There was really almost nothing to say at that time, since so little was known. This remains the posture of most Darwinists. Dramatically varient rates of speciation are a major conundrum.

Although offshore marine life around the Galapagos Islands is generally a cornucopia except in El Niño years, onshore is a different story. The landside climate is one of the most severe to be found. The annual rainfall at the lower elevations varies from one inch to fifty-five inches in El Niño years. In an El Niño year the islands change form a desert environment to a jungle environment overnight. Reproduction quickly follows any rainfall that might occur. Molting will come to a halt, hormones will be affected, and genes will be suppressed or activated by the change in the weather. In short, survival is a lively and dramatic challenge on the Galapagos Islands. Food is such an issue that it has led to such unfinch-like, specialized diets as that of the vegetarian finch and the so-called vampire finch. *Geospiza difficilis* has acquired the habit eating ticks off of reptiles and drinking the blood from the punctured bases of the flight feathers of nesting boobies.

A dramatic contrast is made with the cormorants that are fisher birds. Taking advantage of the wealth of the sea and the absence of predators, the cormorants have become flightless. Having decided to stay, they now must stay. Their population craters during El Niño years when the marine life disappears, but this only occurs every fifth year or so. On the other side of the coin, it is only every fifth year or so that the finches have a bountiful year because of the rain that El Niño brings. The savage variations in the ecosystem of the Galapagos remind us of the steppes of Central Asia that prompted Petr Kropotkin to observe that those animals that know best how to combine have the greatest chance to survive and evolve.

Among Darwin's finches other behavioral innovations beyond creative dietary innovations and timing of courtship are important. When courtship begins, females will mate with males only when they show the capacity and motivation to build elaborate nest structures that the females will line to suit themselves. The males must also show a potential for parental behavior by feeding the female during courtship. Males will do all food gathering for the female and the chicks. Without the participation of both sexes in the nesting and rearing process, these birds would not have survived on the Galapagos Islands. The sexes of this bland, unpretentious bird are very similar. From this it is safe to assume that the females have been doing the selection. They are selecting for mates that are like themselves.

I am reminded of the contrast between the California scrub jay and the Florida scrub jay. The California scrub jay is not challenged by food shortages. When the offspring are fully fledged and capable of taking care of themselves, the adults chase them away. This keeps their territory from becoming overpopulated. The Florida scrub jay experiences a failure in their diet just at the time when the eggs hatch. Under these circumstances the offspring remains close to their parents to act as helpers, giving up or delaying the chance to advance their own genes with families of their own. A similar sort of generational cooperation is found among the Galapagos mockingbirds for the same reason.

In the fat environment of the wet tropics where the male has little to do but implant sperm, trivial secondary-sexual characteristics are often selected for and fought over by competing males. One of the most amazing results of this is the polygamous, vainglorious peacock. The rather more drab, retiring peahens may be mildly interested in the male to be sure, but the peacock's main preoccupation is with other males of the species. Coincidently the male draws attention away from the females during the vulnerable circumstances connected with nesting and birthing. Males chose mates that do not look anything like their same-gender opponents. The Galapagos Islands do not support much in the way of high cockalorum in its birds, although the male Darwin's finch still defends a territory and sings a family song.

A most interesting experiment has been conducted in the wild contrasting Darwin's finches on the Galapagos Islands and Darwin's finches on Cocos Island. Cocos lies several hundred miles off the coast of Costa Rica and several hundred miles north of the Galapagos. Considering the tonnage of material that has been manufactured about the Galapagos Islands and its finches, there is next to nothing on the Cocos Island and its finch. The Cocos Islands Friends Foundation doesn't even mention the bird. There is really no possibility for a tourist industry, for one thing. The island has sheer cliffs all around. Most of the people who go there go for the diving and lie offshore. Another reason there is little interest in the Cocos finch is that they don't support the Central Dogma of Darwinism. Fortunately Tracey K. Werner of the Department of Zoology, University of Massachusetts, and Thomas W. Sherry of the Department of Biological Sciences, Dartmouth College, have done a very detailed study of these finches, published in 1987.

Cocos Island is rare among the islands of the west coast of Central and South America in that it has heavy year-round rainfall. This tropical island has a lush rainforest of complex structure. According to the research published by Werner and Sherry, the Cocos finches eat diverse arthropods including crustacea, nectar, fruit, seeds, small mollusks, and possibly small lizards. The foraging behavior of these birds spans those of many different genera of birds on the mainland. Individual families of Cocos finches consistently use a particular foraging behavior; other families of birds consistently use other foraging behaviors, but the beak morphology is a preferred one-size-fits-all. The feeding method was unrelated to age, gender, or any slight morphological differences in the beaks that might exist, and there was a large overlap in feeding territories. They build the same spherical nest as their Galapagos cousins. Curiously the Cocos finches struck their observers as awkward, even clumsy, feeders compared to other birds. It was as if they were still learning their trade.

In deference to the Darwinian monopoly the authors say, "Feeding behavior specializations could be genetic in origin," but what they observed is juveniles following the parents and learning from them, or even following other birds including sandpipers and learning from them. Their conclusion was, "the Cocos Finches exploit diverse resources with behavioral means."[1] I suggest that the Cocos finches have reincarnated the Adam and Eve situation that existed on the Galapagos Islands after the extinction event. Recent DNA comparisons with Galapagos finches point to a relatively recent arrival from the Galapagos Islands.

The Galapagos Islands does have one finch that reflects the inventive behavior of the Cocos finches. It actually looks like a Cocos finch. The bird I am speaking of is one of the most interesting characters among Darwin's finches, the woodpecker finch. The woodpecker finch occupies a niche for which it does not have the requisite bodily equipment. There is more to a woodpecker than a heavy beak, but the woodpecker finch doesn't even have that. The woodpecker finch does not have a heavy beak, nor does it have the long tongue or the reinforced skull of the woodpecker. In order for the woodpecker finch to get grubs out from behind bark or out of holes in the wood, it uses small twigs or cactus spines. It uses tools. This is cited as a rare use of tools among birds. The woodpecker finch is a woodpecker only in the sense of the food that it eats.

The observation about the woodpecker finch being unique because of its use of tools reveals the continued dependence on descartian-level ethology. Beaks themselves are actually tools that must be realized as such and used with intention. Every new generation must learn the use of their bodily tools by watching and mimicking their parents. The conceptual manipulation of the body is a preadaptation that ultimately leads to the use of externalized tools by animals, but the basis for tool use is as old as the nervous system. Ectopic tool usage is not a favored venue for biological survival for an obvious reason to all but those who have become dependent upon them. Once a tool becomes an article of survival, it must be widely available or forever carried around. This is why body parts make the best tools.

In his book *Galapagos*, published in 1961, Irenäus Eibl-Eibesfeldt casually mentions the New Zealand bird, *Heterolochia acutirostris*, in connection with Darwin's finches. The huias were a dramatic case of a difference in bills, expressed in the two genders. The two genders were totally alike except for their bills, which they use as tools. The male huia had a short, stout, woodpecker-like bill; the female a long, slender, scimitar-shaped bill. The genders worked in concert. The male excavated with his short beak. He was followed up by the female, who probed for the larvae of boring insects with her long, curved beak. This had become such a mutual dependency that young huias had to be mated while they were still being feed by their parents. The huias practiced a proscriptive state of monogamy. Widows and widowers had an immediate problem. They gave out a special a distress call that attracted neighboring couples, who would assist the helpless bird. This revealed a wider social network that was otherwise invisible.

What finished off the huias was another sort of behavior practiced by humans in their environment. The Maori people so highly appreciated and valued huias that only high chiefs were allowed to wear their tail feathers. At the turn of the twentieth century, the Duke and Duchess of York visited New Zealand. One of their Maori guides removed a huia feather from her hair and put it into the band of the bowler hat worn by the Duke. This created a European fashion for huia tail feathers. Soon enough they were extinct. The huias could not change behavior fast enough, assuming it was even possible. A species was lost in the search for a feather.

Huias give us a dramatic example that different beaks can exist within the domain of a single genome, cycling through the differences in gender expression. The huias' high degree of behavioral dependency is a bit extreme, but not out of the ordinary. We have seen it in endosymbiotic and ectobiotic relationships up and down the ladder of creation. We have seen it also in the caste differences in the social insects. It is not all that different from what we see in Darwin's finches divided up among their different ecological niches on the Galapagos Islands. We see a vertically inherited genomic tradition that is elaborated by a broad horizon of epigenomic factors: everything from the temperature and humidity, to hormones, to songs, and to visual images correlated by a memory bank of past experiences and passed along by neurological mimicking.

As much as Darwinists preen and crow about Darwin's finches as different genomic species created by competition, whether these different groups of finches are actually different species is still an open question even among Darwinists. The variation in the beaks of Darwin's finches is certainly not as great as the differences in the morphological differences in the different castes of ants that come from the same genetic tradition. They hardly hold a candle to the different varieties of dogs that all belong to the same species. Darwinists are rarely led off the beaten path of Aristotle and his rule of the excluded middle: all gray areas erased from consideration in order to establish a clear-cut opposition, light versus dark. Speciation is at the heart of the Lamarckian/Darwinian dialectic and the Creationism/Darwinism dialectic. On one occasion Charles Darwin himself dismissed the idea of species as "arbitrarily given, for the sake of convenience."[2] For just a moment we see Darwin broach the age-old question of universals: whether genera and species actually exist or are found in the mind alone. For a moment he leaned toward Plato and away from Aristotle. In short, the definition of a species is an ancient work in process.

The work done by Dolph Schluter from the University of Guelph in Canada attempted to put the theory to test. He did a close-up study of the small-beak, G*eospiza fuliginosa* finches and the sharp-beak, *Geospiza difficilis* finches on one of the islands. He was predisposed to the thesis that finch distribution would be random and that there would be territorial skirmishes along the boundaries between gene-driven individuals attempting to establish their access to genetic immortality. Schluter found nothing of the sort. Each group occupied a particular territory, and where the territories overlapped there was no competition. They hardly recognized each other's presence. In some places where one type of finch had gone missing, the remaining finch would broaden its diet to include the diet of the missing type.

The most intensely focused attempt to find evolution in Darwin's finches by the single method of change allowed in Darwinism is a study led by Rosemary and Peter Grant at Harvard University that has been under way for decades. The site of much of their work is Daphne Major, an almost inaccessible small island north of Santa Cruz Island. Like the classical nymph of old, Daphne has vowed to remain a virgin, but the Grants are bound to know her and her resident birds. It was the El Niño event in 1982 that really got the ball rolling.

Lisle Gibbs, a graduate student working for the Grants, was on Daphne making observations when El Niño hit. It was the biggest El Niño in memory. The seas were hotter than normal. Mushrooms clouds grew right out of the ocean. Vine seeds that had lain dormant at the foot of tent poles for years exploded up the lines. Croton trees flowered not twice, but seven times that year. The total mass of seeds was a dozen times what it had been in the previous drought year. It was too wet for the cactus on the other hand, and the creeping vines also smothered the caltrop plants. Caltrop had provided large highly rewarding but difficult to get to spiny seeds that were the specialty of *fortis* ground finches. A small seed crop was in great abundance and this was the specialty food of *fortis* ground finches with slightly smaller beaks. Would the miniscule difference in beak size in a single species be the wedge that would be the cause of a speciation event?

The birds went crazy, Gibbs recalled. The year before there was no breeding at all because of a shortage of food. This year females laid up to forty eggs and fledged twenty-five youngsters apiece. In the steamy heat of this reproductive frenzy, some females became bigamous, others polygamous. To his amazement he found that finches as young as three months old were fertile, laying, and successfully fledging young. This had never been seen before among passerine birds. In most cases birds are not fertile until the following year.

When he returned to his computer and entered the data, he discovered that the *fortis* finches with the slightly larger beaks, the ones that specialized in caltrop, were crashing in population. It was only the *fortis* finches with the slightly smaller beaks that were thriving. This was the opposite of what had been happening in the drought years that preceded the El Niño event. All across the islands the same pattern prevailed, with the exception of Genovesa. There the cold and the wet caused sitting parents to give up on their eggs and young. Branches heavy with growth and rain snapped and broke, causing further mayhem. The mockingbirds on the Genovesa had their own problem. The adults developed a pox sitting in the poor weather. Many died, leaving immature male helpers at loose ends. Normally the needs of reproduction had these mockers occupied with their tight family groups. Cut free, the male birds banded into roving gangs. These adolescent bands attacked finch nests, breaking eggs and eating young. When it rains it pours. In this difficult environment, none of finches on Genovesa did well.

It was established by the Grant research team that small differences in beak were sufficient to influence survival success, but the sudden nature of the change was a shock. In his review of this research in the book *The Beak of the Finch*, published in 1995, Jonathan Weiner said, "Having seen even this much of the view from the rim of Daphne Major, we can no longer picture the story of life as slow and almost static, a world view for which the chief emblem of evolutionary change is a fossil in stone. What we must picture instead is an emblem of life in motion. For all species, including our own, the true figure of life is a perching bird, a passerine, alert and nervous in every part, ready to dart off in an instant. Life is always poised for a flight."[3] While I am always interested in the Uniformitarian analysis of the life histories of our fellow travelers, it is obvious that Weiner was not familiar with the discipline of impact science and its role in eliminating genomes and cleaning up the survivors. His classical, award-winning, conventional point of view occurred naturally without any censorship being applied. A reporter was simply interviewing a group of scientific specialists whose studies were restricted to conventional issues.

For a moment it looked like the results would be a Darwinian success story, yet the arrow of selection flipped back again as soon as the weather changed. Over the longer haul things averaged out, and it looked like nothing much was actually happening. The shift in beak size in close coherence to yearly changes in the environment could not possibly be the result of the rare positive random mutation. But something else also happened during the flood of 1983 that led the research in a different direction. It was something that happened between a *Geospiza fortis* ground finch and a *Geospiza scandens* cactus finch. It became the greatest surprise of all for the Grants.

During the mating frenzy that occurred during the super El Niño, a *scandens* male courted a *fortis* female. There had been observations of hybrid pairings in the past, but they had nearly all come to grief. Ornithologists had attempted to cross different species of Darwin's finches without success. This fit the prejudiced Darwinian view on hybridization. As I have already mentioned, Charles Darwin followed the Protestant Christian tradition on breeding. The normal and correct rule was no breeding in nature closer than second cousins. Much effort has been put into discovering natural factors that block species from breeding closer than second cousins. But there was also a cultural tradition about out-breeding as well. Darwin used the example of the horse crossed with a donkey, a well-known hybrid whose offspring are sterile.

The hybridization of a *fortis* with a *scandens* would seem to suit the argument about increasing variability that is used to rationalize the justification for out-breeding. Of course if unbridled hybridization of that sort were to go on unchecked among Darwin's finches, the family branch of thirteen species or so would be reduced to a single twig. The same argument used against inbreeding too closely is also used against out-breeding too far: it reduces variability. Amazing! Darwinism narrowly threads the eye of the needle of Anglo-Saxon, Protestant morality on breeding: no closer than second cousins, no further than fourth cousins. That was the basic underlayment to the Victorian class system. In Protestant America this has shifted slightly and opened up a bit: no closer than first cousins, no further than sixth cousins.

Before the flood, Darwin's finches generally did follow the Protestant rule. Having seen the fickle arrow of fate even out the chances of first or second cousins to effect permanent change, however, the Grants were now primed to follow a new lead in order to finally nail down the first observed origin of a species in the wild. They eagerly followed the fate of the potentially star-crossed Romeo and Juliet finches as well as several other hybrid matches. After the flood, hybrid pairs were not only successful, in many cases they were doing better than purebred pairs. On the island of Genovesa the hybrid families were turning out fledgling after fledgling. Over the years they have become royalty on the island. On Daphne Major, hybrid crosses are now doing better than purebred families overall.

As Weiner describes it, "Something has changed since the flood. Something is happening. Strange as it seems, these hybrids are the fittest finches on the island. If the Grants' hunch is right, this is a missing piece of a puzzle, a puzzle they have been struggling with for the last twenty years: 'that mystery of mysteries,' the origin of species."[4] This kind of hybridization is not a case of mass horizontal genetic transfer such as that discovered by Lynn Margulis or Donald Williamson. A change of that magnitude is not even on the computer screen of the Grants or of Weiner. It is a profound violation of the Central Dogma. Of course for Darwin the subject of microbiology did not even exist.

The Grants have pursued the cause of behavioral innovation of the finches into the genome itself. The Grants and their colleagues have shown that calmodulin is the controller of the frontonasal prominence in birds in embryological development. Calmodulin regulates calcium uptake. Calmodulin is a simple protein that has two parts with a flexible hinge region that makes it a very effective binder of calcium. Being Darwinists, they imply that it was calmodulin that influenced behavior, not the other way around, however we know that calmodulin influences and is influenced by processes as diverse as inflammation, metabolism, muscle contraction, intracellular movement, short-term and long-term memory, nerve growth, and the immune response. So what has actually changed here, the genome or the epigenome?

I am reminded of juvenile hormone in the ants that, in combination with special diet, leads to dramatic phenotypical changes. Are the different beaks of Darwin's finches simply a reflection of the Swiss Army Knife range of possibilities influenced by calmodulin, with the exception of a woodpecker beak? Calmodulin can also be influenced by methylation that shapes DNA expression. This may be the method of change. We now know that methylation of the genome can begin in the womb as a result of an environmental shock, especially by way of diet. We know methylation can occur shortly after birth during neurological and hormonal maturation as a result of many environmental influences. We know that methylation can be inherited for at least several generations without any basic change in the genome.

Another source of genetic input that might apply here has been demonstrated by Heidi Parker at the National Human Genome Research Institute. She found that a retrogene is responsible for short-legged dogs such as the dachshund. This gene came from their wolf ancestors. Short-legged dogs were hiding in the wolf genome all along. This retrogene appears to turn on growth mechanisms at the wrong time in fetal development if a long-legged dog is the desired end point. Retrogenes have turned on something else in the general discussion of genetic physiology. Parker states that they may have played a more important role in evolution than previously thought—a Lamarckian form of evolution.

These sorts of genomic changes are a much more coherent explanation for the beaks of Darwin's finches than the shotgun wedding of biological accidents imposed by dogma central. Only the nervous/hormonal system is flexible enough to respond with a change in behavior on such short notice. I propose that the birds were stunned by the real collapse of the world that their behavior had helped to mold them into. They responded to the environmental changes with behavioral intention and innovation. In addition the mild hybridization of *Geospiza fortis* and *Geospiza scandens* suggests hybrid vigor involving the cleaning up of genetic damage in genomes. We have already seen that inbreeding is an inevitable consequence coming out of a population bottleneck; I can now add that mild out-breeding (hybridization) can also be a part of the scenario coming out of following catastrophic events. What would the shelf life be for a genome that didn't go through such perturbations regularly? Could we even ask that question?

Is behavior the primary vector of evolution at this level of creation for creatures with nervous systems? The academic specialty of artificial intelligence has essentially junked the genetically deterministic, tinker toy theory of behavior for all non-human forms of life proposed by Descartes and sponsored by the Church. It simply couldn't work in practice. Artificial intelligence specialists now copy human intelligence as closely as it is possible to do to even approach the simplest forms of organismic behavior in robots. Can theories of evolution do any less? The academic reaction against anthromorphism in popular culture has gone overboard. We are hopelessly behind the eight ball in terms of understanding the evolution of consciousness and the evolution brought about by consciousness.

Darwin's finches can tell us more about creation than just evolution. They can also give us a clue about devolution. Going back to the Age of the Dinosaurs, we remember that it was long enough to be composed of two large chapters. The first chapter was dominated by the archosaurs that included the distant cousins of today's crocodiles. The dinosaurs appeared around 230 million years ago, and for 30 million years they lived with the archosaurs before they became the dominant group. At first the Darwinian mantra of variety prevailed as the explanation for the rise of the dinosaurs: the dinosaurs were the superior animals, more varied, more adaptable. Steve Brusatte at Columbia University and the American Museum of Natural History in New York has led a team that has shown that this was not the case. The archosaurs actually had the larger range of body types, diets, and lifestyles.

The archosaurs came out of the Permian extinction event. Darwin's finches came out of the Typhon tsunami event. They both radiated out to occupy and specialize in a wide range of unoccupied biological niches. This was a successful approach to survival for a while. One small, hardly noticed branch of the family migrated to Cocos Island, where it retained its generalized, less efficient life style. Variety came to its natural end, overspecialization, in the archosaurs. Will the Cocos finches be the ones to occupy the brave new world after the next Second Coming? As the Book of Ecclesiastes put it: To everything there is a season.

In summary, the work of the Grants has completed the deconstruction of Darwinism. There is no indication at all that a random genetic mutation is involved with the coherent changes expressed by Darwin's finches. The selection of a particular strength over another in the group reflects a normal pattern of adaptation out of a common genome. The extinction of a species follows from overspecialization to an ecological niche that is radically altered for some reason or another. Success writes the formula for failure. The wild card here is behavior. Behavior is the least predictable part of who shall move on and who shall not.

The only thing left of Charles Darwin's utopian theory of evolution is that species do indeed change, which comes with the old-Earth theory of geology.

The only thesis that Darwinism can really continue to contend with anymore is Creationism, which is locked into its young-Earth theory of geology. While Darwinists and the Creationists quarrel over their bone of contention, I will proceed with more interesting matters. Paleontologists report that 95 percent of all species that have ever lived are now extinct. In our own little microcosm of creation—evolution/devolution—we have a well-documented, but not well-understood case of extinction involving one of the last of the feathered flying dinosaurs, the passenger pigeon.

The passenger pigeon once existed in America in huge flocks. When the pilgrims first came it is estimated that there were nine billion passenger pigeons. This is probably a high estimate for reasons that we shall see. By the time of John James Audubon, however, they were certainly present in huge numbers. When they migrated to new feeding grounds a river of birds would darken the skies for days. Arriving at a new area, they would occupy a forested area ten to twenty miles long, a mile across. They completely would fill it. Every tree branch would hold countless birds. Smaller branches and younger trees would break off under the weight crashing down, killing birds below them. During their stay the area would be encrusted with excrement so thick it would kill much of the plant life and reduce the variety of animal life. It would take years for the area to recover.

Like locusts these birds would strip woodland and cropland for miles around the roost. Audubon estimated that a billion-bird flock required nine million bushels of fruits, acorns, chestnuts, or grain a day. When a flock would arrive, predation would commence around the periphery of the roost on a major scale, including animals that ordinarily were not even meat eaters. The cost to the flock of a general alarm and precipitate flight from such an attack on their social body would have cost many more in lives than would have been saved. Therefore the birds stayed put at close quarters. Most wild pigeons will brutally attack one another if caged together. Their sense of social distance does not allow for close encounters. The passenger pigeon was unique in its ability to live at close quarters. Predation occurred around the perimeter of the flock, but within the body of the roost they were carefree. Their fertility rates were so high it made up the difference due to predation and then some.

The mob-like behavior of the passenger pigeon got them into real trouble for other reasons on occasion. A Pennsylvania flock got lost offshore in 1740 as reported by Peter Kalm. Incoming ships observed its tens of millions of members floating on the waters of the Atlantic over a three-mile area. The group behavior of this bird was so fixed that an error made by the lead bird or birds was sufficient to condemn the whole flock to a watery grave, but it took only a decade to repair the damage from other flocks in neighboring states.

The pigeon family, Columbidae, occupies all land habitats on the planet except in the extreme high latitudes. They are humorless compared to the towhee; they are at the opposite end of the spectrum of curiosity when compared to the jay or the crow. Unlike any other bird, pigeons and their smaller relatives the doves can swallow water without the need to tip back their heads. Pigeons have no gall bladder; they are vegetarian and are therefore good to eat. Pigeons, as the father of ethology Konrad Lorenz pointed out, are normally very effective in avoiding predation by their rapid takeoff and flight. Predators like the red tailed hawk are too big and have almost no chance catching them on the wing even though one will occasionally give it a try, indicating the desire is still there.

There is a rule of thumb in biology concerning size and climate. As an animal species move poleward, its size increases. Larger size reduces the ratio of surface area to body volume; therefore the potential for heat loss is less. Obviously there is an outer limit to usefulness of increasing size, especially in birds. Size put the passenger pigeon under severe survival stress as a result of another family characteristic. Pigeons have flat feet rather than grasping claws. While this enables the pigeon to walk on the ground rather than hop, the pigeon family is noted for its crude insufficiency in nest building. Even Mother Goose made note of this: "The dove says, coo, coo, what shall I do? I can scarce maintain two. Pooh! Pooh! says the wren, I've got ten, and keep them like gentlemen." The difference was in nest-building capacity. A high level of pigeon mortality comes from eggs and chicks falling out of their inadequate nests. The passenger pigeon was a large bird; it was as large as pigeons ever get. Large size puts a premium on nest building; its nests were a shambles.

Most tropical pigeons are dark-colored birds and stay hidden in the forest canopy. An exception is New Guinea's crowned pigeon, the male of which is a fine-feathered rooster of a bird. The male is simply a sperm in the height of fashion, nothing more. Outside of the tropics, nearly all birds are monogamous and the male helps the female to some degree. A large bird like the passenger pigeon brought the family shortcoming in nest-building to a head. Unlike the huia, whose survival strategy involved a profound interlocking specialization between the genders, the passenger pigeon solved their quandary by the male joining its female mate in behavior following courtship. It was hard to distinguish between the male and the female passenger pigeon either in appearance or in behavior. As soon as one parent returned from foraging, it replaced the other parent sitting on the nest.

The remarkable fertility of this bird was a great success, but they could not have foreseen what would happen when they adapted to the newly arrived presence of Indo-Europeans. The beginning of the end would come when we converted this plague of birds into a feathered cornucopia with gunpowder-propelled nets. Trainloads full of barrels of live pigeons were regularly trapped and shipped into the new American cities to feed the poor at a penny a bird. Peter Alden informs me that the railroad and the telegraph also conspired to reduce the numbers of these birds. As lumbering destroyed their habitat they moved west. Nevertheless, as the railroads pushed into the most inaccessible areas of the country the telegraph lines that accompanied them send forth the news of newly discovered flocks of birds. Professional gunners and amateur sportsmen boarded the trains to continued the slaughter.

The passenger pigeon flocks were reduced from billions to millions, and then to hundreds of thousands, yet this by itself did not bring the bird into extinction as is generally supposed. When the numbers of birds had been reduced to mere tens of thousands, they were no longer economical to hunt. The hunting ceased, and once again they were on their own. In a letter from Dr. Bachman to Audubon, it was observed that the birds no longer bred in communities. The nests were now scattered throughout the woods in northern Michigan, seldom near each other. The nests were still inadequate to do their job, and eggs or young birds were frequently found at the foot of the tree.

Overpopulation, the pinnacle of success for the passenger pigeon or any animal, created a variation on the theme of the loss of an ecological niche. A species may become extinct, for instance, if the tropical environment that it is adapted to cools down and dries out; a species may become extinct if its niche is rudely destroyed by some form of planetary violence; and a species may become extinct if it is so successful that it destroys itself by overburdening its natural resources and by putting itself on the menu of every predator in creation. Overpopulation itself is not the proximal cause of extinction; overpopulation sets up the situation for extinction when the inevitable crash occurs. It is secondary and tertiary factors that weigh in after the fact that make the difference. Once the passenger pigeon was punished for their social success, they were thrown back into the jaws of the original dilemma: their size and their nest-building shortcomings. There was one other important factor that is revealed by examining their cousins.

The band-tailed pigeon is as large as the passenger pigeon. It lives discreetly out West. The genetic difference between the passenger pigeon and the band-tailed pigeon is probably insignificant. These pigeons prefer low- to mid-level conifer or mixed conifer forests, and they eat berries, acorns, seeds, and nuts. Two times a year they congregate around mineral springs to ingest salts. They form loose colonies of fifty birds. The nest is a rudimentary platform of twigs that the male gathers and the female puts in place. They both sit on the eggs; they both nurture the young with pigeon milk that is produced in the crop. They migrate as a flock to find new sources of food.

It seems that the most important difference between the band-tailed pigeon and the passenger pigeon was their original association with Native American peoples. The Algonquin-speaking peoples of the East practiced high-level farming of corn, beans, and squash, among other things. This advanced agriculture allowed these people to reach a population density of 287 people per square hundred miles as apposed to forty-one for hunting and gathering peoples. The passenger pigeon thrived as well. Unlike the Native Americans who did so poorly when Europeans arrived, it is likely that the passenger pigeon population got even greater at first because of the widespread success of European agriculture powered by the horse or the ox pulling the iron plowshare. In the West the Native American peoples did little farming. Tobacco is all that has been noted. The band-tailed pigeon was originally observed in large flocks, but with the arrival of Europeans with their guns they were not prone to any sort of close association. They were able to survive by dispersing.

When the passenger pigeon was forced to disperse they had to return to the challenges of reduced birthing efficiency. The passenger pigeon had to face the challenges of reduced birthing efficiency as well as a radically changed diet. No new behavioral innovation was adopted by them in time to forestall extinction. The replacement rate was below what was necessary to sustain the species, and they died out. Their devolution was complete. The band-tailed pigeon had always stayed with its native forest diet of seeds and acorns. The old habitat of the passenger pigeon is now being filled in by that well-traveled migrant, the collared dove, which has been introduced from Europe. A domesticated variant, the ringed turtledove, lives in dovecotes on our roofs and in our backyard aviaries. It does not do well in the wild. It lets hawks, such as the Cooper's hawk, get too close.

The Victorian theory of evolution was a celebration of change that was no real change. At the pinnacle of empire, devolution was already clearly on the horizon. The island nation had already gone past the point where it could feed itself without the empire. It reached its peak in the age of steam. The first working internal combustion engine made its appearance in London in the middle of the nineteenth century. It was ignored. It was well behind the other industrialized civilizations on the European continent in both the arts and sciences. The savage rate of cultural evolution that had already occurred was staggering enough for this aging culture. The capital city of the great empire was an astonishing cesspool of rot and corruption. *The Communist Manifesto* documented the moral imperative for the devolution of the Victorian empire. *The Origin of Species* offered a scientific justification for no change. Other cultures would lead the next stage of social revolution. The chickens would come home to roost a century later following two pyrrhic victories against a younger, intellectually advanced empire. The bankrupt empire collapsed, leaving 50 million people on a small island that will have grave difficulty supporting 5 million after the age of oil goes up in smoke. The Malthusian hammer will eventually come down. For those who read history it is all too familiar. As Zen monk Alan Watts observed, at least it proves that we have free will.

CHAPTER 13

The Age of the Mammals

WE ARE DEEPLY PERPLEXED by the paleontological record that shows cases where life forms change almost simultaneously throughout the world. We wish to believe that evolution will be slow and that the variability of each species is independent of all others. The past does not predict the future. In reality the slow, erosive processes of wind, rain, and sun over billions of years would have reduced the planet to a sterile, lifeless desert long ago, assuming that creation could have occurred in the first place by that formula. There is an unpopular point of view called the Shiva Hypothesis that holds that mass extinctions occur with startling regularity as the movement of our solar system takes us through the galactic plane, a passage that is thought to perturb millions of Oort cloud comets, some of which cross the path of Earth. According to the Shiva Hypothesis comets, along with the mass extinctions they cause, are the piston that drives earth's biological processes.

It is certainly the case that the basic building blocks of life rode in on the cosmic express. In the nursery of life on Earth, meteor and comet impacts were the very stuff of evolution. In energetic terms the Second Law of Thermodynamics is temporarily reversed by a major cosmic impact: time is reversed, the energy state is raised, the primitive flux of the molten core of Earth pours forth. This is the devil in the details that we wish to ignore. Life is not random and orderly; life is chaotic and symbiotic. What is allowed in terms of biological complexity is predetermined by the ambient energy level.

As the energy level goes down, greater complexity is allowed, up to a point. It is impossible to separate evolution from devolution.

Early on cosmic, impacts were not untoward events. The Earth cooled as the impacts became less frequent. They also began to come out of the background and take on a more defined role of triggers and filters to more complex levels of evolution extinction. Much would depend upon the size of the impact, whether it occurred on the land or the water, whether it occurred at an angle or straight on, and of course, where it occurred. Ironically, the most random event for the more orderly, higher states of biological evolution are the very events we have outlawed as too depressing to think about: cosmic impacts.

The most primitive levels of life remained unchanged as if nothing had happened at the K-T extinction, but many of the varieties of single-celled life that lived in the first few inches of the surface of the water became extinct at that time. Through the bottleneck of survival would come a very reduced range of flowering plants. These would recover more quickly than animal forms and flourish, having very effective means to take a year, or a decade, or even more off if necessary before the vegetative part of the cycle needed to be played out to restore the species. The big dinosaurs had been dependent on a relatively high level of ambient energy and a lush landscape. The K-T extinction event was so momentous that it is estimated that no land animal over twenty-five pounds in weight survived. The mammals that lived underground or in caves were among the survivors.

Innovative intelligence is not of much value compared to a well-adapted instinct in a stable environment. When habitat is suddenly lost, however, habit may well lose the day. The new singers and dancers of higher complexity would be those who had already laid the groundwork for more energetic metabolisms and more highly developed nervous systems. The crash of the ecosystems with the impact event brought on inbreeding followed by hybridization. In some cases, this would have cleaned up some of the genetic damage that had accumulated for one reason or another. In other cases the level of genetic damage would have been so great that inbreeding would have simply finished off the job of devolution. Since we are going through our own bottleneck of extinctions, it is now possible to observe this in the wild.

In a 1978 document, *Wildlife and America*, sponsored by the U. S. Fish and Wildlife Service, S. Dillon Ripley and Thomas E. Lovejoy contributed an article on crashing and remnant populations as well as colonizing populations where gene flow is naturally limited, to put it politely. They mention several cases of reduced populations that, nevertheless, had shown no problems from inbreeding or homozygosity. They also mention the case of the Polish Wisent, reduced to sixteen individuals that had been free of lethal genes but that had reduced breeding success when a male Caucasian Wisent was added to the mix. Their conclusion was that, "The deleterious results of inbreeding are not well documented in wild populations, and natural situations, at least for the moment, remain mostly in the realm of speculation."[1]

After a massive extinction event, the surviving species with their small numbers were in the position of immigrants to a New World: they were free of their normal predators, pests, and plagues. For quite a long time the oceans were warmer, the atmosphere restored, and the soils enriched, and the tropical planet would return. The snakes, frogs, alligators, and the birds, would be the very reduced relatives from the Age of the Dinosaurs, although the birds would exceed all in their numbers. Mammals would not show up again in the fossil record until 10 to 15 million years after the K-T event, but the warm-blooded mammals would now have their chance.

Among the stranger transvestites of the early Eocene epoch were the Mesonychids. They were carnivorous ungulates that looked like a wolf and had tiny hooves instead of claws. They preyed upon other ungulates such as the browsing Chalicothere, which had an equine head, moved most of their weight on short, powerful rear legs, achieving balance with long forelimbs with curved claws. They apparently knuckle-walked like the modern anteater. An explosion of rodents in this age would be followed by an explosion of snakes. These were more familiar in appearance. On the whole the Cenozoic era would become the Garden of Eden for the Age of the Mammals.

The internal fertilization practiced by all mammals involves the ingestion of the male sexual gametes by the female, who fails to release her eggs. At first glance this would seem to be a dead end, the reflection of some sort bad timing carrying with it the potential for serious problems. Internal fertilization can be seen in some insects. It completely changes their behavior. The mother of such predacious insects has to find an appropriate host for the eggs in time or be eaten by the hatched larva herself. This does not turn out to be a critical problem for a species since the mother does not provide any assistance to the young and will soon die anyway, but the mothers involved seem to have no inclination to simply acquiesce to this final solution. As we have seen, the egg-laying reptiles survived the hazards of retaining the fertilized egg. The egg was surrounded by a calcified shell as if it might be perceived as alien, and in fact the eggshell may have given it protection from the mother's immune system. Getting out of this enclosure even after being successfully laid was a challenge, but specialized methods of escape evolved. As we see in some frogs today, internal fertilization became a solution to surviving in a drier, hotter, and cooler environment. Frogs with internal fertilization did not need to return to the water to reproduce, but the numbers of offspring were smaller.

What would distinguish the marsupials and the mammals from the insects, the reptiles, and the duckbilled platypus is that the egg, rather than being isolated within the female body, was joined to it. Although the immune system that produces antibodies has to be neutralized in order for the sperm to survive in the female's body, it is the inflammation of the endometrial lining by the immune system that is necessary for implantation to occur. Joined to the maternal blood stream by a new organ called the placenta, the need for yolk was dispensed with. Perhaps, like the orchid, it had gone missing, or like the hymenoptera after the sawfly, it was insufficient.

The gender of the mature embryo-bearing adult as a female was now as firmly established as it can be in a complicated life. Since the fertilized egg has DNA and protein from a foreign body, one can imagine the possibility of an attack by the immune system is still a possibility. This can happen when a pregnant woman has Rh-negative blood and the baby in her womb has Rh-positive blood. Unraveling how two sexual beings, which are in some ways as different as two species, remain different phases of a single, founding, non-sexual organism is a fascinating story.

Unraveling Turner's syndrome made a major contribution to the deconstruction of the conventional definition of gender in biology. Turner's syndrome is a behavioral disorder associated with the loss of all or part of an X chromosome in women. One third of Turner's girls have autism-like social and communication difficulties despite normal verbal IQ. Normally autism is ten times more common in males than females. Since it is has become possible to identify whether it is the paternal X chromosome that is missing or the maternal X chromosome that is missing, David Skuse decided to identify the origin of the X chromosomes the two types of Turner's girls. The researchers were surprised to discover that the Turner's girls with their father's X chromosome were significantly better-adjusted and more empathic in character. The Turner's girls with the autistic profile, identified by tests to measure the ability to recognize emotions as conveyed by facial expressions, had their mother's X chromosome. For Anglo-American biologists marinated in Thomas Hunt Morgan's strict dogma that males are males and females are females, this was a totally unexpected and unexplainable result.

Japanese biologist Yoh Iwara at Kyushu University was able to offer a reasonable solution. He proposed that it is only the father's chromosomes that normally determine sex. If the egg receives the paternal X, the result is a female. If the egg receives the paternal Y, the result is a male. Sperm, therefore, are either male sperm or female sperm. If characteristic female behavior were to be expressed by the maternal X chromosome in combination with the male Y it would tend to feminize a male; if it were to be expressed by the maternal X in combination with the father's X chromosome, it would tend to show over-expression in a female. Both of the female's X chromosomes are normally blocked. It turns out that the normally blocked X chromosome in the mother carries a tendency for autism. The absence of the male X chromosome allows the mother's autistic X to express itself. As Skuse et al. put it in a recent paper, "X-monosomy may 'unmask' social disability; for example, mild features of autism, such as failure to understand nonverbal social cues, poor maintenance of eye contact and limited 'Theory of Mind' skills ..."[2]

Further enlightenment would come when another parentally imprinted gene called the Mest gene was discovered in mice. Female mice that lack a working copy of this gene make terrible mothers. Only the gene inherited from the father is functional. The mother's gene is switched off. Effective mothering is determined by a gene that females have received from the male in these mice. Other clever experiments have been done to produce viable eggs with two mothers but no father, and with two fathers but no mother. In both cases embryos failed to develop properly and soon died in the womb. However, as reviewed by Matt Ridley in *Genome*, in the two-mothers case, the embryo itself was properly organized, but it could not make a placenta with which to sustain itself. In the two-fathers case, the embryo grew a large and healthy placenta and most of the membranes that surround the fetus. Inside, where the embryo should be, there was a disorganized blob of cells with no discernible head. The placental genes were expressed in the two-fathered eggs; the embryonic genes were expressed in the two-mothered eggs. There is a clear separation of function along the male line for the development of the placenta and along the female line for the development of the fetus.

David Haig at Oxford attempted a typical sort of explanation for this as follows: the placenta that comes from the male is designed to parasitize the maternal blood supply, forcing its way into the vessels, and then to produce hormones raising the mother's blood pressure and blood sugar. The mother counters these moves et cetera, et cetra. His reasoning is a variation on sperm competition. Haig's theory of gender warfare predicted that parentally imprinted genes would be rapidly evolving, because sexual antagonism would drive a molecular arms race in which each benefited from temporarily gaining the upper hand. A species by species comparison did not bear this out. In fact, parentally imprinted genes seem to evolve quite slowly if at all. As Yogi Berra would have put it: in theory practice and theory are the same; in practice they aren't. The fact that Haig could conjure up such an interpretation is not surprising in the present environment. His only opposition is Charismatic Christianity. The fact that Haig's one prediction turned out to be dead wrong does not seem to matter much. He is singing to the choir.

We have traced the origin of sexual reproduction or meiosis to protozoa where the separation of the two strands of DNA acted as a metabolic switch. The reduction of metabolic activity by meiosis allowed simple organisms to survive states of resource impoverishment or severe climate. Correction for genetic damage occurred during recombination of the two strands of DNA. This was another, equally important function of meiosis. The basic nature of gender specialization occurred long before it was embedded in elaborate metazoan phenotypes with what we think of as sexual motivations.

Originally genetic variety was accomplished in life by the sharing of genes by protozoans. This was followed by the formation of hybrid organisms by endosymbiosis or ectosymbiosis. The doubling of the genes (polyploidy) opened up other possibilities for variation. We are now beginning to see the co-evolution of the simplest organisms (viruses) with the most complex (mammals) under our very noses. Virologist Keizo Tomonaga at Osaka University has found that our own DNA has kidnapped viruses in its vicinity and incorporated them into our genetic self. Thierry Heidmann of the Gustave Roussy Institute in France has shown that one such captive virus is responsible for making syncytin. Syncytin is the protein coat made by the virus that affords it protection. Secondarily this protein allowed the virus to fuse with cells as a first step in invading them. Found in all mammals, syncytin is the protein that allows the placenta of the mammalian embryo to fuse with the mother. Without syncytin, without the capture of the syncytin-producing virus, no mammals. Evolution of this sort was beyond the wildest dreams of Charles Darwin and is a nightmare for Darwinists to try to explain.

Another key to mammalian evolution is oxytocin. Oxytocin is a small peptide hormone that is celebrated for stimulating and orchestrating maternal behavior in metazoans. This molecule goes back to the earliest stages of Animalia and to a gene duplication event that is dated to be about 500 million years ago. It is produced in the hypothalamus that sits on top of the brainstem. The hypothalamus is well connected in the nervous system, and it responds to the circadian rhythm of day to night, to pheromones from the outside, and to hormones from within, as well as to neurological input from the stomach, heart, and the reproductive tract.

The connection between the nervous system and the hormone system is mediated through the pituitary gland of the hypothalamus, where oxytocin is produced. Oxytocin's slightly altered genetic twin is vasopressin, which helps to control water regulation and raises blood pressure. Released in the brain, vasopressin stimulates pair bonding and, paradoxically, induces the male to become aggressive toward other males. Herein we have the chemical inducements that allow the continued specialization of the egg and the sperm into male and female, all stemming from a common ancestor. The expert in behavioral neuroendocrinology at the University of Illinois at Chicago, C. Sue Carter, has nominated oxytocin and vasopressin as the Eve and Adam of

Mammalia, further suggesting that an appropriate cocktail of the two celebrates the possibility of monogamy between two very different organisms.

It turns out that it is not the amount of vasopressin that is so important; it is the pattern of vasopressin receptor sites in the body. Many vasopressin sites enable prairie voles to be monogamous. Mountain and meadow voles have few vasopressin sites and are not monogamous. Hasse Walum at Karolinska Institute has found a similar correlation in human males. Men with a certain gene variant that produced fewer vasopressin receptor sites had exacerbated autistic personalities and were markedly less successful in sustaining the Judeo-Christian social preference for monogamy.

In the early stages of metazoan evolution hybridization occurred due, apparently, to the miscarriages in the method and the madness of the specialized genders. The complications related to integrating evolving metabolic systems were daunting, and the possibilities for genetic confusion great. Specific gender identities associated with meiosis developed new dimensions of variability and of eggish behavior and spermish behavior. The evolution of gender has become a textbook example for the dimensions of genetic change that are still possible in the genome.

Like the Hox genes, sex specialization in the chromosomes is a deeply embedded genetic tradition, but the most elaborate expressions of gender are a more recent development. The latest theory about how gender including the secondary sexual characteristics came to be is that an allele or variation occurred on what had been a pair of identical chromosomes in an otherwise non-sexual organism. The recombined chromosome failed to correct the variation. This new gene, at least in the case of marsupials and mammals, belonged to a group that encoded for a type of protein whose function was and is to aid in transcription, replication, recombination, and DNA repair called the high-mobility group of proteins (HMG). In short a repair gene had became a repair problem. Although people with autoimmune diseases can have antibodies to HMG proteins, normally they are seen as friendly, so no correction of the botched correction was forthcoming.

The malfunctioning HMG gene created an area that was sheltered from correction. Once this sheltered area was established, it would evolve into the sex-determining area (SRY). Other genes migrated or translocated to the SRY region to be preserved under the umbrella of non-correction. The normal chromosome of the pair continued to reflect the normal, non-sexual genetic tradition of the organism. As we recall from above, the two mothers XX embryo developed a proper embryo, an ancient sacred tradition. The formation and insertion of the placenta into the body of the female—a sort of secondary version of the male fertilization of the egg—came along with the evolution of the X chromosome into the male Y chromosome.

The SRY area was a foreign embassy that would tolerate foreign bodies not subject to ancient traditions. In marsupials and mammals the sheltered SRY gene produces a protein that regulates other parts of the genome to create the male testis. From the testis comes testosterone. Testosterone is a slightly altered, which is to say non-corrected, form of the steroid compound estrogen. Estrogen and testosterone are major factors in the determination of sexuality from beards to breasts. In mice the male brain develops in the very early stages of embryological development under the influence of testosterone that is converted by an enzyme into estrogen. The female brain is the default state developing when there is no testosterone or estrogen present. From the testis also comes antimullerian hormone (AMH). AMH blocks the development of the normal female reproductive organs in the male. Sexual reproduction, particularly in the male of the species, is a revolutionary biological evolution layered over a well-established biological evolution.

In birds, some fish, and some insects there is a ZW sex-determination system rather than an XY system. In this case it is the female's chromosomes that are heterogametic (ZW). It is the ovum that determines the sex of the offspring, and the male's chromosomes (ZZ) are turned off. It has been found that the Z chromosome in birds is similar to non-sexual chromosome number 9 in humans, so the non-correction SRY incident in the genomic histories of the two were independent events. It seems not to matter which chromosome will be involved, which gene will be corrupted, or which gender will carry the active mismatched pair. We recall that the platypus has both the XY and the ZW systems separated by three other paired sex chromosomes. This means that it is an organism that has come after the original foundation of the separate sex-determination lines of descent and must, therefore be entered as a candidate on the list of hybrid organisms to explain how the platypus acquired both bird-like characteristics and marsupial-like characteristics. I would love to hear Haig's explanation for the platypus.

Since recombination with a normal X is impossible, correction eventually does become a problem for the Y chromosome. A secondary mode of local gene self-correction called palindrome correction has been discovered by David C. Page of the Whitehead Institute. Even so, errors in duplication are still an Achilles' heel of the Y chromosome. Retrogenes, such as the retrogene that created the short-legged dog, often hid in the Y chromosome. It is also the source of a wide variety of sexual disorders. Depending upon where the error occurs in the chromosome, a complete range of sexual intermediates from male to female are possible. As reported by Nicholas Wade in the *New York Times*, Page describes this as a fantastic experiment in nature. This confirms the work of Alfred Charles Kinsey done over fifty years ago, and it also suggests the what we label as a sexual disorder depends to great degree upon the social organization that it is born into.

Whole sections of the Y chromosome have been turned off because of harmful mutations. Compared to the 1500 working genes on the X chromosome, there are only seventy-eight working genes on the Y chromosome in humans. It is small and getting smaller. In the rodent-like marsupial called the dunnart, there are only four working genes left on its Y chromosome. Not much Y chromosome is needed for a functional marsupial male, nevertheless I suggest that sexual specialization, so creative in its theme and variations, does has a shelf life. The devolution of DNA in organisms that have adopted sexual reproduction can be profound as we saw in the ants. In their case the damage to the male sex is so profound that reproduction has returned to parthenogenesis or cloning, in some cases with the aid of a bacteria. The male sex, where it is still involved, is merely a spermish timing device. Highly developed SRY sexual reproduction is a high-risk approach that offers the possibility of variety. However, species extinction by spermacide probably occurs when the burden of error becomes too great. It has probably already happened to many species even through the proximal cause of extinction was the result of some other event.

As we have seen, sexual specialization has affected the brain itself. Simon Baron-Cohen at Cambridge is well known for his theory that autism is an extreme form of the normal "male brain." The empathic tendency of the "female brain" reflected in its capacity to recognize emotions in others, especially fear, is a necessary component for the parent responsible for child care in marsupials and mammals. It is also necessary for dealing with the emotionally closed-off and potentially irritable male who may be suitable for the defense of territory. In this we recognize a postmodern interpretation of gender, replacing the heroic definition of the male that replaced the divine definition of the male created by our Judeo-Christian tradition.

For practical purposes we have to reduce our understanding of creation to slick formula and the social fashion of the moment, but we have seen the most rudimentary dialectic of gender expression explode into a star burst pattern of overlapping yins and yangs that has synchronistically created new patterns and different perceptions with a rainbow of results. It is one of the main themes of evolution in the animal kingdom. The capacity to see the genders as a dialectic, as in the war of the sexes, is an attractive illusion. Fundamentally we are talking about a single organism with a generational phase change that must negotiate with variety in order to survive. It is a broad reflection of genetic, embryological, and all manner of epigenetic influences within and between species all the way up and down the line.

It was a very simple distinction that would create the separation between the marsupials and the mammals. The distinction was a simple difference in gestation time. The marsupials are characterized by short gestation periods. They are also capable of suspending embryological development with the fetus hibernating in the womb when external conditions are harsh or there is a shortage of food. Marsupial neonates are outrageously premature when birthed from a mammalian perspective. In comparison to the adult they look like larva in their level of development. Marsupial neonates are protected in their mother's pouch, where they develop rapidly. Marsupials produce large numbers of offspring. A kangaroo has a life span of only four to six years. They represent the continuation of an older norm with their solution to reproduction emphasizing huge numbers a short life span. The marsupial way of life favors genetic correction over sexual innovation and the challenges of character development.

The placental mammals are characterized by long gestation periods. This allows for greater neurological development, but with lower numbers of offspring. This imposes a greater requirement for maternal involvement. Long gestation period would go so far in the mammals that a sojourn of nine months in the womb would be followed by a decade and a half of continued maturation before sexual maturity was reached in *Homo sapiens*. The retention of infantile characteristics for such a long period from the womb into adulthood is called neoteny. It is a major theme in the evolution in the mammals.

In the oxygen environment of earth the corrosive effect of free radicals created by oxygen reactions is a major factor in biological aging. Most of the mutations that cause longevity, at least in flies and worms, seem to be in genes that block the formation of free radicals. As we have already seen, the somatic cells in the multi-cellular organism have their telomerase gene switched off, therefore the genes of somatic cells age and eventually die when they are no longer able to produce coherent chemical information. The replication and growth of somatic cells in the multi-cellular body have to be highly coordinated in order for the body to function as a whole. Cancer, primarily a disease of aging, is an example of uncoordinated growth. It appears that the original size of a telomere is roughly correlated with the expected productive life span of a multi-cellular organism. At some point the accumulation of genetic damage is simply too great, and we all return to the basic genetic cell line for gene correction and refreshment. Anything that is learned by a particular generation must be passed on behaviorally.

At first, among the small forerunners of the marsupials and the mammals the difference between them would be insignificant. The marsupial opossums have changed little since their fossils were first identified in the Cretaceous period. The lack of change for such a long time suggests that DNA correction at meiotic recombination has been greatly enabled by the very short lifetime in the marsupials. The relatively rapid genetic review and repair of the germ cell line keeps the genome well conserved and clean. Among the opossums, whose life span is short, whose broods are large, and whose parenting demands are limited, aging begins in the second year of life. Opossums that have been isolated from the normal onslaught of predation on offshore islands live longer and begin to age later, but not much longer and not much later. In general the spermish approach epitomized by the marsupials fills the environment with seed, to hell with the mortality rates.

Among animals, especially the mammals, where the hazards of internal birthing are great and the needs for effective parenting are also great, ten to a hundred years or more is the greatest age for an adult to live. The eggish approach epitomized by *Homo sapiens* conserves resources and emphasizes intelligence, and there is a longer time to integrate and embed genetic variations. Among the larger mammals, greater intelligence would be needed in the adult just to be able to survive the severe conditions of birthing concurrent with long gestation, conditions that would make the mother and offspring vulnerable to predation. Yet a higher neurological development was a pure advantage, and the mammals would radiate and become the dominant strain even before the beginning of the Cenozoic era. First there was organic evolution, then biochemical evolution, then embryological evolution, and finally would come social evolution based upon the evolution of the nervous system.

As one would expect from cell lines that have been diverging one from the other for a long time from a common ancestor, the kangaroos and the koalas look and act somewhat differently than the bears and the beavers. In spite of a genetic separation of millions of years, however, there have been some striking convergences. Given the skulls of the marsupial wolf of Australia and the mammalian wolf, only a taxonomic expert can tell the difference, yet gestation in the marsupial wolf was measured in weeks; gestation in the mammalian wolf is measured in months, and the mother suffers travail at birth. A small genetic difference is magnified by the changed conditions of embryological development in the mammalian wolf. This, in turn, is reflected in ethology. The marsupial matured quickly and developed no social skills. The marsupial wolf hunted alone.

There was a remarkable similarity between the marsupial *Thylacosmilus* of South America and the saber-tooth cats of North America. *Thylacosmilus* became extinct when North and South America were joined during the Pliocene epoch and the saber-tooth cats went south. By outward appearance they looked the same, and they were both predators of herbivores. Underlying their phenotypical similarities was a relatively small genetic difference stemming from their common ancestry. It is when we step away from the basic genetic similarities, the Hox genes for instance, that the real differences begin to emerge. It is likely that the saber-tooth cats were social animals, as are some of the modern big cats. This gave them the upper hand over the non-social, marsupial *Thylacosmilus*. The contrast between the two strains shows how genetic strains can meet similar environmental imperatives and end up looking the same, but how a very small adjustment in embryological development can lead to quite different behavior and extinction or survival. The saber-tooth ate *Thylocosmilus's* lunch.

The largest of the mammals are among those that returned to the water, the cetaceans. Norihiro Okada of the Tokyo Institute of Technology believes that the hippopotamus and the whales share a common ancestor. The whale story begins with a small, furry, meat-eater that walked on hooves around 50 million years ago called Pakicetus that has been unearthed in the Himalayan foothills. A closer relative of the whale called *Ambulocetus natans*, the walking, swimming whale, apparently took up life at the edge of the sea. In the conversion from landlubbers to seafarers the nostrils of *Ambulocetus* had drifted from the end of their snouts to the top of the head. The four-legged *Ambulocetus* still had well-developed rear legs and a pelvic girdle to support them. A toothed whale fossil *Maiacetus inuus* dated at around 47 million years ago has been found in Pakistan. It was a pregnant female. The fetus was oriented to be delivered head first, which is common for land animals in order to allow breathing to begin as soon as possible. Modern cetaceans normally delivery flipper first in order to promote swimming first. Apparently *Maiacetus* was returning to land to mate, for parturition, and perhaps to rest. This whale had four flipper-like legs that could support the animal for short distances.

The modern whales and dolphins have their pelvis, legs, and hind feet reduced to mere vestiges. The vestigial hind legs of the whale now act as an anchor for the muscles of the genitalia that need to be big, very big. The front legs and feet have become flippers. The cetaceans have conveniently combined the big brain of the mammals with the ease of parturition of the marsupials. Because of their anatomy, with no pelvic bottleneck to traverse, the whales and dolphins could maximize the period of gestation and still bring to birth a fully formed neonate without having to traverse the pelvic girdle. Along with the buoyancy provided by the water, the lack of a bony birth canal allowed the body and the brain of the whale to evolve to great size: up to twice the size of the largest dinosaur, up to twenty times the largest elephant.

The cetaceans seem to have more brain than they actually need. While cetaceans all use echolocation by sound to find their way around in the marine world, the humpbacks go well beyond pure utility and use their large brain and well-developed auditory sense for the creation of long, complicated, seasonally-changing songs. Why they do this is beyond the ken of those of us with a need for a Puritanical concept of life. What would be the survival value? Any sort of sound-making capacity is sufficient to the purpose of communicating presence, territory, and sexual state, and it serves to keep the group together. Creativity is the main theme in humpback communication. The true measure of the intelligence of the cetaceans can be seen in the symbiotic hunting relationship between humans and orcas in Eden, Australia, where the prey is the baleen whale.

As recently as fifteen years ago the transformation of the whales was poorly understood and quiet miraculous. Creationists held them up as examples of species that could not possibly have come about by natural selection. They must have been the bemused creation of a good-natured god of the New Testament. Now we can see, as Hans Thewissen suggests, that whales underwent the most dramatic and complete transformations of any mammal. Whales are offered as one of the better-understood examples of evolution, but which theory of evolution? Occasionally a modern whale is born having sprouted a leg or two. I think it is safe to assume that the Hox genes that played a role in fin development in the paddlefish, and were the same blueprint for limbs in land dwellers with a similar neurological pattern of activity, are still in play as the land animal returns to the sea. This is called an atavism. Giving it a name doesn't explain, it however. It only explains it away.

The fact that animals can be remodeled from ancient blueprints does not fit our need for a theory that a uniform rate or microscopic genetic damage is the method of evolution. In the same genome where the Hox genes are shouldering their way forward changing not at all, there are areas of the genome that are changing more rapidly than theory allows. The major histocompatibility complex is such an area. This area plays an important role in the immune system. The astounding diversity of genes in this area is created by gene duplication and promoted by animals seeking out mates that smell different from themselves in their histocompatibility profile. This builds a library of pathogen protection that is much broader than what an inbred family might carry. It creates a moving target for pathogen evolution to keep up with. With our nearsighted attention to the species level of change, we miss the insight that it is the coherence of a cell line that is the main consideration of metazoan evolution. Species are just fashions of the moment. A coherent cell line is one that can change with the flu season as well as preserve billion-year-old metabolic rituals.

The biggest climatic disturbance in the Cenozoic after the K-T extinction event that marked its beginning occurred at the Eocene-Oligocene boundary. The falling temperature was correlated with the decline in carbon dioxide levels. The carbon dioxide levels were about the same place they are now. Halfway through the Cenozoic era the first permanent ice sheets formed in the Antarctic. The planet that had been tropically forested from pole to pole began to open up. Savannahs appeared. The grasses began to expand, shaping the birds and animals that fed on them. It was getting hotter during the day, cooler during the night, and drier. The first blossoming of mammals was replaced by a new more familiar set. As the Oligocene and the Miocene proceeded to the Pliocene epoch around 2.3 million years ago we would begin to recognize dogs, raccoons, horses, beaver, deer, camels, whales, and the crow in the fossil record. Coming out of the tropical jungle about the same time as the hominids was the cat around 7 million years ago.

The male lion is the king of the jungle in human fantasy. The male lion kills off all competitors, including all offspring in his pride that are not his own, thus making him a poster child for patriarchal culture. In reality the male African lion is now largely out of the running as a predator at 450 pounds, except for large, slow, but still dangerous mammals such as the buffalo. The male generally leaves the hunting to the females, who cooperate as a pride using deception and ambush. In fact, it was probably the need for females to do cooperative hunting as well as the convenience of reducing the number of quarrelsome males that allowed the lion to be the only sexually dimorphic cat. The male lion's mane is a luxury like peacock feathers and it has no real survival value. The mane of the male is a vanity in the eye of the lioness.

E. O. Wilson speaks of the dangers of excessive homozygosity (inbreeding) in *Sociobiology* and looks for behavior in the wild that would avoid incest. "Virtually all young lions, for example, leave the pride of their birth and wander as nomads before joining the lionesses of another pride."[3] But then he admits this may be due to the low rank that they occupy in relation to the lionesses of their birth pride. It is hard to become a lion with your mother around.

In the harem-like setting of three or so females to one male, it is more than size that keeps the male out-of-shape for hunting. Cats are induced to ovulate by the physical act of copulation, and it may take several attempts. It is estimated that it takes 3,000 copulations for one cub that survives. In this polygamous relationship, an active sperm-carrying male is all that is required, and if the male can drag down the occasional aged wildebeest, so much the better. The male lion loses his virility in two to four years and is replaced. The female lion looks like she is fully prepared to extend the race by parthenogenesis, but there is one problem. All primary and secondary gender distinctions come from the male, as we have seen with *Homo sapiens*.

The African lion is now protected in preserves. The Asiatic lion that once roamed the hills around Haran threatening Jacob's flocks is now down to about 300 members in the Gir Forest in India. The DNA prints from a few individuals look like prints from identical twins. This is a result of extensive inbreeding. Although genetic abnormalities are not observed in the inbred lions of the Gir Forest, high levels of immobile sperm are found, bringing up the problem of infertility. As we have already seen, sperm vigor is a timeless, unvarying necessity to keep the races going, but it has no other evolutionary program of its own. It is not connected with the body-building genome. In the case of the Gir lions the low sperm counts are doubtless associated with the overcrowding within the sanctuary as well as pollutants from the farms and factories that surround the park. Added to these factors is the depressing circumstance of existence: the lion's normal physical and psychological character is simply now longer matched to their present zoo-like environment. Under such depressive conditions, sperm count will also be low.

The big cats are generally disappearing in the Old World, especially the cheetah. It is said that the cheetah is in trouble because its genome is relatively damage-free and homozygous as a result of inbreeding. Of the fifty specimens observed, every one is identical with every other. Furthermore, the conventional genetic abnormalities that are assumed to follow from the inbreeding—things such as hip displasia, blindness, intestines outside of the body cavity, two heads, and the like—are not observed. Their problem is not genetic.

The cheetah founding fathers launched themselves on the million-year-movable-feast of grass-fed ungulates in the Age of the Mammals. Over time their genome and physiology became superbly fine tuned to this activity. As the prey became faster, the cats also had to become bigger, faster, and more specialized. The cheetah at 130 pounds can do bursts of up to fifty miles an hour. Now that that prey is virtually gone, they are still dragging around a genetic body adapted and augmented to one narrow, specialized profession. Neither breeding nor the wisdom of a saint can save them from the loss of habitat.

The big cats of the New World, by comparison, are not in immediate threat of extinction even though, in the case of the jaguar, it has lost two-thirds of its original range in the last few centuries. The reason the jaguar is not a threatened species is that these New World cats have a much broader diet. While cheetahs largely limit themselves to animals on the hoof, the jaguars take other small mammals from monkeys to rodents, as well as fish, iguanas, turtles, and birds. In the Old World the cheetah is in trouble because they moved out of the protection of the tropical rain forest onto the exposed, open African savanna where there was an abundance of grazing animals. Such an enticement did not exist in South America in the first place. The bulk of South America lies along the equator. The area of dry tropics that could support large, savanna-grazing fauna is small. The jaguar is still largely a jungle cat.

The small Old World cats, especially *Felis chaus* and *Felis silvestris lybica*, have a wide diet for a carnivore, rather more like the jaguar. They are doing quite well in the wild, thank you, and the domesticated house cat is exploding in population. The house cat is doing well because it has learned to work around its deleterious genetic mutations and to live in a symbiotic relationship with *Homo sapiens*. The house cat has around 150 mutations compared to *Homo sapiens* that carry over 6,000. In a general collapse of *Homo sapiens* culture the cat, which still has most of its wild-type genome still intact, will probably be able to go it alone.

Bad genes should be weeded out, we all agree. Probably even God agrees if he would only listen. It is quite apparent that the weeding-out mechanism falls short of being completely efficient, but how could it be otherwise? How do we even know what is inefficient until after the fact, sometimes well after the fact? Surviving and sustaining a clear-cut genetic deficiency that has no survival value is well illustrated by achromatopia, which is an absence of cone vision. Achromatopia is the completely normal state for the common house cat. The genetic loss of cone vision in the eye is also listed as a rare congenital disease in the general human population. Nobody points to it as an advantage. In the main, achromatopes are treated as dependents and hidden from public view. Yet achromatopia is an accepted variety in insular populations, such as the people on the Pacific atoll of Pingelap where it is found in up to 20 percent of the population. As Oliver Sacks so engagingly described it, Pingelap achromatopes perform unique, specialized tasks such as fishing underwater at dusk, or on a moonlight night, or weaving indoors in the near dark. Achromatopes on Pingelap are still clearly dependent upon the larger population, but in this social setting it begins to take on the appearance of normality.

The house cat has only rod vision. Rod vision is movement-sensitive and is generally useful in darkened conditions. This makes the cat very lively to quick movements, but in the full sun these receptors are bleached out and almost completely ineffective. This is why the house cat pupil is a slit that can close down the light to almost nothing even at the expense of optical clarity. A slitted pupil is only an advantage when combined with another genetic disadvantage, achromatopia. Because of its achromatopia the house cat prefers to be active at dawn and dusk, and to sleep away the day.

The cat also compensates for the absence of cone vision by intelligent cross correlation in the brain between the other senses, from sensitive footpads, active whiskers, and a lively attention to the aural and olfactory environment. When we observe a cat looking at something, the cat may actually be paying at least as much attention to what he or she is hearing or smelling. In fact, olfaction, the phylogenetically oldest sensory system, is the most important sensory system for the cat. It is with its olfactory brain that the cat poses the question to itself, "What is this world coming to?" The result of their sensory cross-correlation is so successful that most people, including many cat specialists, actually believe that cats have excellent vision. The cat has survived a fundamental genetic loss through cross-modality sensitivity funneled through an active memory of practical experience that is adjusted by high intelligence.

Some absences of genetic information for essential biochemistry reflect dependent relationships with microbial life from time before the multi-cellular cell lines were established. Achromatopia is a deficiency that has to be solved with appropriate behavior in the generation that gets the bad gene. Other losses of genetic information may be sustained and accumulated without being immediately recognized. The loss of the genetic information needed to make amino acids may be masked by a diet that already includes them. The house cat has lost the genetic capacity to make the organic acid taurine and to a somewhat lesser extent the amino acid arginine. These are an absolute necessity for the cat to survive. These losses were almost certainly a result of a point mutation. Cats may have survived the loss of the capacity to make taurine and arginine through their carnivorous diet. Other animals produce these lost amino acids for them. Cats produce their own Vitamin C. The loss of the genetic information to produce this in a new brood is unlikely to be sustained since Vitamin C is not a part of their normal diet.

Dietary science is very low on the academic totem pole. Medicine was notorious for being behind the eight ball on issues of diet and its role in good health. If medicine was slow to see that diet is rather more than some instinctive reflex that can be ignored, evolutionary biology has been even slower. Rats, which are an essential part of the wild cat's diet, are unable to synthesize eleven of the twenty amino acids present in their protein structure; their food supply must contain substantial amounts of these molecules. Rats are omnivorous. It is possible that the wider range of their diet has allowed them to lose more of their genetic integrity through mutation, or to never have the capacity from the start. Their omnivorous diet places them lower on the ladder of prey and preyed upon in comparison to the cat, but it is a more sustainable lifestyle. The domesticated house cat is learning to eat an increasingly vegetarian diet, with taurine and arginine added. It has a longer intestine as a result.

Central to the evolution of the mammals has been the pronounced evolution of the limbic system in the brain. The limbic system surrounds the brain stem like a belt as it arises from the spine to enter the base of the neocortex. The limbic system is central to the operations of the vertebrate nervous system. The body is a sea of fluids bathing a civilization of cells. Cells release into the bodily fluids proteins called neuropeptides that influence the activity of other cells in the body. There are upwards of 300 of these neuropeptides. The limbic system is richly imbedded with receptors for these proteins. The limbic system synthesizes the metabolic needs of the body, in conjunction with sensory and hormonal input that has been ordered and cross correlated in the cerebral cortex and with its own patterns of neural activity that constitute memory. Walter J. Freeman has spent a lifetime doing research on the common house cat. His battle-scarred alley cats, abandoned by their owners, gave the last days of their lives so we could apprehend the secrets of life. The conclusion that he drew from decades of work was that each animal brain has its own private language, self-constructed, that in some respects resembles the chemical labeling of the self that the immune system gives to the body. Each individual animal acts in the world with intention.

Like the dog, the cat finds no difficulty in linking an arbitrary sound, like the ringing of a bell, to a real or imagined event. Such capacities lead to auditory and visual communication between members of the same species and to the evolution of social groups from pairs to prides, packs, and tribes. In fact the basic signs of life—warning calls, the sounds of rut, the sounds of satisfaction, the cries of infancy—are known across species. These signals are more stable than some parts of the genome. The concentration of irritable tissue in the brain came with a cost of course. In the cat the brain makes up less than a fiftieth of the cat's weight, but it receives a fifth of the entire blood flow directly from the heart; as long as the cat's ecology in life is doing well, the cost seems to be well worth it.

The large land mammals dominated the latter part of the Cenozoic era, the Miocene epoch, 24 million to five million years ago. This was when the greatest variety and numbers of mammals existed. Elephants and elephant-like mammals inhabited the world in prodigious numbers and in a bewildering array of genera and species. Even up until the end of Pleistocene, roughly 10,000 years ago, a few elephant species were still present in great numbers. The last of the mammoth population survived on Wrangle Island north of Siberia until around 3,500 years ago. The African elephant does not like or cannot survive the cold, and they disappeared from Africa during the Pleistocene era. This apparent disappearance has recently been explained by supposing that they moved into the rain forest, or perhaps a variety of them had always been in the rain forest, but because the modern rain forest does not produce fossils, the paleontological record is blank.

The African elephant that we are familiar with in the wild state has a matriarchal social organization. The genders of the elephants prefer their own company and divide into adult male herds and matriarchal herds that include the young. This makes the greatest use of the elephant's large brain and long life. It is an effective way to utilize the possibilities for maternal intelligence, passing on tradition by example. Elephant families with older matriarchs successfully produce more calves per female-reproductive year.

Although elephants seem to be limited to grunts and groans, rumbles and roars, their communication capacity is quite extensive. Many of the rudimentary facets of complex communication that opened up for land animals are available to the elephant. The complex vocal-auditory channel is well developed with a potential for directional reception. In air sound has a rapidly fading signal that allows for rapid communication. Elephants will turn on a dime to immediate threats in response to vocalizations. And with the ability to repeat what is heard, to hear what is said, learning by feedback can easily occur. Joyce Poole, who has studied elephants for decades in the wild, finds that they have seventy kinds of vocal sounds as well as 160 different visual and tactile signals, expressions, and gestures that they use in day-to-day communication.

Richard Lair and David Sulzer have directed these abilities to communicate to a high level in the Indian elephant. They have created the Thai Elephant Orchestra for out-of-work Indian elephants. They use special elephant-sized instruments, including slit drums, a gong hammered from a sawmill blade, a diddly-bow bass, a xylophone-like renats, a thundersheet, and harmonicas. Some of the players improvise distinct meters and melodic lines, and vary and repeat them. Some players are efficient but bland; others are experimental and enthusiastic by nature. The learned musical abilities of elephants show that some of the features of communication that Europeans have thought to be unique to humans have been just been poorly developed or latent, simply waiting for the opportunity to come forth.

The elephants have almost completely died out and will likely do so soon enough with or without human assistance. Fortunately they have already left us indelible evidence for their high intelligence and creativity. Had we only the bones and tusks of the elephant to guide us, there would be absolutely no way that we would have imagined the character, the social structure, and the intelligence of this magnificent creature. This example highlights a huge bias of misinterpretation that overhangs all of paleontology. This is the White Man's Burden of paleontological science that we are only beginning to dig our way out of with ethology. In Indian culture this prejudice has not been manifested. In the Hindu pantheon Ganesha, the elephant-headed god, is the god of intellect and wisdom.

The great age of the mammals appears to be winding down by the regular succession shortcomings, overspecialization, loss of habitat, and the like, punctuated by the occasional big bang. Overall the Second Law of Thermodynamics is still in full effect. The Second Law of Evolution temporarily damns the flow. Most of the extinctions over the million years or so of the Pleistocene have occurred among the large mammals and the remaining large marsupials as well. A million years from now, looking back on our era, the three-impact sequence at 13,000, 9,000, and 3,500 years ago will look like a single event. The sudden disappearance of the mammals in the fossil record, along with many other species, will appear to be caused by this event. The most successful mammals at the present time in terms of numbers and varieties are house cats and roof rats. They are fast breeders. They refresh and restore their DNA frequently and efficiently, they have learned to live with *Homo sapiens*, and they have far fewer problems giving birth than the large mammals generally do. They will probably join the snakes and lizards as sturdy reminders of their profligate past in the next new age.

There will be some interesting backwaters of antique perseverance, some unexpected eruptions, and some profound reversals in the story of the mammal, however. The family Ursa is likely to survive beyond the Age of the Mammals. This Komodo dragon of mammals survives with a rare combination of a remarkably experimental omnivorous diet and a very short gestation period for a mammal of its size. The 750- to 1,000-pound female gives birth to a two-pound neonate. She does this during hibernation. There is no need for socialization in the polar bear, no need for extra brain, no need for the male to do anything but impregnate the female. The only shortcoming of this social arrangement is that the male is a mortal threat to the female and her cubs. Our Ice Age ancestors set up bear skull altars in caves. They were probably deeply impressed with the ease with which Ursa goes through parturition. Some believe that Ursa was the first domesticated animal during that time.

For another possible survivor into the new age we offer the case of the family Capra, the goat. The Capra ibex male has a most impressive set of horns. In the female the horns are insignificant. The two genders are nearly the same size. The horny males gather groups of females into harems during courtship season. Anyone untrained in animal psychology might well assume that the impressive male is the alpha or leading animal in ibex society. The zoo man H. Hediger corrects this misperception stating, "The truth, however, is that the he-ibex is the perfect hen-pecked husband, and it is the female that wears the trousers, so to speak."[4] The female is considerably more nimble and will have delivered several stomach prods before the male can get his heavy ceremonial horns at the ready. It is the female ibex that is the alpha gender. The only competition for the male is another male. Perhaps the female tolerates this spermish behavior of the male in order to check the health of her potential sperm carrier. Otherwise the male is largely expendable. After courtship season the males retire to the relative calm of all-male herds. The male is not called upon to show much intelligence, the female does not suffer travail at birth, and the ibex has never been domesticated. The breeding season of the ibex was noted on the calendar bones kept by our Ice Age ancestors. Their image was marked with an X. It occurred in the spring. Presumably this indicated a season for abstention for *Homo sapiens*. *Homo sapiens* did better if copulation occurred in the June/July season so that parturition occurred in the following spring.

The case of the family Bos is also of particular interest because of its close relationship with the family *Homo*. The remains of wild cattle around the Mediterranean basin at the end of the last Ice Age, when so many large mammals became extinct, perplexed paleontologists. There seemed to be two species: a large species and a pygmy species. It turns out that the size difference was a gender difference. The male was large and horny. This is a reflection of a full-blown harem society with the male being actively present and constantly on guard. The bull to this day has never been domesticated by anything short of castration. As was known anciently by those who lived with the herds, but apparently not to academics, female heifers grow to a larger size where there are no bulls around.

The ruminants on the whole are eminently polygynous, many females to one male, and the males express secondary-sexual differences such as horns more commonly than any other group. According to Alison Jolly at Princeton University the higher the degree of polygyny, the lower the intelligence of the species in general. Harem-type culture demonstrated by Bos stems from males being hostile to each during courtship other while the females look on in stunned amazement. Cattle are unique, however, in that both the sexes have horns of the same size. It appears that the males not only competed against each other for the chance to plant their seed, but that they also selected against large, threatening, horny females. The horns didn't go away, but the female did become much smaller.

Evolution in the direction of greater complexity has required much variation on the theme of mistaken, or tolerated, or suppressed identity. This was necessary in order for there to have been sexual determination in the germ cells, male and female, and for them to have acquired quite different identities. Accommodation and tolerance was necessary in order for internal fertilization to evolve in the mammals. Sometimes the genders have confused self-interest with survival, to the detriment of the species. As we have seen polygyny is often an opportunity for the female to be able to choose a healthy male after she has lost the capacity to choose healthy sperm in the first place. In the case of Bos, the male was choosing the female. I suppose that all survival solutions have had or will have their fatal conclusion and final comeuppance when pursued too far and that there have to be constant adjustments to survival, not final solutions to survival. In the family Bos we have a remarkable chance to see the shortcomings of an approach to survival when it is taken too far.

The out-of-control spermishness in the male Bos put the species into a deadly spiral. I suspect that the selection for smaller female brought with it increased birthing stress due to the reduced size of the pelvic girdle. This further increased the need for the male to be a paranoid protector of the family. We may well suppose that Bos was on the verge of extinction because of the bull's aggressive rejection of secondary-sexual characteristics in the heifer. The trigger for extinction occurred following the impact events at 13,000 and 9,000 years ago, until our ancestors intervened to disrupt the consequences of the bull's deadly survival-of-the-fittest strategy.

The full implications of the out-of-control spermish behavior of the bull are obviously not one of the things that we care to talk about. Nor is the following anecdote. Richard Irving Dodge made some fascinating observations in his *The Plains of the Great West and Their Inhabitants*. Lieutenant Colonel Dodge was in command of Fort Dodge, Kansas, in 1872–3. He recorded numerous unique and invaluable observations on the life of the American plains before they were overrun by civilized by Europeans, before they could be carefully catalogued by professional naturalists. "In writing these pages," he says in the preface, "I have carefully abstained from consulting 'authorities,' and have treated the different subjects from my own standpoint. Whether valuable or otherwise, the ideas are my own; and the beliefs expressed are the natural growth of long and varied experiences."[5]

He observed things that naturalists rarely ever get to see in the wild. Most of the time naturalists only observe well-adapted set pieces in nature. These are behavioral routines that passed along to the young who mirror the behaviors of their parents: what the parents do, they do; what the parents do not do, they don't do. Mirror cells in the brain that fire in response to what the subject is seeing in the environment rather than what the subject is doing is a great excitement in neurology, but this reflects the process of passing on habitual behavior in the main, not the process of solving basic behavioral adaptations in the first place. At the same time we are also fixated on the figure over the ground in the same way that we have overemphasized DNA over the epigenetic environment. The whole brain is a simulacrum of the environment making sense out of localized activity in ways that are still quite beyond our understanding. Once behavior is routinized there is almost no activity in the brain. Our initial attempts to fashion theories of behavior, both in ourselves and in nature at large, always begin with routinized behaviors.

In any case Dodge had the unique opportunity to observe the behavior of the long-horned, Spanish cattle that had gone feral and were adapting to an entirely new environment. The feral bull did not display harem-type behavior as in the past, but it was still a dangerous animal, always on its muscle. Without the bull defending his harem, the wild cow had become a daring and desperate defender against predators such as the cougar. The female heifer was now nearly the same size as the male and still had a full set of horns. Because of their changed circumstances, the female even had to defend her family from the bull on occasion, rather like the female bear.

Colonel Dodge also observed the native buffalo. The buffalo bull was "gregarious, inoffensive, seldom or never fighting, and truly fatherly in his care for his progeny."[6] The buffalo bulls would form defensive rings around their collected families when under attack from wolves or cougars. The buffalo cow appeared to have little or no natural instinct to protect her progeny; that was a function performed by the male. Dodge also reported in the initial interactions between domesticated cattle and wild buffalo. He observed that buffalo and domestic cattle interbred, but only successfully when the buffalo cow was the mother of the hybrid. The domesticated cow would receive the amorous attentions of the buffalo bull, allow impregnation, embryological development occurred, but the cow would invariably die, being unable to bring forth the calf. There was no basic genetic problem with crossing these dissimilar species; it was the way the genes were expressed in embryological development that was a problem.

We know from the contemporary circumstances of rustic cattle ranching wherein the modern accouterments to cattle ranching are too expensive to implement, that cows are loath to mate with large bulls and much prefer bulls their own size. In modern, high-tech cattle herds the bull is still generally bred for size, but insemination is performed manually by a vet. The female is chosen for factors such as milk production. The female is delivered by the full array of modern veterinarian devices from drugs, to chains and winches. There are varieties of cattle that cannot deliver at all except by Cesarean section. Although present in very large numbers today, the family Bos will probably not survive any longer than the human civilization nurtures and supports it.

Canis is another mammal in the grips of a feedback loop that favors increasing intelligence, in part because of the travails of birthing. Wolves are sexually dimorphic, with the females being 20 percent smaller than the male. While this certainly adds to the travail of birth, the male is an intelligent and enthusiastic family member. He helps to guard the pups, and he brings back food to the den. As observed by L. David Mech, wolf packs are not hierarchic communities ruled by a dominant "alpha" breeding couple, a group of subordinate "beta" individuals, and a scapegoat "omega" wolf at the bottom end, as they are commonly portrayed. This is something that occurs with the helter-skelter wolf packs that form in captivity. Wolf packs in their present state in the wild are generally nuclear families led by a monogamous couple. Young wolves are certainly submissive or playful in relationship to their parents, but when they reach sexual maturity they disperse and form their own families. The wolf was domesticated by, or it domesticated itself to, our Ice Age ancestors. It has always seemed to be a close friend and was not exalted, as was Ursa. The family Canis will probably survive into the next new age through intelligence as reflected in their social cooperation.

The whole subject of the decline of the mammals and the shortcomings of internal fertilization is one of the most tabooed subjects in human culture. It is the most profound indictment of the gods we worship, be they religious or secular. Our much beloved theory of evolution is contrived so that we, the chosen people, are at the top of the heap and that with us creation comes to an end. It is a suitable reflection of Capitalism or, with a few minor variations, Socialism. The kind of social metamorphosis that would be necessary to understand evolution as we are now capable of doing is hard to imagine.

CHAPTER 14

Human Evolution

FROM ANNE CAMERON WE have the story of the *Daughters of Copper Woman*. This is a collection of tales from the female oral tradition of the Nootka living on the Northwest coast of North America, where they have been living for 9,000 years. In this story of origination, Copper Woman is alone, the first inhabitant of the island. One day a bowl-shaped boat brought three sisters so old and withered it could not be believed they were women. Copper woman nursed them and kept them alive. The sisters emptied their shell of wisdom and teaching. They made the runes and said the words around the fire so that the smoke and sparks went straight to heaven. Before she gave up her meat and bones, the last old woman delayed her own passing over to share the final secret in a great outpouring of love. With this knowledge Copper Woman successfully raised snot boy, watching as he became something like the horse clam, something like the sea urchin, something like the crab, and something like the sea lion. As it had been told she would never be alone again. Yet as she soothed snot boy, who cried like Qui-na the gull, she wondered if her loneliness would ever totally go. Snot boy would grow up to become raven to tell the Bear Mother story.

In the Linnaean system, humans are categorized first in the kingdom of Animalia; then the phylum Chordata because we have a backbone; the class Mammalia because we have hair and suckle our young; the order Primates because we share with apes, monkeys, and lemurs certain morphological characteristics; the superfamily Hominoidae because we are not lemurs, nor are we monkeys; the family Hominidae because we, along with the other apes except the gibbons, have lost our tails; the subfamily Homininae that includes hominids, gorillas, chimpanzees, and humans, but excludes orangutans; the tribe Hominini that comprises two species, chimps and humans; the genus *Homo* that began with *Homo habilis* who used tools (now extinct); and the species *Homo sapiens* the wise.

By the rough estimations provided by genetic analysis a monkey became an ape, losing its tail around 25 million years ago. This is the estimated date of the separation between the line that would become *Homo sapiens* and the line that would become the Rhesus monkey, *Macaca mulatta*. There is a 93 percent identity between the two genomes at the present time. The loss of the tail was certainly due to an alteration in the most posterior of the Hox genes that orchestrate the linear and bilateral aspect of embryological evolution in all vertebrates. Occasionally a tail-like appendage does show up in humans, suggesting that the change was due to blocking or to changes in embryological timing rather than to genetic damage to the Hox genes. What was lost with the loss of the tail? In many animals the tail is a particularly sensitive expresser of emotions. It is also the case that the loss of the tail meant the loss of a balance pole, but considering the number of bone fractures found in the aerial gibbons, it also meant the loss of an inherently dangerous sport.

The Chimpanzee Genome Project places the divergence between *Homo sapiens* and *Pan troglodytes* from a common ancestor at around 6 million years ago with a 99 percent identity at the present time. The most obvious genetic difference between *Homo* and *Pan* is visible to the human eye under a microscope however. *Homo* has twenty-three pairs of chromosomes; *Pan* has twenty-four pairs of chromosomes. Actually *Homo sapiens* have two pairs of chromosomes fused into a single large pair. In *Homo* this fused chromosome, chromosome number 2, contains one of the four sets of Hox genes created by polyploidy long ago. Therefore there may be a 99 percent identity between the compiled list of *Pan* and the *Homo* genes, but there is a considerably larger difference between the chromosomal landscapes of the two animals. This leads to significant differences in embryological development. It is a difference we can see with our eyes. The problem here is that this change fits under the heading of "hopeful monsters of evolution." The dogmatic rule is that only point mutations can be transferred to the next generation. But there is simply no way that such a massive mutation could have been recreated point by point so the whole issue is ignored. It is an irreducible complexity for both Darwinists and Creationists.

Along with the irreducible complexity of the fused chromosomes came a duplicated piece of chromosome number 9 that includes a gene called forkhead box D4. This gene is a type of gene that codes for a transcription control protein that blocks the expression of other genes. It is a change such as this that could have influenced the architecture of the spine, especially the two balance points at each end, the head and the tail. In the later stages of embryological development, the remaining tail portion of the spine in all apes, except in *Homo*, recurves outward. In *Homo sapiens* the remains of the tail stay tucked back under the pelvis. This non-event could be due to the blocking the action of a Hox gene in embryological development.

The coccyx bones are the residual remains of an ancestral tail in *Homo*. Coccyx means "cuckoo" and refers to the shape of the bones. They are shaped like a cuckoo's beak. They are part of the pelvic cavity. The pelvic cavity encloses the female reproductive organs; it carries the fertilized egg and the developing fetus. It is through the pelvic opening that the neonate must be delivered to the waiting world. The pelvis also forms the massive bony structure that anchors the legs by way of the hipbones to carry the weight of the body in bipedal posture. The coccyx is an atavistic reminder of past evolutionary stages like legs on a whale, but our vestigial tail is also a fundamental part of what it means to be a human.

In the Hermetic tradition the ontological discontinuity embedded in human existence was associated with the spinal bones that are fused with the pelvic girdle. It is from that tradition that we get the os sacrum. The modern medical fraternity pin still shows the os sacrum or pelvis surmounted by a skull, indicating that birthing is traditionally its primary consideration. In Slavic languages and in German this bone is called the "cross bone," a reference to the travail at birth, and by analogy to the sacrifice of the savior. The tradition of the twined snakes has been abominated or generally ignored by the Judeo-Christian tradition, but it has never been able to do without it in practice. A popular variant of the Hermetic medical icon is the skull and crossed thighbones meaning, "poison": this subject is not open for public consumption.

At the other end of the *Homo* spine, the head would retain its ninety-degree, fetal angle to the spine so that the head would face forward rather than up to the sky for someone who is going to stand on two legs. Like the crack of a whip in slow motion, a movement in the base of the spine ripples up to the brain. Any local shift that is not harmonically compensated for above and below will not work. The brain, as well as the coccyx bones at the other end, continues to mature in *Homo* until after the age of twenty. If this were an insect, we would be talking about a juvenile hormone causing a delay in maturation of adult features. I suspect a delay in maturation over some earlier primate ancestor may have been caused by the fusing of two sets of chromosomes. Whether this alone was sufficient to extend the embryological process until long after parturition, or if it was also accompanied by juvenile-type hormones or other factors is not known at this point. The os sacrum reflects the reality of a comfortable bipedal hominid.

Our oldest supposed ancestor, the 6-million-year-old *Orrorin tugenensis*, shows both tree climbing and bipedal characteristics. It is assumed that our primeval ancestor was a creature high up in the trees, a creature with the depth perception of a flyer, with a perspective of both heaven and earth, and a well-developed cerebellum. The cerebellum provides proprioceptive feedback on the position of the body in space. From this comes good balance. Four strong, well-articulated hands are the universal constant for the arboreal monkeys. The shape of the spine and the presence of a tail, even a prehensile tail, are less important than four hands. Even the most grounded ape still shows their arboreal past with feet that are like hands that can use tools. We continue to see the residue of our Primate four-handedness in the clutching of our infant feet, the Babinski Reflex, and even in the destruction of our elders' feet when we belatedly realize that walking on our rear hands for more than five decades is still not our wont, especially at the size that we have become.

With the exception of our close cousins the family of *Pan*, other apes have probably come out of the trees too late to be obligate bipedalists, even with the appropriate mutation. They were too big to take advantage of the sort of genetic changes that were sustained by the hominids. They use their front hands to help bear the brunt of gravity. Their spine retains the ancient and well-tested role of ridge beam for the ribs in the quadrupedal vertebrate. The chimps and the gorilla probably came out of the trees in the manner that their distant cousin the orangutan is now doing. The orangutan is still largely arboreal out of fear of the tiger. On the ground they can walk on their rear hands, with a clumsy gait, with front arms held over their head for balance. To move rapidly they clench their fists and use their arms as crutches to swing across the ground. Gorilla mothers look on fondly as their young males struggle to their feet and pound their chests like dad, but standing on their rear hands with bowed legs and recurved spines, they are easily unbalanced.

The chimps were probably candidates for obligate bipedalism, but lacking the appropriate mutation they have been more comfortable moving in a horizontal plane most of the time. Chimps have been observed doing a droll imitation of *Homo sapiens*, a drunken sailor's stroll, with comic effect as far as the youngsters are concerned, but they can only bipedal for short distances. Situational bipedalism such as this is surely an example of creative use of the body as a tool. It is yet another example of how creative behavior sets the stage for biological evolution.

There is no evidence from the earliest fossil hominid bones so far discovered of knuckle walking such as we see in the terrestrial chimp and gorilla, or fist walking as we see in the orangutan. There is a rare human genetic mutation that results in a greatly reduced cerebellum. Such people are easily unbalanced. Without special training, such people remain on all fours from youth. They use their palms like *Orrorin tugenensis* might have done rather than their knuckles. The gracile *Australopiths* of our ancient past were in the range of four and a half feet tall, around fifty pounds, and had a life span from fifteen to thirty years at the outside. They were probably more like the modern macaques in their physique and in their behavior. The macaques spend much of the day on the ground, but return to the trees at night. The macaques are generally upright, sitting on their haunches when at rest whether they are in the trees or on the ground. Although they are not obligate bipedalists, they may walk or run on their rear hands, particularly if they need to carry something. They don't knuckle walk like chimps. The classical image of *Homo* evolution, the stooped, knuckle-walking ape rising up to civilized rectitude, is unsupported by the evidence. Leaning over is a sign of aging in the individual and in the culture as well. It is a sign that we are getting bigger, heavier, and less mobile. We are those who walk on four hands in the morning, two feet in the afternoon, and in the evening two feet with the aid of a stick used as a walking tool.

We fall all over ourselves extolling the virtues of bipedalism to the applause of the choir. Yet we see that standing on two hands is not new or unique. Many dinosaurs were bipedal. The infamous velociraptor stood on its hind legs. Its hands were completely unsuited to walking. Theropods such as the velociraptor were the predecessors of birds and had many characteristics that we associate with birds, including feathers. Birds had to be feathered as well as bipedal in order to fly. Wings are almost completely useless for ground support duties, although the chicks of the Hoatzin still have claws on the thumb and first finger of their wings that they use to climb back into their nest before they are able to fly. Pigeons have become flat-footed ground-walkers. Their skill at nest building has declined through the loss of strong, grasping claws. Fish had to give up the use of fins for other than walking when they came on land. Kangaroos, with a relatively intact Hox gene genome, are permanently bipedal. In their case balance comes on the heels of the large, counterbalancing tail. Technically they are tripedal. This has allowed for the selection of feet in the rear and hands in front. They use their forepaws to box with each other in play or in competition for mates. Dogs that have lost their front legs to genetic mutation can learn to be bipedal within the protective society of *Homo sapiens*. As we grope around blindly looking for insight, we feel the monumental columns of the elephant and assume a large and powerful animal. If we only had the bones to examine, little would we expect an articulated nose used as a tool for tool manipulation by the elephant.

The evolution of bipedalism is another one of those irreducible complexities of life that is difficult to explain. We stretch reason thin looking for a cause that is not a purpose following the rule of random change and blind justice producing gradual progress. The attempt to give bipedalism in hominids an immediate survival-of-the-fittest utility has been source of continuous argument. Sherwood L. Washburn suggested that striding long distances with a minimum of effort was an asset to the hunter, but such a result requires a multitude of unproductive adjustments ahead of time. Hominid skeletons such as *Australopithecus* have shown that they must have covered ground with quick steps, a sort of jog trot. They were still walking on hands like crows. John Napier appealed to a paradoxical resolution, suggesting that it was the precisely the inefficiency of this jog trot that motivated the human family to adopt a hunter's high-energy diet. Taken together, the circularity almost makes sense. A little of this, a little of that—who knows, the cart might pull the horse.

The fact of the matter is that even the first tool user at the beginning of the Pleistocene, *Homo habilis*, was probably not a hunter. Quite the reverse. As they moved out into the savannah, they were probably scavengers, and their remains have been found in the diet of a large, predatory saber-toothed cat. However they probably remained largely in the middle: neither hunted, nor hunter be. The sense of smell in both hunter and hunted is usually highly developed. Olfaction is a very poorly developed sense in *Homo*. What is very highly developed in *Homo* is the olfactory cortex, so much so that it takes on a much larger role in the brain. The huge size of the human brain comes largely from the explosion of the olfactory cortex into the frontal cortex.

With the aid of the frontal cortex *Homo* was of a mind to adopt group defensive tactics such as hurling stones and clubs. This was probably very effective against the cats. Similar group tactics have given the otherwise defenseless African wild dog a widespread reputation that rarely needs to be proven. The dog can run, so its group behavior also sponsors a hunting career. Even when *Homo* finally was able to add hunting to their repertoire with bow and arrows, they still appealed to tradition in the land of the big cats. We see this today with the San people, who live with the lions in South Africa. According to Elizabeth Marshall Thomas, when a San hunting group approaches lions around a kill that they think is theirs they give deference to the big cat, but they insist that the trophy is theirs. They do not use their bows and arrows, which would just enflame the king of the jungle. To remind them of their ancient relationship, they pick up a few pebbles or a clod of dirt and toss them in the cat's direction. The lions retire from the scene. Another thing the San people do, or don't do, is run in an encounter with a lion. More than anything it is running that triggers a lion attack. The nearby agricultural peoples have lost this relationship with the lion and must protect themselves and their animals with *kaarals* and guns. The San people find this very funny. In general we don't pay much attention to these ethnographic discoveries if they don't fit our favorite armchair theories.

Occasionally a reality check on our heroic commemoration of bipedalism occurs, such as W. Krogman's article in *Scientific American*, 1951, entitled "The Scars of Human Evolution." Krogman listed the deficiencies that accrued to the permanently vertical posture from a body that had evolved for hundreds of millions of years on the horizontal plane. They are scars that are still with us: hernias, hemorrhoids, multiple spinal problems, increased abdominal musculature and with it increased birthing problems are among the burdens accruing to bipedalism. Gravity is the unexpected consequence and the villain of this story.

Gravity is the force that keeps us from falling off the planet. Gravity is what does the sorting in the plant cell that will determine what will be root, stem, and leaf. In fact, an animal weighing thirty or forty pounds that is not walking great distances, and does not live longer than a few decades, will not suffer from most of the problems of gravity listed above. Take falling as an example of the differences due to size in respect to gravity. A squirrel will get up, dust itself off, and run away from a fall of ten feet, but horse will be crushed by the same fall. There is no need to jump the gun at this point about the drawbacks of bipedalism, but it does remind us of a certain reality about creation that is willfully overlooked—devolution.

One of the prescient observers of nature, Eugène Marais, described it long ago, "No specializations in nature are without an attendant number of disadvantages. Whether the specialization continues and the attendant disadvantages are eliminated by selection of course depends upon their relative values in the struggle for existence. But even the elimination of a disadvantage can only be effected by fresh specialization. There is no such thing in nature as an organism in perfect accord with its environment."[1] Marais was a genius before his time, and his time has yet to arrive. When *Homo* got too large to return to the trees, well after we had adopted the gait of a hunter and were totally dependent upon tools, this is when we also began to carry the burden of proscriptive bipedalism.

Elaine Morgan suggested one of the immediate problems connected to bipedalism. She suggests that the areas of pleasure associated with copulation seem to be caught mid-stride between vaginal sensitivity, good for rear-entry quadrapedal copulation, and clitoral sensitivity, good for ventral-ventral copulation. This has left the female with a tendency to be less than entranced with the mechanical aspects of insemination, according to Morgan, and the male is in the unfortunate position of tending to have to force the issue. I wonder if Morgan is envious of the hyena female who has an enlarged clitoris, has complete control over who will mate with her, and lives in a matriarchal social structure. The male under these circumstances provides no assistance in raising the cubs. In the *Naked Ape*, Desmond Morris cogently points out that ventral-ventral or missionary position is a personalized form of sex. It tends to moderate the raw, unsociable tendencies in the male whether it is the best position for the female or not.

The issue of sexuality and the role of the male has provided a way to close the conventional, circular argument about why inefficient walkers were motivated to overcome their limitations to become hunters. C. Owen Lovejoy closed the circularity with the formula: sex for meat. The more meat you can carry, the more sex you get. The more bipedal you are, the more meat you can carry, the more sex you can get. This is so compelling that it is almost a waste of time going to the ethnographic literature, but if we do we find the bonobo chimpanzee. This gracile chimpanzee lives south of the Congo River. The bonobo chimpanzee practices ventral-ventral copulation like the hominids with shocking frequency and widespread application. Copulation in the bonobos appears to be a pleasurable activity. Seen from the outside it is almost comic. However, free-for-all sex such as this doesn't suit the Lovejoy thesis of meat for sex, as any good Capitalist can tell you. Instead of the territorial conflicts observed in the common chimpanzee, bonobos gather in groups of a hundred for the night and then split up into small groups during the day for the purpose of gathering food. Bonobo chimpanzees are largely frugivorous, but when they occasionally hunt monkeys they do it as a group.

Among our Stone Age relatives there is considerable variation among the hunters and gatherers as to who does what and how, but one thing stands out: gathering provides the lion's share of the food that is eaten, and is almost always done by the women. In the Ituri Forest where the pygmies live, they gather fruits, nuts, seeds, roots, leaves, and flowers as well as scavenging for fish dug out of the mud, termites, insect larva, eggs, and the occasional dead animal. Pygmy men are completely responsible for gathering honey. They do this by climbing up trees with an astonishing, ape-like facility. The whole tribe, including children, is involved in the trapping of small animals.

The only circumstance that bears any resemblance to the Lovejoy thesis is the case of the individual male elephant hunter. One or two males in a pygmy extended family might aspire to this highly skilled, extremely dangerous endeavor. He hunts alone. The danger to the group is too great, but a single hunter is expendable. As with honey-gathering the reward is great, but the necessity is low compared to the cost. The successful elephant hunter may kill only one animal in his lifetime. Considering activity such as this, clearly marginal to the economy of the group, recalls Wheeler's description of the opulent circumstances often achieved by ant culture. He observed a large margin of vitality that allowed for extravagances that had little or no survival value.

Morris points to the unusual breast development in the bipedal human female as the inducement for the frontal approach to copulation. Perhaps the engrossing sight of clitoromegaly in the male—the human male has a considerably larger penis than other primates—is a reflection of and contributes to the tendency for upstanding posture. The measurement of the pupillary response by human males to pictures of nude females shows that the male frequently doesn't even bother to take glance above the neck. In his unguarded and instinctive way he is looking for the place to plant his organ. Human females look first and longest at the head and face of nude photos of males, and then to the rest of the body. The female perspective is looking for a well-formed, healthy, and intelligent mate.

Genetic instincts do nothing more than activate certain tendencies; individuals have to reverse-engineer what they are designed to do, after a false start or two. Copulation in the wild by most animals most of the time is a highly ritualized, short, and violent moment. Pleasure and intercourse are not two words that one would necessarily put together very often. For the *Homo* male there are the few seconds of intense excitement as he launches his wiggly microbes from his testosteronally enlarged clitoris, followed by a release of tension. For the female, sustaining the species is a much more holistic affair. A painful interlude here and there, a moment of satisfaction when things go well, are all part of the program. From this comes the potential for strong pair relationships.

The social relationship between the hominid genders was almost certainly not like the male-dominated harem of the vegetarian gorillas, where the females leave the social group and have little status or power in their adopted group. Although I have seen gorilla culture through the sympathetic eyes of Dian Fossey, George Shaller, and Penny Patterson, I cannot imagine any way in which the impressive silverback gorilla culture could have ever have been the vehicle for social innovation that we see in *Homo*. The social arrangement of the hominids was not likely to have been like the gender-separated troops of the common chimpanzee, where the male and the female are about the same size. Through the eyes of Jane Goodall I can certainly identify with the anger and psychosis of the chimps as they respond to the species-crushing changes that are now going on all around them—the loss of habitat—but gender-separation encourages male group behavior like hunting. It is not suited to nurturing behavior like child rearing.

Morris's observation that the missionary position is a personalized form of sex tending to build strong pair bonds works to a certain extent for the bonobo chimpanzees. Females also engage in sexual contact. This supports strong female bonding precluding the formation of monogamous relationships. The bonobos are matriarchal, more peaceable than their common cousins, and egalitarian. This is a good start, but everywhere we look we find gender bonding and specialization in human culture. Free sex seems to erode tendencies for gender specialization.

Vitus Dröscher made a fascinating study of gender bonding in nature. He describes two degrees of monogamy in the animal kingdom in his book *They Love and Kill*, strict monogamy and weak monogamy. An example of strict monogamy is found in the gibbons. Gibbons are apes that still have their tails, and they still fly like the monkeys. They do walk bipedally with their arms raised above them for balance, but their preferred mode of locomotion is brachiation through the forest canopy. In gibbon society the offspring of both genders are dispersed from the nuclear family, creating strictly isolated family groups. When the male of a couple dies he is replaced by a son; when a female dies she is replaced by a daughter. Their form of inbred monogamy supports a strong sense of territoriality with many local tribes. For them there is no impetus to go beyond the most minimal of social aggregations, the pair bond. Although Taoists believed the gibbons were able to live a thousand years and then to turn into humans, there is little evidence of much change in their physiology or behavior over a much longer period of time.

Dröscher points to weak monogamy as the situation that favors extensive social development. In weak monogamy there can be strong nuclear families, but there is also plenty of opportunity for switching partners. This sort of monogamy undermines the tendency for complete domination by either gender as we see in the gorillas. It sanctions a certain amount of partner exchange and encourages the development a broad social network. Pair bonds are necessary, but in the early stages of a relationship new pair bonds can be formed. The constancy of a relationship increases as their time together increases. After a few years a relationship may even become lifelong, and not even disease, sexual impotence, or mutual irritation are grounds for divorce.

We have already seen that the macaques have the size and general behavior that could be the best model for hominids becoming bipedal. They are, in addition, second only to *Homo sapiens* in range of communication. In spite of the fact that they have a significantly smaller brain than any of the apes, they are also second only to *Homo sapiens* in adaptation to different environmental niches. And, like us, they are omnivorous. Fortunately the most thoroughly researched of all the primates in the wild is the Japanese macaque or snow monkey. Sadly our ideological propensity for drawing family trees and looking for models based strictly upon genetic relationships, along with our fixation on brain capacity, has caused us to overlook these animals as a model for hominid culture.

The snow monkey is about one third size of the earliest *Homo*, male and the female are similar in appearance, and the male is about 15 percent larger than the female. Neither the males nor the females are notably antagonistic to their own gender; in fact homosexuality has been observed in both sexes. The snow monkeys are basically a female-bonded society. The males generally disperse from their birth troop. This means that they generally have less influence in a group. The snow monkey troops are composed of several tightly knit female kinship groups loosely surrounded and filled in by imported males. There are no exclusive gender troops. There is plenty of genetic exchange between troops of snow monkeys; therefore there is no appearance of local varieties. Nor do there appear to be problems stemming from inbreeding even though the breeding populations are very small. There appears to be a nice balance between inbreeding and out-breeding. The snow monkey is an example of a weak monogamy.

The females are promiscuous, mating with any and all males. Many suitors raise the probability of conception. Females outnumber males three to one, so we may well suspect some form of natural selection for gender after birth. Although there is a dominance hierarchy, the troop is really more like an extended family, and this is important since offspring take four to five years to mature. All adults participate in nurturing the young. Unlike tropical macaques like the Rhesus monkey that has a menstrual cycle of around twenty-eight days and a year-round mating season, the snow monkey female is fertile for only a short period of time. Snow monkeys are remarkable in that they time their conception so that birthing occurs in the summer months. This is an important part of the puzzle that allowed them to move into climates where there is a winter. Another well-known innovation is their habit of occupying thermal hot springs to survive the winter.

Snow monkey females are affectionate mothers forming a strong bond with their offspring. Snow monkey infants are dependent on their mothers for two years, and weaning is very stressful for the mother. Females will occasionally favor the particularly well-buffed, healthy male, who will then become the vain autocrat of the female core group and never leave home. These pashas revel in an overdose of maternal affection. It produces a self-centered, adolescent adult male whose only skill is manipulating others. Some observers describe these troops as a male-dominated harem society. Such a male may dominate sexual access to the females. In captivity pashas depend on underlings they can dominate. When isolated they may refuse to eat, become ill, and die.

Snow monkeys are fruit eaters, but they also eat seeds, cereals, insects, young leaves, flowers, bark, and roots. The most well-known story associated with the snow monkey is the story of the adoption of the sweet potato to their diet. In 1953 Imo began to wash the dirt off her sweet potatoes in a stream. It was the dirt that had made the sweet potato distasteful. Imo was a member of the Koshima troop that had a pasha. By 1956 half of the troop of twenty-two were washing and eating sweet potatoes. It was through juveniles that this innovation was propagated from Imo. By 1958 all infants were learning this behavior from their mothers in the Koshima troop.

In 1956 Imo began another experiment. She learned to separate wheat from sand by washing it in the ocean. This was so successful that washing sweet potatoes in salt water became the preferred method. One female, Nami, did not pass on this new behavior, and the observers thought she must be mentally deficient. By 1962 most members of Koshima troop had taken to bathing in the ocean, although females did so more often than males. In the same year members of this troop began to solicit food from humans. Individuals from the Nami lineage refused to participate. Now members of Nami's lineage are being seen as conservative rather than mentally deficient.

An adult male introduced from another troop into the Koshima troop learned the innovation of washing the sand out of the wheat in the ocean. When returned to his old troop he introduced the new idea like a returning Columbus, and it spread like wildfire. It was adopted by one and all within four hours. Back in the Koshima troop, washing food in the ocean led to swimming in the ocean, and within a few years snow monkeys were swimming out to collect rock-dwelling shellfish. Shellfish were added to the diet, but at no time was the reigning pasha of the Koshima troop involved in any of these innovations.

In their normal social context the macaques are superior in communication to the apes as reflected by their remarkable cultural achievements. Imo used mental representation to categorize, differentiate, and to extrapolate by causal reasoning. The successful result was passed on by mimicry, first to those most open to learning, then to infants, and finally up the chain of the social hierarchy to the pasha. The pasha was valued by the group for his physical presence. He was an icon for conservation, for minimal change. Removed from their social context and put into zoos or research labs macaques become village idiots—nasty, angry village idiots. Captive monkeys are no exemplars of evolution or survival. Japanese culture that has lived with the wild snow monkey uses them in the Koshin folk tradition of the three monkeys: see no evil, speak no evil, hear no evil. It is their icon for civility. The macaques demonstrate a remarkable tolerance for a range of personalities. This is the basis for specialization and high civilization.

A challenge for this sort of arrangement is when a sexually active male is brought into the bosom of the family. All human tribes have complex social mores to control the potential for abuse by an overly spermish patriarch.

Some of these mores are conventionally misconstrued as prohibitions to inbreeding. Inbreeding can be a problem, but disruption of the family by an uncontrollable sexually active male is a more immediate and tangible problem. It's a problem we have to the present day.

The social arrangement of the snow monkey gives us an excellent idea of what it must have taken for hominids to move out of the tropical forest, but there was another challenge unique to hominids to deal with before we get to that—travail at birth. Morgan deals with the travail of human birthing in the following manner: "During pregnancy," she says, "the muscles which supported the weight of her unborn child were all slung from the spine, which was fine for a quadruped, but when she began walking upright everything slipped sideways like an upended line of washing. She would have profited from an entirely new arrangement with the muscles attached to the shoulder bones instead, but though she complained sporadically about backache and prolapse and varicose veins and other feminine disorders, nothing was ever done about it."[2] Desmond Morris in *The Naked Ape* states, "The demands of vertical locomotion have not been kind to the female of our species: the penalty for this progressive step is a sentence of several hours 'hard labour.'"[3]

These are modern upgrades to the nineteenth-century intellectual posture on the subject, i.e., silence. The default implication was that birthing and everything connected with it was a simple reflex with no particular impact on evolution. For the religious, travail in birth is a sentence handed out to Eve by God in Genesis 3:16, "Unto the woman God said: I will greatly multiply thy sorrow and thy conception, in sorrow thou shalt bring forth children; and thy desire shall be to they husband, and he shall rule over thee." The Christian tradition also stayed out of the birthing chamber, by and large leaving that domain to whatever ancient pagan traditions could survive under the circumstances, until modern medicine took over with the diminution of the power of the church.

Of all the scars of evolution accumulated from becoming permanently bipedal, it was the increase of birthing stress that was immediately the most troublesome. Bipedalism requires that the legs be close enough together so that the person can walk without a waddle, but this requires the pelvic girdle to be well developed and the birth canal to be small. Rapid brain enlargement, especially the frontal cortex, was being selected for by those hominids. Wenda Trevathan is a physical anthropologist at the Department of Sociology and Anthropology, New Mexico State University. She is the daughter of an obstetrician. After studying the pelvises of humans and humanoids from as far back as we can go, she stated that the need for midwifery began at the separation between the hominids and the apes 6 million years ago. By the time of *Homo erectus* midwifery, in her opinion, was obligate.

Trevathan's thesis is dramatically highlighted by aging data. Until recently women lived an average of ten years less than men as a group. When the mortality rate of mothers dying in childbirth was reduced to a few percent for modern women, not only do they live as long as men, they live as much as ten years longer as a group. This confirms the heavier wear and tear on women due to birthing. It suggests that the length of the telomeres that determine aging are naturally longer in females than in males in partial compensation for the much higher level of stress, and this is something long-buried in the human genome.

As more and more females enter the field of anthropology we find many taboos that were designed to protect the fetus and ease delivery for the mother. These had not been asked about by male anthropologists or made available by female informants if there were any. In reality the high level of mortality that is connected with birthing has driven all human cultures to use midwives and pursue rational solutions to the challenge of birthing. In addition to numerous spurious inventions, such as the black cat crossing the road, some herbs, some painkillers, some blood thickeners and the like have been discovered that have helped.

Morgan and Morris were on the right track that bipedal posture was an important contributing factor of travail in birth. Trevathan adds the collision course between bipedal posture and increasing encephalization. We should note here that the X chromosome has been very active during all this period of time. It also appears that sexual dimorphism prevailed in *Australopithecus*. The male was larger than the female. This is often a sign of a close relationship of some kind between the genders, but it is another factor that puts the female under increased birthing stress. I think it is safe to assume that this dimorphism was not like the male-dominated social structure of Bos the bull with its extreme dimorphism; I think this dimorphism was more like the social structure of the Japanese macaque where matriarchy prevails. This would have been the natural social setting for midwifery to flourish.

God's supposition that Eve's curiosity was closely related to travail at birth is not without merit. I suppose that Eve was selecting for increased knowledge in the tree of life or anywhere she could find it. I assume the snake in Genesis refers to the Hermitic tradition of medicine that was pagan and matriarchal, and therefore profane. It was the male contribution to midwifery. In the time when the principles of Genesis were conceived, poor diet, miscarriage, death at birth, low fertility, and profound social disorder were major issues for humanity. Blaming the victim is a standard device to escape an imponderable moral jungle hopelessly out of synch with nature. We had been thrown out of the Garden of Eden. The same thing happened all around the planet after the impact of Typhon.

Evolution of the hominid species was certainly being provoked by travail at birth, and coupled with it was a form of selection that we also prefer not to know about. Ironically this was exactly the form of selection that shaped our secularized theories of evolution in the first place: the selection for desired traits by animal breeders or farmers. We had to modify it to accommodate the sensibilities of a culture that had and has an uncomfortable memory of abandonment. We deferred to the relatively harmless botanical metaphor, Parable of the Sower casting his seeds where intentionality and responsibility are erased. This is the blinded god of justice administering the invisible hand of evolution.

Among all animals, selection at birth against damaged offspring has been important to maintain genetic integrity, but it is only among mammals where parents are closely involved in the care of the newborn that choice really becomes an issue. It is a choice the mammalian mother makes either to adopt or abandon her newly born infant or infants. Among mammals in general anything odd about the newborn is potential cause for abandonment. The initial experience of birth by the mammalian mother is a great surprise and tends to be an experiment. She uses the experience to gain practice for more successful attempts in the future. On occasion males are observed making the selection, killing physically injured or genetically damaged neonates. Twins may be abandoned among hunting and gathering humans, one or both, because of the impossible burden they would represent to the mother.

Among Stone Age humans, twins were either a symbol of a miracle or a curse. If successfully delivered it was a miracle. More often it was a curse. Among some African tribes Adam and Eve are the twin offspring of the founding creatrix, which makes a much better story than the masculinized fable involving a rib with which we are familiar. In Greek mythology, Apollo was the twin of Artemis. Artemis was born first and became the mid-wife to Leto, the mother, in order to help her to give birth to Apollo. This was a monumentally traumatic birthing story. All of these birthing issues are still taboo in our culture, even though our revived Hermetic tradition is now 95 percent successful at delivering babies, including twins.

Having brought this subject up we might well ask why it is any neonate so dependent, so thoroughly underdeveloped would have been allowed to live, let alone be nourished and cared for among mammals. It appears there is a biochemical adjuvant involved that goes back to the roots of internal fertilization. The neuropeptide oxytocin seems to be part of an orchestration of neuropeptides, released during courtship and intercourse, that lead to profound changes in emotion and behavior. Oxytocin closes the social distance of dissimilar genders so that rough business can be attended to. Another neurohumoral bath of oxytocin created during the act of birthing causes a meltdown of the personal identity by a sense of euphoria and a decrease in blood pressure, but without the loss of declarative and procedural memories: the what and how of things. Here oxytocin has the paradoxical or a least the unrelated role of promoting uterine contractions that will eject the fetus from its mother.

The shock of birth triggers adrenaline that counters the positive effects of oxytocin and can have serious consequences on the birthing process. A mother thoroughly exhausted by the struggle to give birth and with insufficient oxytocin may well be of a mind to abandon a neonate with fingers missing, or limbs missing, or other mutations shortly after birth. In a fairly short time after birth, imprinting is promoted with another onrush of oxytocin along with lactation. This is the cause of imprinting. Imprinting is a powerful bond that is formed between parent and newborn. Psychologist Ruth Feldman at Bar-Ilan University in Israel has shown that bonding between infant and mother is correlated to levels of oxytocin starting with the first trimester. Normally the lactation rush seals the bond between mother and newborn no matter what the newborn looks like. By this point in the astonishing oratorio of rebirth it will be hard to separate mother from infant no matter what the burden may be. The relationship between mother and child recapitulates the theme of courtship and can easily be interpreted as infidelity by the father.

Early on there is a registration in the infant between the bodily genome and its immune system. This is when potentially foreign foods and bacteria are being introduced into its diet through mother's milk for the first time. During this open season, potentially foreign proteins are labeled as friends. During this period the infant will neurologically imprint upon whoever is acting as parent. Husbandry began here with human females nursing the neonates of other mammals. Males can play the role of nurse, but only if they have the gene that gives them a high number of vasopressin sites. Vasopressin itself comes on with sexual activity. Even then the male is usually well outside the birth chamber, defending the family from perceived threats including other sexual males.

I suppose that the high levels of oxytocin involved with human birth influence and extend a very subtle form of selective evolution. All animals have a body image mapped out on their cortex of the brain. It is called a homunculus. Located in the motor cortex of the brain, this identity contains an epitome of the body. It is here that movement of the body is coordinated with the sense of space and time and self. The homunculus image was originally invented by neurologists to describe the pattern of neurological activity in the motor cortex and the sensory cortex of the brain that was devoted to different areas of the body. It is not a mirror image. The image exaggerates hands, mouth, and eyes, but it under represents the rest of the body. The homunculus image is, in a word, infantile.

I suppose that without the aid of photographic self-reflection this is how we feel ourselves to be, how we "mirror" ourselves to be. We do not automatically recognize a mirror reflection of our self. It is something that we have to learn. I suppose that with oxytocin to set the mood, human mothers would be particularly attracted to an infant that tended to reflect her own homuncular self image. In modern females, continued attraction to infantile images in nature and art goes back to fundamental homuncular self-identification. The mirror image is connected to a sexual concept of self. I suggest that there was a high degree of directed neurological co-adaptation or feedback going on with this complex mother-infant interaction. It occurs in all animals, but the greater the role played by the mother and the longer the dependency of the offspring, the greater the effect.

The most active area of change in the *Homo* genome, the HAR1 area, plays a crucial role in the formation of the brain from seven to nineteen months of gestation. The multiplication of brain cells only occurs within the anoxic environment of the womb. The extended length of neurological maturation following birth also profoundly affected what *Homo sapiens* could become. This is when the nervous system is shaped. Whereas the typical monkey is born with 70 percent of its brain having achieved its adult size, *Homo sapiens* has only 23 percent of its adult-size brain at birth. Whereas a chimp brain is mature six years before sexual maturity, the *Homo sapiens* brain is not fully mature until ten years after sexual maturity. It is speculated that if full gestation could have been allowed to occur in the womb, it would have taken twenty-one months rather than nine.

In the ants the larval stage is exposed and is capable of being adjusted by externalized epigenetic influences to produce different physical types. In *Homo* it is the adolescent brain that is exposed to the environment. It is during this time that language will be learned. Just as different racial types in birds can be correlated to different songs, human racial types can be identified by a mutual adaptation of language with the environment. The window of time during which this can occur—before the sexual hormones close down neural development to its largely fixed adult role—is much greater than in any other animal. This situation has created a unique adolescent subculture of the species. It has led to the possibility of playful, curious, childlike attributes being extended into adulthood in *Homo sapiens* to a very high degree. Neotenous selection has possibly returned us to a recrudescence of some marine facilities with marginal value at first, and it has certainly led us to the first *Homo*, *Homo habilis*.

As the savannah expanded during the Oligocene and the Miocene we would begin to recognize dogs, cats, horses, and the crow in the fossil record. In the Miocene there were one hundred species of apes in what remained of the shrinking wet tropics. There were as many apes as we have monkeys today. With the Pleistocene epoch the tropics were a narrow belt around the equator. On either side were two dry cold/hot desert areas. While the rest of the apes continued to practice a survival strategy that led to little change, the hominids followed a path fraught with travail. Descending out of the tree of life was accomplished with a considerable degree of profane curiosity. Under the increasing pressure of loss of habitat, the apes were down to a mere handful of species. They are still there. *Homo habilis* left the Garden of Eden at the beginning of the Pleistocene epoch around 2 million years ago. It was a matter of survival.

As provident as the Ituri Forest can be for its native species, it is not an environment one can easily adopt. As far as I know the only way to introduce new species into the areas occupied by the wet tropics is by destroying it. The jungle of recent times is quite different from the birthplace of terrestrial life hundreds of millions of years ago, where the great cycad forests fell in on themselves generation after generation, creating the vast coal seams that would be pressurized and distilled down for our tiny little Age of Fossil Fuels. The jungles of more recent times are swarming with microbes, funguses, flies, termites, and ants constantly cleaning up, recycling, looking for an opening. The very narrow range of temperature (around 88°F) with an 85 percent average humidity keeps things humming. There is no humus on the jungle floor to speak of. Nearly all the organic matter is tied up in the living substance of the forest. The humans that patrol the ground will quickly be in a state of protein deficiency if they don't hunt, scavenge, or fish, and they must gorge fresh kill immediately to keep ahead of the ants and the fly larva. The efficient recycling capacity of this environment is the reason why there are no early hominid or *Homo* fossils preserved inside the wet, hot cradle of creation.

As far as we can tell migration has been one way: out of the Ituri Forest, not back in. The conventional view of the ancestry of *Homo* is warped by the discipline of paleontology that always attempts to construct a linear line of descent with its various fossils. Rather than seeing the first-order evolution of the hominids and of *Homo* as it was expressed in the jungle, we are only seeing those that have decamped. Donald Johanson, professor of Human Origins at Arizona State University, comments on this taxonomical dilemma, stating that the various *Homos* could all just be trial decampments from the single jungle rootstock of *Homo* that have been forced into a straight line. Perhaps all of the fossils carefully preserved in museums were dead ends rather than missing links.

The chimpanzee has never gotten out of the Ituri Forest or of any other areas of tropical jungle nearby. The Congo River has created boundaries within the jungle that have resulted in two races that never or rarely ever interbreed: a race north of the Congo and a race south of the Congo. The genetic differences between these two groups is two to three times greater than what is found between any human races, yet it is hard to see much physiological difference between them. I suggest that this offers a comparison for inbreeding within two races where deleterious mutations are efficiently weeded out, but largely trivial point mutations and simple inversions, accounting for nearly all of the genetic change between the two groups, goes uncorrected. The essential genome remains unchanged. As we have already seen it is not the genome, not the compilation of individual genes that make the difference between *Pan* and *Homo*, it is the chromosomal landscape and how it is expressed in the embryo.

There is little genetic difference between the tribes of pygmies within the Ituri Forest, reflecting the fact that they represent a single breeding group. Pygmies cross the Congo; chimpanzees do not. There is a much higher degree of genetic variability between *Homo sapiens* races outside of the jungle than there is within the jungle. This is a reflection of the one-way filter that *Homo sapiens* is able to cross to leave the Ituri, but not return. Individual females do occasionally move into savannah societies to vary their genetic constitution, and we even find hybrid pygmy tribes that have decamped wholesale and adopted outsiders. It is safe to presume that this has been going on for millions of years. There is much greater genetic variability from village to village just outside of the Ituri Forest than there is in the rest of the world. In fact, Stephan Schuster of Pennsylvania State University has found that any two Bushmen who spoke different languages were more genetically diverse than a European compared to an Asian even if the Bushmen were within walking distance from one another. Cultural differences reflected in language difference tend to maintain inbreeding rather than out-breeding.

The variety found in African genomes is not a result of mutations to the basic genome. This was predicted thirty years ago when Mary-Claire King and Allan Wilson suggested that the difference between humans and chimpanzees might be found in the way they express their genes. While the difference between humans and chimpanzees is how the genes are expressed after a major rearrangement of the chromosome landscape; the difference between culturally identifiable human groups is almost entirely due to how the genes are expressed epigenetically. All humans can interbreed. Transcription factors that promote or block the basic function of different gene variations are what have made us what we are. Many active regions of the human genome reflect adaptation to new environments; genes that effect skin color and hair texture.

Even with these factors, factors that have determined the different races, there is more genetic variety within races than between them. The concentrated and highly varied populations with social hierarchies are the hallmark of high civilization. These circumstances add a significant variable to the definition of ecological niche. A life span difference of ten years between high and low class is normal. Out-breeding and a high level of conserved deleterious genetic damage is also one of the results of civilization. As I have suggested already, behavior is leading the way with genes being adjusted by the epigenome where they are able. With the major exception of the fusion of chimp 2a and 2b chromosomes that I suggested made us more comfortable on two legs and contributed to a big brain, we are basically chimpanzees in drag genomically speaking: a 1% difference. On the whole, however, we are very different: there doesn't seem to be any interbreeding between the two. Alas, poor Yorick. Where art thou indeed? The only active role that point mutation plays in this drama is to create damage that we must learn how to survive if we can.

Homo habilis, our earliest ancestor according to conventional taxonomy, had a more flattened face, a larger brain with bigger eyes, shorter jaw, and smaller teeth. *Habilis* brain capacity was 50 percent larger than the hominids, and about half the size of our own. There is a bulge in the cast of one skull at Broca's area, which indicates that *habilis* had a rudimentary power of speech. I will leave the subject of language evolution to my book *A Modern Theory of Language Evolution* but I will say that language was an important key for *Homo* to able to leave the jungle and occupy the surrounding savannah. Language was the tool for conveying the likeness of things. It was the tool used for identifying the substitutions that were necessary upon moving into Terra Incognita.

One of the most important unexpected consequences that had to be attended to by habilis was the challenge of a new diet. The hominids, along with the other apes, had lost the genetic information necessary to produce their own vitamin C. In the wet, tropical jungle fruits come ripe all year round. Flora and fauna go into flower or heat as soon as they have the stored energy reserves to do so in the Ituri Forest. Replacement of vitamin C is relatively easy. Outside the jungle where there are seasons, a dry and a wet season, this was no longer the case. A special effort had to be made to insure that foods that provided vitamin C were available. The pygmy, still living in the Ituri Forest, is the only member of our genus that doesn't require a ritualized knowledge of the seasons in order to anticipate a vitamin C shortage. This was a problem the macaques did not have to solve.

Another challenge involved the estrus cycle. The adult female hominid, like the female tropical macaque, was probably in constant state of estrus. At the complex level of hormonal interaction, the hominid line was missing a timed reproductive cycle. Females don't return to a state of fertility until they stop breastfeeding, but that is all. Outside of the Ituri Forest, both sexes of *Homo* have been adjusted by cultural mores one way or another, often dramatically, as a substitute for a reproductive cycle hormonally triggered by the seasons. A reproductive cycle controlled by hormones that we see in most temperate climate mammals has never evolved in *Homo*. Only modern man with his control over the environment gets away with breeding in all seasons.

Having undergone so many changes in the past few million years it is extremely difficult for us to understand who we are and where we came from. One of the most relentless themes in evolutionary stories is how consistently dangerous life must have been outside our protective corral. In our language culture this prejudice goes back to Thomas Hobbes who supposed that hunter-gatherers as well as country folk in general lived lives, "solitary, poor, nasty, brutish, and short." The psychiatrist Branko Bokun supposed that humans lost their hair because of the constant anxiety. Elaine Morgan has our forbearers running into the water to escape predators, so much so that we became marine animals for a while. Desmond Morris and Robert Ardrey focus the whole of human early evolution on the male-dominated hunter-killer pack. Raymond Dart found what he supposed was an *Australopithecine* tool kit, the tool kit of our nasty, cannibal ancestors. Bob Brain, however, would return to the evidence and show that it was simply the leftovers from leopard and hyena meals that were then further shaped by the gradual processes of weathering.

Ethnocentricity is a constant bias to our understanding even among world travelers such as that implied in the question asked a century ago by Paul Du Chaillu of a pygmy: why did they not build villages and settle down following the example of the tall agriculturists that surround the forest. His interlocutor responded that they loved to move, they loved to be free like the antelopes and gazelles. They were sitting at the edge of the forest where pygmies go to trade and do menial labor for the odd tin pot. Why do you not plant food as other people do, and work iron, and build bigger cabins, persisted Chaillu. Too much work was the reply. These psychological projections give us a way to feel better about our achievements and to ameliorate our own anxieties, but fundamentally it is hard to understand any other life purely through language.

Very few people have actually lived in the Ituri with the pygmies. Those who have, such as Colin Turnbull, describe the most musical, clean, carefree, and natural lifestyle on the planet. The pygmies, he observed, gave constant praise and love to the forest that gives them life. Pygmies sing about chimps and consider them the smartest animals in the forest. They mimic chimpanzees when it comes to diet as well as what plants are good for medicinal purposes. In fact, they use the forest to know the forest. Except for the shrinking environment and the unnerving reality of travail at parturition, the jungle is still a bountiful environment for its native people. Marshall Sahlins raised a few eyebrows in 1966 when he offered ethnographic data that confirmed that hunter-gatherers worked far fewer hours and enjoyed more leisure time than most members of an industrial society.

Our discovery of the bonobos south of the Congo River was cause for great excitement, and rightfully so. There is a small, fervid clique of bonobo fans who have gone so far as to advocate prolific copulation as the key to blunting the war-like tendencies of our anthill civilizations. Unfortunately when bonobos are shipped out of their preferred habitat to zoos and other ghettos, they became considerably less sexy, considerably more psychotic. Whatever they were doing in the jungle is not exportable. Our *Homo* ancestry got its larger breasts and penises in the jungle. To a limited extent we can refer to the bonobos for inspiration, but it was also upright posture, the gender relationship, and the larger brain, especially our larger frontal cortex, that was the difference between us and our primate cousins, not sex, not hunting, not hunting for sex, not sex for hunting.

One of our leading neurologists, Walter J. Freeman, describes the function of the frontal cortex in his *Societies of Brains*. "The immense frontal cortex provides for the neural machinery for constructing patterns of behavior shaped by understanding the personalities of others (insight) as the basis for socialization, and foresight, the ability to plan. Here is located the capacity for imagination and also for anticipating pain and death."[4] During the long development of the brain in childhood, exposure to stress can cause lesions in the frontal cortex that are traceable to adrenaline. People with such lesions exhibit poor attention regulation, disorganized and impulsive behavior, and hyperactivity. Humans who get frontal-cortex brain tumors often engage in impulsive and unnatural sexual behaviors that they had never done before the tumor. When the tumor is removed the behavior disappears. The frontal cortex provides the ability to delay gratification.

The frontal cortex is an outgrowth of the primitive olfactory cortex. Smell was an augmentation to the sense of taste in our aquatic ancestors. Located outside the mouth, this sense could anticipate sweet suggesting food or bitter suggesting poison at a distance, before actual ingestion. The evolution from marine to land animal only gave smell greater importance. In fact, there are now many more types of smell detectors than there are taste detectors. A most unique aspect of this sense is that it operates relatively independently of the other senses. The other sensory systems—auditory, somatic, visceral, gustatory, and visual—are all funneled through the thalamus at the top of the brain stem up to the primary sensory cortices. There is a feedback loop back from the sensory cortices back down to the thalamus that influences input. These sensory systems are basically conservative. What we are taught to perceive is what we will perceive. Growing out of the olfactory cortex the frontal cortex, operating on its own, is the Wizard of Oz of the brain.

At first we thought the frontal cortex did nothing at all. Damage to this area often seems to do nothing, or at least nothing too different from what we can see in "normal" people. It became part of the myth that we use only 10 percent of our cortex. As we have continued to learn more about the frontal cortex, we realize that this area doesn't have a narrowly defined sensory or motor function. It is the area where we construct our cultural personality with the tool of the language that we are born into. The capacity for foresight takes empathy to a higher level, beyond pair bonding. It allows for the formation of social organizations that individuals fall in love with, social organizations that can be brutally simplistic and narrow-minded in relation to other social organizations. It is our equivalent to the social stomach of the ants that also relies on taste and smell. But in the first case it is unlikely that we would have been capable of moving out of the jungle into the savannah without a functional frontal cortex and the complex social group that it reflects. And it is certainly the case that we would never have been even interested to write books about evolution without the frontal cortex.

The pop science theory of the gender-divided brain or bicameral brain was based on the work of Roger Sperry. While there is a division left and right of basic sensory areas in the cortex, the enthusiasm for one side of the brain as rational, analytical, and more intelligent, the other side as emotional, creative, and feminine is not the biggest reality of brain function. The greater reality is that there is a gender difference in brains and it involves the neural commissures that connect the two halves of the brain together. Females, on average, have 33 percent more fibers in their commissures than men. This means females have more bandwidth. On average they can use more of their brain with greater speed. This is the reality of gender difference in brain function. They are more able to handle the most difficult intellectual challenge we generally have to face, the cocktail party. It is also the case that a large brain can be filled with trivia.

Another variation on pop neurology was elaborated by P. D. Maclean in his book *The Triune Brain*. He states that the reptilian brain is fundamental for such genetically constituted forms of behavior as selecting home sites, establishing territory, engaging in various types of display, hunting, homing, mating, breeding, imprinting, forming social hierarchies, and selecting leaders. Maclean's hopeful thesis is that the great cauliflower that sits atop our brain stem—the civilizing cortex that controls and directs the primitive emotional lower brain—is what keeps the Freudian self in control.

Freeman criticizes this hypothesis, stating, "Maclean's theory carries substantial nineteenth-century baggage. The hierarchal control of lower centers by inhibition from higher centers derives from Hughlings Jackson's concept (1884) of successive levels of control, which is revealed by 'devolution' as deterioration of cortex releases lower centers to autonomy and escape from reason. But reason and its quale, emotion, rise together to the heights of musical, poetic, or dramatic composition, and fall together in postconsummatory and postcopulatory torpors. Rational acts can be passionately engaged in or coldly executed with unspeakable consequences ..."[5] In short, the anthill-like brutalities conducted by high civilizations of recent experience were not anomalies that we can sweep away by evoking demons or casting racial epithets. They are simply a product of what our civilization is capable of under stress.

CHAPTER 15

LAST THOUGHTS

AS THE GREAT TROPICAL forest continues to die with the cooling and drying of the planet during the Pleistocene epoch, the chimpanzee and the gorilla will become extinct. The pygmies will look on from the little villages that they have already begun occupy on the edge of the forest where they support themselves by singing and dancing for their distant relatives that left the forest long ago. When this ecology is gone—along with our cousins that know how to live in it—we will have lost our primary and most forgiving survival refuge against the travail of major impact events. Another thing we are losing is brainpower. Parallel with the evolution of human social organization beginning with the impact events around 10,000 years ago, our brains have shrunk by about ten percent. This has also happened to our best friend the dog. Apparently we don't need as much brain now that we have civilization to organize for us.

The social aggregations of Stone Age peoples that set the stage for high civilization were found in places rich in resources the year round, such as the Queen Charlotte Islands south of Alaska occupied by the Haida. In such places a hierarchy can evolve with the male elevated to and occupying the role of chief, reminding us of the snow monkey pasha. It is under these circumstances that we see a higher level of craft, the concentration of wealth, and the use of slaves. Among the Northwest Native American tribes opulent ceremonial feasts were held, during which the raven stories were told and possessions were given away or burned as a display of wealth and power.

These tendencies would eventually be magnified by high civilization into environment-crushing bonfires of the vanities.

The Aryan invasion of the Indian subcontinent under the banner of the thunder god followed the impact of Typhon. This was the beginning of the Iron Age. The Aryans found a devastated native culture. Priests were worshipping the lingam, an icon of Typhon. Shiva the "Feared One" dominated early Hinduism the way Jehovah commanded attention from his terrified, rustic Hebrews. Recovery from that impact event occurred to our Indian cousins first. While our own ancestors were living in wattle and daub huts in northern Europe, India was practicing a high civilization. When our Persian cousins took Babylon under Cyrus the Great, he released the Jews from captivity. By that time Jehova had been retired to a distant heaven, replaced by the inscrutable Yahweh. In Hindu culture Shiva had retired from center stage to become part of a trinity: Brahma the creator, Vishnu the maintainer or preserver, and Shiva the transformer. A rarified part of the high culture was in the middle of an existential quandary about the meaning of life.

In the "Song of God," the Bhagavad Gita, Arjuna is standing on the sacred battleground. The Pandavas, his own family, are endowed with righteousness, self-control, and nobility. His cousins the Kauravas are cruel, unscrupulous, greedy, and lustful. War is inevitable, but he is having second thoughts. Who is the enemy? Year after year, war after war, who is the enemy? Both warring families invoke the support of Krishna, the wisest teacher and greatest yogi.

Krishna offers Arjuna either the services of his great army or his own personal support as charioteer and counselor. Arjuna chooses to have Krishna as his driver. Krishna proceeds to give Arjuna a profound recitation on the yogic practices that lead to understanding the true nature of the life and to recognizing the delusional personas we invent in order to play our role in this endless melodrama. He shows how our language uses us in this endeavor to reason rather than we using it. As a final boon he gives Arjuna a transcendental vision revealing that he is the ultimate essence of Being and therefore all of its material expressions. Called the Vishvarupa, this vision shows that all the gods and goddesses, all the saints and devils, all the heroes and villains, are simply embodiments or avatars we invent to locate our special place in creation. Essentially the Vishvarupa is the story of evolution and what we have done is add more of the missing pages. A basic knowledge of the human condition is one of our oldest gifts. It is the greatest story that can ever be told because it includes all stories.

Paul Ehrlich detailed the challenges we now face in the *Population Bomb* published in 1968. He made predictions about when the collapse would occur that turned out to be wrong; we have always been notoriously bad at predicting second comings. Yet the issues he raised are still before us including the greenhouse problem. The Firestone group gives our overpopulation situation an evolutionary context. "Today, most of humanity's current problems, including starvation, warfare, pollution, global warming, and AIDS, result from or relate to the severe overpopulation that has occurred in the aftermath of the Event [the impact events of 13,000 and 9,000 years ago]. Past evidence tells us that after a catastrophe has disturbed the ecosystem, some species go extinct long after the initiating events are over. That threat of extinction still looms over humanity today …"[1] We should also remember that our rate of evolution over the past 10,000 years has been savage. If we can no longer keep up with the need for change, we did our best. No one is to blame, but one thing is for sure. The odds of our species surviving will be dramatically reduced if we finish off the destruction of the Ituri Forest and the people who live in it, or we exterminate the few remaining Stone Age level human groups around the planet.

It is impossible to understand who we are, where we came from, or the magnitude of our achievement from the perspective of a single tradition. Personally, a lifetime of pursuing those aims has been a rich reward. Riding the leading edge of scientific discovery has been breathtaking. And all of it has turned around to sharpen the appreciation of my own cultural tradition, specifically, at this moment, the great one who lived in the very midst of the last great ecological collapse. This was Shakespeare's final resolution voiced by Prospero in *The Tempest*:

> *Our revels now are ended. These our actors, as I foretold you, were all spirits and are melted into air, into thin air; and—like the baseless fabric of this vision—the cloud-capped towers, the gorgeous palaces, the solemn temples, the great globe itself, yea, all which it inherit, shall dissolve, and like this insubstantial pageant faded, leave not a rack behind. We are such stuff as dreams are made on, and our little life is rounded with sleep.*

Notes

Chapter 2: The Little Ice Age and the Outrageous Darwin

1 B. Woolley, *The Queen's Conjurer*, p. 133.
2 D. Sobel, *Galileo's Daughter*, p. 47.
3 Ibid., p. 256.
4 B. Woolley, *The Queen's Conjurer*, p. 279.
5 D. King-Hele, *Doctor of Revolution*, p. 31.
6 Ibid., p. 44.
7 Ibid., p. 179.
8 Ibid., p. 238.
9 Ibid., p. 244.
10 Aeschylus, *The Eumenides*, p. 260.
11 Aristotle, *Generation of Animals*, p. 165.
12 J. Farley, *Gametes & Spores*, p. 23.
13 W. Paley, *Natural Theology*, p. 63.

Chapter 3: Charles Darwin and Victorian Culture

1 D. King-Hele, *Doctor of Revolution*, p. 309.
2 A. E. E. McKenzie, *The Major Achievements of Science*, p. 251.

3 A. Bunn, "On the Religion of Charles Darwin", p. 1.

4 Ibid.

5 C. Darwin, *The Origin of Species*, p. 6.

6 D. King-Hele, Doctor of Revolution, p. 310.

7 J. Farley, *Gametes & Spores*, p. 108.

8 C. Darwin, *The Various Contrivances by Which Orchids are Fertilized by Insects*, p. 292.

9 C. D. Yonge, *The Works of Philo*, p. 321.

10 J. Farley, *Gametes & Spores*, p. 108.

11 Ibid., p. 108.

12 D. King-Hele, Doctor of Revolution, p. 308.

13 P. Deane, The *First Industrial Revolution*, p. 261.

14 Ibid., p. 214.

15 H. Mayhew, *London Labour and the London Poor*, p. v.

16 H. Spenser, *The Study of Sociology*, p. 327.

17 Ibid., p. 256.

18 Ibid., p. 313.

19 C. Darwin, *Darwin on Earthworms*, p. 56.

Chapter 4: Biology on the Continent

1 E. Ludwig, *Goethe*, p. 186.

2 Ibid., p. 187.

3 L. Orr, *Jules Michelet*, p. 19.

4 Ibid.

5 Ibid., p. 17.

6 Ibid., p. 18.

7 Ibid., p. 22.

8 Lucretius, *Of the Nature of Things*, p. 181.

9 J. Farley, *Gametes & Spores*, p. 68.

10 Ibid., p. 128.

11 Ibid., p. 254.

12 Ibid., p. 257.

13 P. Kropotkin, *Mutual Aid, a Factor in Evolution*, p. 57.

14 A. Vucinich, *Science in Russian Culture*, p. 275.

Chapter 5: Evolution Comes to America

1 D. J. Struik, *Yankee Science in the Making*, p. 24.

2 D. Boorstin, *The Lost World of Thomas Jefferson*, p. 36.

3 D. J. Struik, *Yankee Science in the Making*, p. 300.

4 Ibid., p. 305.

5 Ibid., p. 306.

6 Ibid., p. 309.

7 Ibid., p. 311.

8 J. Farley, *Gametes & Spores*, p. 112.

9 D. J. Struik, *Yankee Science in the Making*, p. 317.

10 J. D. Watson, *Molecular Biology of the Gene*, p. 233.

Chapter 6: Modern Geology

1 C. E. P. Brooks, *Climate Through the Ages*, p. 7.

2 I. Velikovsky, *Worlds in Collision*, p. vii.

3 F. Hitching, *The Mysterious World*, p. 26.

4 N. D. Newell, *Scientific American*, p. 37.

5 J. Gribbin, *Climate change*, p. 3.

6 Ibid., p. 249.

7 Ibid., p. 64.

8 N. Calder, *The Comet is Coming*, p. 135.

9 Ibid., p. 150.

10 J. and M. Gribbin, *Ice Age*, p. 76.

11 Ibid.

12 C. Darwin, *The Origin of Species*, p. 262.

Chapter 7: The Fitness of Darwinism

1 M. M. Nice, *Studies in the Life History of the Song Sparrow*, p. 90.

2 R. Dawkins, *The Selfish Gene*, p. 172.

3 M. R. Rosenzweig and A. L. Leiman, *Physiological Psychology*, p. 596.

4 C. Bernard, *An Introduction to the Study of Experimental Medicine*, p. 135.

5 G. R. Taylor, *The Great Evolution Mystery*, p. 233.

6 D. R. Griffin, *The Question of Animal Awareness*.

Chapter 8: In the Beginning ...

1 F. Capra, *The Tao of Physics*, p. 186.

2 J. Horgan, *The End of Science*, p. 110.

3 M. Eigen and R. Winkler, *Laws of the Game*, p. 124.

4 A. I. Oparin, Life: Its Nature, *Origin, and Development*, p. 21.

5 Ibid.

6 E. Schrödinger, *What is Life?*, p. 89.

7 D. King-Hele, *Doctor of Revolution*, p. 289.

Chapter 9: Microbiological Evolution

1 L. Margulis and D. Sagan, *Microcosmos*, p. 113.

2 J. Farley, *Gametes & Spores*, p. 262.

3 L. Margulis and D. Sagan, *Microcosmos*, p. 165.

4 J. Farley, *Gametes & Spores*, p. 108.

5 R. D. Manwell, *Introduction to Protozoology*, p. 369.

Chapter 10: The Shapes of Things to Come

1 L. Margulis and D. Sagan, *Acquiring Genomes*, p. 204.

2 J. D. Watson, *Molecular Biology of the Gene*, p. 541.

3 J. T. Bonner, *Size & Cycle*, p. 47.

4 Ibid., p. 3.

5 M. Ridley, *Genome*, p. 178.

6 J. D. Watson, *Molecular Biology of the Gene*, p. 546.

7 C. Patterson, *Evolution*, p. 89.

8 C. Patterson, *Evolution*, p. 65.

Chapter 11: Plant/Animal Evolution

1 C. Darwin, *The Various Contrivances by Which Orchids are Fertilized by Insects*, p. 292.

2 J. Monod, *Chance & Necessity*, p. 120.

3 W. M. Wheeler, *Ants*, p. 246.

4 Ibid.

5 T. Price and N. Wedell, "Selfish genetic elements and sexual selection: their impact on male fertility", 2008.

6 V. B. Wigglesworth, *Insect Hormones*, p. 62.

7 C. Darwin, *The Origin of Species*, p. 205.

8 W. M. Wheeler, *Ants*, p. 405.

9 Ibid., p. 411.

Chapter 12: The Last of the Dinosaurs

1 T. K. Werner and T. W. Sherry, "Behavioral feeding specialization in *Pinaroloxias inornata*, the "Darwin's Finch" of Cocos Island", 1987.

2 C. Darwin, *The Origin of Species*, p. 46.

3 J. Weiner, *The Beak of the Finch*, p. 112.

4 Ibid., p. 125.

Chapter 13: The Age of Mammals

1 H. P. Brokaw, *Wildlife and America*, p. 369.

2 L. A. Weiss et al., "Identification of EFHC2 as a Quantitative Trait Locus for Fear Recognition in Turner Syndrome".

3 E. O. Wilson, *Sociobiology*, p. 79.

4 H. Hediger, *The Psychology and Behavior of Animals in Zoos and Circuses*, p. 66.

5 R. I. Dodge, *The Plains of the Great West*, preface.

6 Ibid., p. 150.

Chapter 14: Human Evolution

1 E. Marais, *The Soul of the Ape*, p. 105.

2 E. Morgan, *The Descent of Woman*, p. 97.

3 D. Morris, *The Naked Ape*, p. 85.

4 W. J. Freeman, *Societies of Brains*, p. 80.

5 Ibid., p. 169.

Chapter 15: Last Thoughts

1 R. Firestone, A. West, and S. Warwick-Smith, *The Cycle of Cosmic Catastrophes*, p. 188.

Bibliography

Achtemeier, Paul J., editor. *Harper's Bible Dictionary*, San Francisco: Harper, 1985.

Aeschylus. *The Oresteia*, New York: Penguin Books, 1979.

Aristotle. *Generation of Animals*, Cambridge: Harvard University Press, 1953.

Becker, Carl J. *A. Modern Theory of Language Evolution*, New York: iUniverse, 2004.

Bernard, Claude. *An Introduction to the Study of Experimental Medicine*, New York: Dover Publications Inc., 1957.

Blackman, Margaret B. *During My Time, Florence Edenshaw Davidson. A Haida Woman*, Seattle: University of Washington Press, 1992.

Bokun, Branko. *Man the Fallen Ape*, New York: Doubleday and Co., 1977.

Boorstin, Daniel. *The Lost World of Thomas Jefferson*, Boston: Beacon Press, 1960.

Brokaw, Howard P. ed. *Wildlife and America*, Washington: U.S. Government Printing Office, 1978.

Brooks, C. E. P. *Climate Through the Ages*, New York: Dover Publications, Inc., 1970.

Brown, Jerram L. *The Evolution of Behavior*, New York: W. W. Norton and Company, 1975.

Bugnion, E. *The Origin of Instinct*, London: Kegan Paul, Trench, Trubner & Co. Ltd., 1927.

Bunn, Andrew. *On the Religion of Charles Darwin: A Brilliant Man's Search for the Divine*, http://www.duke.edu/~agb2/ws.html, 1996.

Cahill, Thomas. *How the Irish Saved Civilization*, New York: Anchor Books, 1995.

Calder, Nigel. *The Comet is Coming*, New York: The Viking Press, 1980.

Cameron, Anne. *Daughters of Copper Woman*, Vancouver, B.C.: Press Gang Publishers, 1981.

Capra, Fritjof. *The Tao of Physics*, Boulder: Shambhala, 1975.

Clube, Victor, and Bill Napier. *The Cosmic Winter*, Oxford: Blackwell Publishers, 1990.

Colbert, Edwin H. *The Evolution of the Vertibrates*, New York: John Wiley & Sons, Inc., 1969.

Corliss, William R. *Unknown Earth: A Handbook of Geological Enigmas*, Glen Arm, MD: The Sourcebook Project, 1980.

Cox, C. B., Ian N. Healey, and Peter D. Moore. *Biogeography*, Oxford: Blackwell Scientific Pub., 1976.

Daly, Michael J., Elena K. Gaidamakova, Vera Y. Matrosova, Alexander Vasilenko, Min Zhi, Richard D. Leapman, Barry Lai, Bruce Ravel, Shu-Mei W. Li, Kenneth M. Kemner, and James K. Fredrickson. "Protein Oxidation Implicated as the Primary Determinant of Bacterial Radioresistance." Public Library of Science, 2007.

Darwin, Charles. *The Origin of Species and The Descent of Man*, New York: The Modern Library.

Darwin, Charles. *The Expressions of the Emotions in Man and Animals*, Chicago: The University of Chicago Press, 1965.

Darwin, Charles. *The Various Contrivances by Which Orchids are Fertilized by Plants*, Chicago: The University of Chicago Press, 1984.

Darwin, Charles. *The Formation of Vegetable Mould*, London: Bookworm Publishing Co., 1976.

Dawkins, Richard. *The Selfish Gene*, Oxford: Oxford University Press, 1976.

Deane, Phyllis. *The First Industrial Revolution*, London: Cambridge University Press, 1965.

Dodge, Richard Irving. *The Plains of the Great West and Their Inhabitants*, New York: Archer House, Inc., 1959.

Dolhinow, Phyllis. *Primate Patterns*, New York: Holt, Rinehart and Winston, Inc., 1972.

Dorst, Jean, and Pierre Dandelot. *A Field Guide to the Larger Mammals of Africa*, Boston: Houghton Mifflin Co., 1969.

Dröscher, Vitus B. *They Love and Kill*, New York: E. P. Dutton & Co., Inc., 1976.

Duplaix, Nicole, and Noel Simon. *World Guide to Mammals*, New York: Crown Publishers, 1976.

Durant, Will. *The Age of Faith*, New York: Simon and Schuster, 1950.

Durant, Will, and Ariel. *The Age of Napoleon*, New York: Simon and Schuster, 1975.

Eibl-Eibesfeldt, Irenäus. *Galapagos*, New York: Doubleday & Company, Inc., 1961.

Eigen, Manfred, and Ruthild Winkler. *Laws of the Game*, New York: Alfred A. Knopf, 1981.

Engle, Phillip. *Far From Equilibrium*, Greensburg, PA: Laurel Highlands Media, 2002.

Farley, John. Gametes & Spores, *Ideas about Sexual Reproduction 1750-1914*, Baltimore: Johns Hopkins University Press, 1982.

Fogle, Bruce. *Natural Cat Care*, New York: DK Publishing, 1999.

Firestone, Richard, Allen West, and Simon Warwick-Smith. *The Cycle of Cosmic Catastrophes*, Rochester: Bear & Company, 2006.

Freeman, Gordon R. *Canada's Stonehenge*, Kingsley Publishing, 2009.

Freeman, Walter J. *Societies of Brains*, Hillsdale, New Jersey: Lawrence Erlbaum Associates, Publishers, 1995.

Gammie, John G., and Leo G. Perdue. *The Sage in Israel and the Ancient Near East*, Winona Lake: Eisenbrauns, 1990.

Gribbin, John, and Mary. *Ice Age*, New York: Barnes & Noble, 2002

Gribbin, John. *Climatic change*, Cambridge: Cambridge University Press, 1978.

Griffin, Donald R. *The Question of Animal Awareness*, New York: The Rockefeller University Press, 1976.

Goodrich, Norma Lorre. *Merlin*, New York: Harper & Row, 1988.

Haggard, Howard W. *Devils, Drugs, and Doctors*, New York: Harper & Brothers Publishers, 1929.

Hall, Manly P. *The Secret Teachings of All the Ages*, Los Angeles: The Philosophical Research Society, Inc., 1988.

Hancock, Graham. *The Sign and the Seal*, New York: Crown Publishers, Inc., 1992.

Hediger, H. *The Psychology and Behavior of Animals in Zoos and Circuses*, New York: Dover Publications Inc., 1968.

Hitching, Francis. *The Mysterious World*, New York: Holt, Rinehart and Winston, 1979.

Hope, Murry. *Atlantis: Myth or Reality?* London: Arkana, 1991.

Horgan John. *The End of Science*, New York: Broadway Books, 1997.

Hoyle, Fred. *Ice*, New York: Continuum, 1981.

Hulse, Frederick S. *The Human Species*, New York: Random House, 1971.

Human Variation and Origins, San Francisco: W. H. Freeman and Company, 1967.

Jelink, J. *The Evolution of Man*, London: Hamlyn, 1975.

King-Hele, Desmond. *Doctor of Revolution, the Life and Genius of Erasmus Darwin*, London: Faber & Faber, 1977.

Keys, David. *Catastrophe: An Investigation into the Origins of Modern Civilization*, New York: Ballantine Books, 2000.

Kraus, Bertram S. *The Basis for Human Evolution*, New York: Harper & Row, Publishers, 1964.

Kropotkin, Peter. *Mutual Aid, a Factor of Evolution*, New York: McClure Phillips & Co., 1904.

Kuhn, Thomas S. *The Structure of Scientific Revolutions*, Chicago: University of Chicago Press, 1970.

Leakey, Richard E., and Lewin, Roger. *Origins*, New York: E. P. Dutton, 1978.

Le Conte, Joseph. *Elements of Geology*, New York: D. Appleton and Company, 1895.

Lorenz, Conrad. *On Aggression*, New York: Harcourt, Brace & World, Inc., 1966.

Lucretius. *Of the Nature of Things*, New York: E. P. Dutton & Co., Inc., 1957.

Ludwig, Emil. *Goethe*, New York: G. P. Putnam's Press, 1928.

Maclean, P. D. *The Triune Brain*, New York: Plenum, 1969.

Manwell, Reginald D. *Introduction to Protozoology*, New York: Dover Publications, Inc, 1968.

Marais, Eugène. *The Soul of the Ape*, New York: Atheneum, 1969.

Margulis, Lynn, and Dorion Sagan. *Microcosmos*, New York: Summit Books, 1986.

Margulis, Lynn, and Dorion Sagan. *Acquiring Genomes: A Theory of the Origins of Species*, New York: Basic Books, 2002.

Marshack, Alexander. *The Roots of Civilization*, New York: McGraw-Hill Book

Company, 1972.

Mayhew, Henry. *London Labour and the London Poor*, New York: Dover Publications, Inc, 1968.

Merrill, Elmer D. *Plant Life of the Pacific World*, New York: The MacMillan Company, 1945.

Mckenzie, A. E. E. *The Major Achievements of Science*, New York: Simon and Schuster, 1973.

Mitchell, James, ed. *The Random House Encyclopedia*, New York: Random House, 1977.

Monod, Jacques. *Chance & Necessity*, New York: Vintage Books, 1972.

Morgan, Elaine. *The Descent of Woman*, New York: Stein and Day, 1972.

Morris, Desmond. *The Naked Ape*, New York: Dell Publishing Co., Inc., 1969.

Nice, Margaret Morse. *Studies in the Life History of the Song Sparrow*, New York: Dover Publications, 1964.

Oparin, A. I. *Life: Its Nature, Origin and Development*, New York: Academic Press, 1964.

Orr, Linda. *Jules Michelet: Nature, History, and Language*, Ithaca: Cornell University Press, 1976.

Paley, William. *Natural Theology: or, Evidences of the Existence and Attributes of the Deity*, J. Faulder, London, 1809.

Patterson, Colin. *Evolution*, Ithaca: Cornell University Press, 1978.

Paine, Thomas. *The Age of Reason*, Buffalo: Prometheus Books, 1985.

Price, Tom A. R., and Nina Wedell. "Selfish genetic elements and sexual selection: their impact on male fertility", Genetica, Volume 132, Number 3, March, 2008.

Rad, Gerhard von. *Old Testament Theology*, Vol II, New York: Harper & Row, 1967.

Ryan, William, and Walter Pitman. *Noah's Flood*, New York: Simon & Schuster, 1998.

Ricketts, Edward, Jack Calvin, and Joel W. Hedgpeth. *Between Pacific Tides*, Stanford: Stanford University Press, 1968.

Ridley, Matt. *Genome*, New York: Harper Collins Publishers, 1999.

Rosenzweig, Mark R., and Leiman, Arnold L. *Physiological Psychology*, Lexington, Mass.: D. C. Heath and Company, 1982.

Sagan, Carl. *The Dragons of Eden*, New York: Random House, 1977.

Schrodinger, Erwin. *What is Life?* Cambridge: Cambridge University Press, 1951.

Smith, Adam. *An Inquiry into the Nature and Causes of the Wealth of Nations*, Indianapolis: Liberty Classics, 1981.

Sobel, Dava. *Galileo's Daughter*, New York: Walker & Company, 1999.

Spanuth, Jürgen. *Atlantis of the North*, London: Sidgwick & Jackson, 1979.

Spencer, Herbert. *The Study of Sociology*, New York: D. Appleton and Company, 1891.

Stent, Gunther S. *Paradoxes of Progress*, San Francisco: W. H. Freeman and Company, 1978.

Stevens, S. S. *Handbook of Experimental Psychology*, New York: John Wiley & Sons, Inc., 1951.

Struik, Dirk J. *Yankee Science in the Making*, Boston: Little, Brown and Company, 1948.

Swan, Lester A., and Charles S. Papp. *The Common Insects of North America*, New York: Harper & Row, Publishers, 1972.

Taylor, Gordon Rattray. *The Great Evolution Mystery*, New York: Harper and Row, Publishers, 1983.

Thomas, Elizabeth Marshall. *The Harmless People*, New York: Alfred A. Knopf, 1959.

Trevathan, Wenda R. *Human Birth*, New York: Aldine de Gruyter, 1987.

Turnbull, Colin M. *The Forest People*, New York: Anchor Books, 1962.

Tyndall, John. *Fragments of Science*, New York: A. L. Burt Company.

Vaneechoutte, Mario. http://users.ugent.be/~mvaneech/Index.html.

Velikovsky, Immanuel, *Worlds in Collision*. New York: The MacMillan Company, 1950.

Velikovsky, Immanuel. *Earth in Upheaval*, New York: Doubleday &Company Inc., 1955.

Vucinich, Alexander. *Science in Russian Culture*, Stanford: Stanford University Press, 1970.

Watson, James D. *Molecular Biology of the Gene*, New York: W. A. Benjamin, Inc. 1970.

Weiner, Jonathan. *The Beak of the Finch*, New York: Vintage Books, 1995.

Weiss, Lauren A., Shaun Purcell, Skye Waggoner, Kate Lawrence, David Spektor, Mark J. Daly, Pamela Sklar, and David Skuse. "Identification of EFHC2 as a Quantitative Trait Locus for Fear Recognition in Turner Syndrome", Human Molecular Genetics, 2007.

Werner, Tracey K. and Thomas W. Sherry. "Behavioral feeding specialization in *Pinaroloxias inornata*, the 'Darwin's Finch' of Cocos Island", Costa Rica: Proc. Natl. Acad. Sci., Vol 84, August 1987.

Wheeler, William Morton. *Ants*, New York: Columbia University Press, 1910.

White, Andrew D. *A History of the Warfare of Science with Theology in Christendom*, New York: George Brasiller, 1955.

Wilson, Edward O. *Sociobiology*, Cambridge: The Belknap Press of Harvard University Press, 1975.

Wilson, Leonard G. *Charles Lyell*, New Haven: Yale University Press, 1972.

Wigglesworth, V. B. *Insect Hormones*, San Francisco: W. H. Freeman and Company, 1970.

Wood, Gaby. *Edison's Eve*, New York: Alfred A. Knopf, 2002.

Wood, J. G.. *The Natural History of Man*, London: George Routledge and Sons, 1868.

Woolley, Benjamin. *The Queen's Conjurer*, New York: Henry Holt and Company, 2001.

Yates, Frances A. *The Rosicrucian Enlightenment*, Boulder: Shambhala, 1978.

www.ingramcontent.com/pod-product-compliance
Lightning Source LLC
Chambersburg PA
CBHW031818170526
45157CB00001B/110